Driving Future Vehicles

Driving Future Vehicles

Edited by

Andrew M. Parkes

HUSAT Research Institute, Loughborough, UK

and

Stig Franzén

HUMTEK hb, Molndal, Sweden

Taylor & Francis
London · Washington, DC

UK   Taylor & Francis Ltd, 4 John St, London WC1N 2ET

USA   Taylor & Francis Inc., 1900 Frost Road, Suite 101, Bristol PA 19007

Copyright © Taylor & Francis Ltd 1993

British Library Cataloguing in Publication Data

A catalogue record for this book is available from the British Library

ISBN 0 7484 0042 7

Library of Congress Cataloging in Publication Data are available

Cover design by Hybert · Design & Type

Typeset by Mathematical Composition Setters Ltd., Salisbury, Wilts SP3 4UF

Printed in Great Britain by Burgess Science Press, Basingstoke on paper which has a specified pH value on final paper manufacture of not less than 7·5 and is therefore 'acid free'.

Contents

Contents

Preface

Although there has been long standing interest in driver behaviour and traffic safety in many countries of the world, the emergence of large scale, multi-institutional, collaborative projects in recent years has accelerated the knowledge gathering and dissemination process. This book has a European perspective and as such it is based on the success of the DRIVE and PROMETHEUS programmes. These programmes have been instrumental in bringing together groups of European researchers from academia and industry in collaborative work, accelerating the two-way knowledge transfer, and making a constructive link between basic and applied research. In both these initiatives a heavy emphasis has been placed on system safety and acceptability in addition to just technical efficiency. 1992 was an appropriate year for writing this book, as it heralded the introduction of the single European market, and an era of increasing openness among member states. It also marked the start of the large scale field trials planned for many of the technologies and services proposed and developed in DRIVE. The movement in PROMETHEUS towards an increasing consideration of MMI problems, and the building of a wide range of demonstration systems, also marked this year as timely for such a collection of review papers and empirical research reports as we have here.

This book cannot hope to cover all the developments that will influence the act of driving future vehicles, nor can it include all the research of interest in the DRIVE, and PROMETHEUS programmes. Instead, this must be seen as a collection of some of the interesting and available material from both programmes. The contributions cover a broad perspective of issues, but are obviously biased towards the personal interests of the editors and the contributors, and are based largely on the network of contacts made during the first phases of the two programmes.

We have contributions from many countries in Europe and have tried to ensure a balance between industry and academia. The backgrounds of the authors can be found in engineering, computer science, traffic safety, psychology, physiology and many other disciplines. These backgrounds influence the approach of the different papers in this collection, but the focus is consistently on the act of driving the vehicle of the future. Although English is not the first language of the majority of the authors, the editors have needed to make only minor typographical or stylistic changes to papers and have not attempted to impose a rigorous conformity in layout or construction.

This book cannot be comprehensive and contains only a sample of the output from the two programmes. It does not address every topic currently under consideration, nor does it deal exhaustively with any single topic. However, it is hoped to give a flavour of the range of work currently being undertaken in Europe.

The papers have been divided into four main parts. Each is provided with an overview which attempts to draw the links between the papers present and point out any salient themes that have emerged. Each part can be dipped into; all papers are self contained.

The editors would like to thank the representatives of DRIVE and PROMETHEUS who gave their blessing for the organization of this book, all of the contributors and publishers who accepted with good grace the many delays that hampered the progress from concept till the final submission of manuscript.

Finally the editors would like to offer a profound apology to all those European researchers who have contributions that merited inclusion in this volume, but who could not be included.

Andrew Parkes
Loughborough, UK

Contributors

Håkan Alm
Swedish Road and Traffic Research
 Institute
S-581 01 Linköping
Sweden
Telephone 46 13 204195
Fax 46 13 141436

Martin C. Ashby
HUSAT Research Institute
The Elms, Elms Grove
Loughborough
Leicestershire LE11 1RG
UK
Telephone 44 509 611088
Fax 44 509 234651

Guro Berge
Institute of Transport Economics
P.O. Box 6110, Etterstad
0602 Oslo
Norway
Telephone 47 2 573800
Fax 47 2 570290

K. A. Brookhuis
Traffic Research Centre
University of Groningen
Rijksstraatweg 76
P.O. Box 69
9750 AB Haren
The Netherlands
Telephone 31 50 636772
Fax 31 50 636784

Carl Brown
SICS
Box 1263
S-164 28 Kista
Sweden

Telephone 46 8 7521500
Fax 46 8 7517230

O. M. J. Carsten
Institute for Transport Studies
The University of Leeds
Leeds LS2 9JT
UK
Telephone 44 532 335348
Fax 44 532 335334

M. Draskóczy
Department of Traffic Planning and
 Engineering
Lund Institute of Technology
Box 118
221 00 Lund
Sweden
Telephone 46 46 104824
Fax 46 46 123272

Martin P. Emberger
PROMETHEUS Office
c/o Daimler-Benz AG
Postfach 80 02 30
D-7000 Stuttgart 80
Germany
Telephone 49 711 1756920
Fax 49 711 1757748

B. Färber
Human Factors Institute, LRT 11
University of the Armed Forces,
 Munich
Werner-Heisenberg-Weg 39
8014 Neubiberg
Germany
Telephone 49 89 60043250
Fax 49 89 60044092

S. Fairclough
The HUSAT Research Institute
The Elms
Elms Grove
Loughborough
Leicestershire LE11 1RG
UK
Telephone 44 509 611088
Fax 44 509 234651

Hélène Fontaine
INRETS
2, avenue du Général Malleret-Joinville
94114 Arcueil-Cedex
France
Telephone 33 1 47407000
Fax 33 1 45475606

Stig Franzén
Franzén Consult AB
Safjallsgatan 4
S-431 39 Molndal
Sweden
Telephone 46 31 864530
Fax 46 31 864530

D. A. Fraser
Department of Electronic and Electrical
    Engineering
King's College
London University
Strand
London WC2R 2LS
UK
Telephone 44 71 8732368
Fax 44 71 8364781

Per Gårder
Department of Traffic and Transport
    Planning
KTH
S-100 44 Stockholm
Sweden
Telephone 46 8 7908008
Fax 46 8 212899

Georg Geiser
Fraunhofer-Institut für Informations-
    und Datenverarbeitung
(IITB)
Fraunhoferstraße 1
D-7500 Karlsruhe 1
Germany
Telephone 49 721 6091240
Fax 49 721 6091413

Detlef Gerhardt
Commission of European Communities
Directorate General
    Telecommunications, Information
    Industries and Innovation
DRIVE Central Office
Rue de Trèves 61
Brussels, Belgium
Telephone 32 2 2363542
Fax 32 2 2362391

Hans Godthelp
TNO Institute for Perception
P.O. Box 23
3769 ZG Soesterberg
The Netherlands
Telephone 31 3463 56447
Fax 31 3463 53977

John A. Groeger
MRC Applied Psychology Unit
15 Chaucer Road
Cambridge CB2 2EF
UK
Telephone 44 223 355294
Fax 44 223 359062

J. Harper
Computer Science Department
St Patrick's College
Maynooth
Co. Kildare
Ireland
Telephone 353 1 6285222
Fax 353 1 6289063

Robert E. Hawken
Department of Electronic and Electrical
    Engineering
King's College
London University
Strand,
London WC2R 2LS
UK
Telephone 44 71 8732368
Fax 44 71 8364781

Kristina Höök
SICS
Box 1263
S-164 28 Kista
Sweden
Telephone 46 8 7521500
Fax 46 8 7517230

Ian Howarth
Department of Psychology
University of Nottingham
Nottingham NG7 2RD
UK
Telephone 44 602 515300
Fax 44 602 515324

C. Hydén
Department of Traffic Planning and
  Engineering
Lund Institute of Technology
Box 118
S-221 00 Lund
Sweden
Telephone 46 46 109129
Fax 46 46 123272

Wiel H. Janssen
TNO Institute for Perception
P.O. Box 23
3769 ZG Soesterberg
The Netherlands
Telephone 31 3463 56325
Fax 31 3463 53977

Robert G. Leiser
Human Factors Group
BAeSEMA
1 Atlantic Quay
Broomielaw
Glasgow G12 0SL
UK
Telephone 44 41 2042737
Fax 44 41 2216435

Christer Lindh
Department of Traffic and Transport
  Planning
KTH
S-100 44 Stockholm
Sweden
Telephone 46 8 7908010
Fax 46 8 212899

K. S. Lorenz
Planungsbüro VIA
Matthias Richter
Kai S Lorenz
Saalmannstrasse 3
D-13403 Berlin
Germany
Telephone 49 30 4122166
Fax 49 30 4121839

Gilles Malaterre
INRETS
2, avenue du Général Malleret-Joinville
94114 Arcueil-Cedex
France
Telephone 33 1 47407000
Fax 31 1 45475606

Lena Nilsson
Swedish Road and Traffic Research
  Institute (VTI)
S-581 01 Linköping
Sweden
Telephone 46 13 204140
Fax 46 13 141436

Günther Nirschl
Fraunhofer-Institut für Informations-
  und Datenverarbeitung
(IITB)
Fraunhoferstraße 1
D-7500 Karlsruhe 1
Germany
Telephone 49 721 6091257
Fax 49 721 6091413

R. Onken
Universität der Bundeswehr München
Fakultät für Luft- und
  Raumfahrttechnik
Institut für Systemdynamik und
  Flugmechanik
D-8014 Neubiberg, Werner-Heisenberg-
  Weg 39
Germany
Telephone 49 89 60043452
Fax 49 89 60043560

A. M. Parkes
HUSAT Research Institute
The Elms, Elms Grove
Loughborough
Leicestershire LE11 1RG
UK
Telephone 44 509 611088
Fax 44 509 234651

Ep H. Piersma
Traffic Research Centre VSC
University of Groningen
P.O. Box 69
9750 AB Haren
The Netherlands
Telephone 31 50 636780
Fax 31 50 636784

Gunter Reichart
BMW AG
Khorrstr. 147
Postfach 40 02 40
W-8000 Munchen 40
Germany
Telephone 49 89 31294984
Fax 49 89 31294988

Ralf Risser
c/o FACTUM
Danhausergasse 6/8
1040 Wien
Austria
Telephone 43 222 50415462
Fax 43 222 5041548

Tracy Ross
HUSAT Research Institute
The Elms
Elms Grove
Loughborough
Leicestershire LE11 1RG
UK
Telephone 44 509 611088
Fax 44 509 234651

J. A. Rothengatter
Traffic Research Centre
University of Groningen
Box 69
NL 9753 AW Haren
The Netherlands
Telephone 31 50 636778
Fax 31 50 636784

K. Rumar
Swedish Road and Traffic Research
 Institute
S-581 01 Linköping
Sweden
Telephone 46 13 204229
Fax 46 13 1411436

Jan Maarten Schraagen
TNO Institute for Perception
P.O. Box 23
3769 ZG Soesterberg
The Netherlands
Telephone 31 3463 56211
Fax 31 3463 53977

J. M. C. Schumann
Human Factors Institute
LRT 11
University of the Armed Forces
Munich Werner-Heisenberg-Weg 39
8014 Neubiberg
Germany
Telephone 49 89 60043170
Fax 49 89 60044092

Victor Sievey
Department of Geography
King's College, London University
Strand, London WC2R 2LS
UK
Telephone 44 71 8732081
Fax 44 71 8732287

Ove Svidén
Sandtorpsgatan 8
S-582 63 Linköping
Sweden
Telephone/Fax 46 13 136020

Willem B. Verwey
TNO Institute for Perception
P.O. Box 23
NL 3769 ZG Soesterberg
The Netherlands
Telephone 31 3463 56211/56439
Fax 31 3463 53977

A. M. Warnes
Department of Geography
King's College
London University
Strand
London WC2R 2LS
UK
Telephone 44 71 8732081
Fax 44 71 8732287

W. van Winsum
Traffic Research Centre, TRC
University of Groningen
P.O. Box 69
9750 AB Haren
The Netherlands
Telephone 31 50 636764/636780
Fax 31 50 636784

PART I
INTRODUCTION

1

Introduction

S. Franzén

New systems are being developed which will address the social objectives of increased safety and efficiency, better economy, more comfort or convenience and environmentally acceptable solutions for future road traffic. To accomplish this the solutions have to comply with the needs and objectives of individual users of the systems. These users can be seen as the potential customers whom the automotive industry and their allies, the electronic and supplier industries, have to meet in the market place. The real human needs and problems have to be in focus of the design work. If they are not actively taken into account in the design process, the new advanced systems designed might have a potential for danger, accidents and even market resistance.

The social and the individual objectives can be seen as both complementary and contradictory at the same time, and an acceptable solution must incorporate both perspectives. The R&D activity in Europe on the future road transport system is emphasizing these two views. The Eureka programme PROMETHEUS has its focus on the interests of the automotive industries and the electronic and supplier industries. They are mainly involved in the product development work trying to meet the individual preferences of the potential customers, i.e. buyers of cars and additional equipment. Some of the features to be provided will, however, need support from infrastructure representatives, e.g. road and telecommunication authorities and other governmental agencies. The CEC programme DRIVE has its focus on the social objectives and how to influence and support the development of future pan-European infrastructure systems. However, these systems must also somewhere meet the users of the systems. The customers of the automotive industries and the users of road traffic service organizations are the same, i.e. the drivers of future vehicles.

In order to tackle the sometimes contradictory interests of society and individuals, systems engineering methodology has to be applied. An ideal systems approach would start with the analyses of both the users and the problems these user groups experience in traffic. Procedures have to guarantee that the results of these analyses are used in the design process itself. The

3

introduction of such a user-oriented perspective has similarities with the introduction of quality assurance procedures in other industrial activities. Finally, both new and old methodologies and tools will be applied to meet the new demands from the design process in order to reach the ultimate goals of systems for road traffic informatics, RTI. However, the ideal process presented above can never be found in reality. Nevertheless, an awareness of the different aspects of the systems approach which have been highlighted will make designers more prepared to handle the challenges and problems of future road transport systems. The structure of the book with its four main parts is in line with this intention.

In Part 1 representatives of the two R&D programmes, PROMETHEUS and DRIVE, will further elaborate on the objectives, already accomplished results and future expectations of the work. The focus of the contributions in the book will be on the driver, and the ideas and concepts of future road traffic informatics. The success of both PROMETHEUS and DRIVE will depend on how the human being, as a driver, will be taken into account from the very early stage of the R&D work, and how the risk of introducing new functions and concepts based only on technological considerations must be avoided. A scenario of a road transport system of the future is also presented. We hope that the contribution will stimulate wide discussions. By giving some examples of future road transport systems in operation, the importance of having links to existing solutions and an open mind accepting new concepts in road traffic are emphasized. The scenario presented is the result of both PROMETHEUS and DRIVE work.

In Part 2 on 'Users' the contributions are examples of some aspects of user analysis work. The focus is limited as the user is seen as the driver of future vehicles. The heterogeneousness of the group of drivers is emphasized and topics like driver characteristics, driver capacity, driving models and task analyses are treated, both explicitly and implicitly. In order to show that there is concern for road users other than drivers in the European work, the cyclist and pedestrian situations are presented. The effects of RTI systems introduction on these road user groups are discussed.

The main track of Part 3 on 'Problems' is the vast area of applications to be implemented as future RTI functions. Concepts behind the applications, as well as the underlying problems, must be analysed as the driver experiences them in traffic. The relevance of a solution to potential user (driver) groups must be clarified. One application in particular, route guidance and navigation, has been chosen to provide the basis for a discussion of the many aspects of concept development and system implementation. The safety aspects of the introduction of RTI, as well as some non-driving tasks of drivers, are also presented. The problems which are addressed are based on both individual and social needs and will certainly be reflected in the solutions a design team has to develop.

Part 4 is devoted to the 'Design' process and what can be expected when many new functions and information sources are introduced. The information flow will increase, and the driver's situation will change dramatically if not met by a systematic approach in the design work. The concept of a co-pilot and the role of the driver when driving future vehicles will be in focus. All the knowledge coming from the work on driver and problem analyses must be incorporated in the design to guarantee a driver-oriented solution, not in conflict with social goals. Much work must be devoted to the investigation of what technologies, old and new, can meet the demands and the ideas behind the specification for new RTI functions. The long term goal that both safe and economic design could be met by standardization is discussed in the last contribution.

Finally, Part 5, 'Methodology' is focused on some of the problems a design team will meet in the necessary experimental work for the analysis, design and evaluation of RTI systems. For instance, the validation of simulator results, the relevance of different measurements, the criteria used are important aspects to be discussed. Some new tools and methodologies based on other ideas, e.g. computer science, are presented.

The purpose of the book is not to cover all possible aspects of the design of future RTI systems. Only a glimpse of the European R&D work is presented. The emphasis is on results available from the first half of the R&D programmes, i.e. until 1991. The structure of the book is related to the ideal systems approach in the design work even if the contributions only cover some aspects. However, we are convinced that the many excellent contributions will bring out a better awareness of the challenges all RTI design teams are facing today and will meet tomorrow. A systems approach in the design work is the only possible way to meet both the driver needs, as well as the goals of society, when developing new RTI systems.

2

Advanced transport telematic systems for the European road infrastructure and human reliability: contributions from the DRIVE I programme

Detlef Gerhardt

Introduction

The European single market will be operational very soon, and the economy will gain a major advantage from this operation resulting in enormous savings of costs and time (Ceccini, 1988). European citizens have great hopes that it will promise them more freedom and wider possibilities in many fields. Integration of the 12 economies will inevitably have, as a consequence, an increased partitioning of production between the countries which will result in an increase of goods and a higher mobility of persons. Transport has to be considered as one of the key elements for the success of this operation in terms of economic and social development.

The transport sector already represents an important factor in the European economy, but it is well known that it is also responsible for a series of problems. So it is not surprising that some forecasts are very pessimistic about the evolution of the numerous transport problems. A recent report of 'The Group Transport 2000 Plus' states that many indicators point to an impending crisis when the single market is operational, and at a time when there will also be a massive increase in the movement of goods and services between the European Community and Eastern Europe. This report forecasts that 'the crisis will paralyse the system, slow down economic progress, provoke serious social tensions, increase damage to the environment and destroy the balance in the central and peripheral regions of the continent. The process of building a unified Europe will be set back severely'. Other documents forecast a so-called transport-infarctus (Verkehrsinfarkt).

The underlying problems are manifold. They are generally related to safety, land allocation and need for competing uses as well as to air and noise pollution.

1. When the problem of road safety is looked at, the scale of the problem becomes immense. Currently 50 000 people are killed each year on European roads and about 1.5 million people are injured, often with serious consequences for the rest of their life and for their family. The costs also are enormous (45–90 billion ECU per year) (Gerondeau *et al.*, 1991). Comparison with other equivalent countries shows how critical the situation actually is in Europe: whereas the number of people killed per million vehicle kilometres in USA is 1.5 and in Japan 2.3, this number goes up for the EC in average to 2.7, varying from 1.4 to 10.6 for the different EC countries (Gerondeau *et al.*, 1991). So effective solutions are needed to make a real improvement in safety for the European road transport user.

2. Road transport has to struggle with congestion problems which become more and more acute:

   - Urban congestion paralyses many cities. Parking space is more and more limited. Public transport does not seem to solve the problems with the efficiency which might be expected. There are at present few other new solutions. This appears to be a local problem, but it requires international solutions: similar problems can be found in different urban areas throughout Europe so solutions for one area could be transferred to another, resulting in a considerable saving of time and money.
   - Congestion problems on access roads to the big cities and ringroads around them create considerable delays for road haulage and present a significant economic loss for European society.
   - On motorways it is nowadays not unusual, especially in holiday or weekend periods, to find queues of more than 200 km, in which foreign drivers are often involved. So a common international concept of solutions is needed.

3. As traffic volume increases, so also does the amount of environmental damage it causes. On a global level the problems of the greenhouse effect and acid rain have to be tackled. On a local level noise and air pollution put into danger health and quality of life, and create more and more, sometimes violent, reactions from the population (Müller, 1989).

In order to find and choose solutions to these road transport problems different major European constraints have to be taken into account:

- There are strict limitations and strong oppositions to increasing the traditional road infrastructure. So additions have not only to be planned carefully, but also to be coordinated very closely inside Europe. The benefits from a European viewpoint should be shown.

- A well developed network of other transport modes exists in the different European countries (rail, air, waterways etc.). For the longer term choice of new solutions for the road transport problem these existing national networks have to be taken into account. This should permit the optimization of the solutions in the overall transport network.

These constraints show that it is necessary to base the response to the existing infrastructure, to render its use more efficient and to find a coherent approach on a European level (CEC, 1990).

New technologies of information and telecommunications have the potential to render the existing infrastructure more effective. They can help to collect, process, combine and dispatch the various data concerning the road environment, status, traffic etc. The isolation of the driver can be overcome by these infrastructure-based technologies. The provision of information, advice or guidance to the driver makes it possible also to optimize traffic flow over space and time.

It is clear that improvements in car performance especially for safety and emissions, while possible and necessary, can by themselves only partly cope with the needs arising from the expected increase of demand. The so-called Advanced Transport Telematics (ATT) will be applied to both road vehicles and road infrastructure. Various ATT information and control systems should be installed in the cities and corridors of European interest.

The infrastructure, the vehicle and the user are three closely inter-related components of the road transport system. The final user plays a major role: it is evident that these systems will only be really effective and accepted by the different road users when they take into account the user's constraints and needs. The systems have to be technically reliable (the system has to be safe), they have to be integrated and adapted to the normal driver tasks (the man-machine interaction has to be optimized), and they have to ensure that driver behaviour will not change with the system in a way that results in a decrease of traffic safety, as has been already noticed in many examples. So the introduction of ATT systems has to be determined by user needs.

The DRIVE (Dedicated Road Infrastructure for Vehicle safety in Europe) programme focusses on the application of information and telecommunication technologies to the road infrastructure, whereas other programmes or projects (e.g. PROMETHEUS in the EUREKA programme) concentrate more on the improvement of the vehicle itself. These programmes can be considered as complementary, supporting themselves mutually on the level of pre-competitive research. The DRIVE I programme represented a financial effort of 60 thousand ECU corresponding to 1.1 per cent of the total amount devoted to the research programmes within the Second Framework Programme of the European Communities (1987–1991)(with 42.3 per cent for information and telecommunication technologies).

d human reliability

developing new systems for road transport management and comm... DRIVE I programme is dedicated to offer a new approach to improve road safety through the applications of ATT. Safety problems have already been addressed in many other domains, e.g. in dealing with accidents or catastrophes in other transport systems (rail, maritime, air, space), or in industries of continuous process (chemical plants, nuclear power plants etc.). In all these systems, beside the technical reliability of equipment, one of the major problems is the reliability of the human operator: human error is often cited as one of the major causes of failures and accidents. It is clear that the contribution of the human operator cannot be isolated from the context of his activity. So the term 'reliability' can be defined with Leplat and de Terssac (1990): 'A domain of knowledge having as object the arrangement of connecting the human and technical components of a system in order to respond in an efficient way to his task.'

Two categories of analysing human reliability can be distinguished (Fadier, 1990): general methods for system analysis, and specific methods for human reliability studies.

1. Methods for system analysis are used both for systems in a normal functioning state and for malfunctioning systems. The objective of the methods in the first case is to describe and to understand the system from the structural and functional point of view. In the second case the objective is on the one hand to know the different conditions and contexts which affect the normal functions and performance of the system, in order to be able to compile a list of the failures, their causes and consequences, affecting the system; and on the other hand, to decrease their frequency of occurrence and the scale of the consequences. This can be done by describing new functions to be performed by both the human operator and the system.
2. Specific methods of human reliability analysis have as a principal goal the precise study of the tasks realized by the human operator and the analysis of errors committed at different levels of perceptive and cognitive functioning (Leplat and de Terssac, 1990).

The application of these methods is expected to provide two types of outcome:

1. The system analysis provides indications on the general problems affecting safety. This should permit identification of the functions, which should be improved in order to increase safety.
2. Specific behavioural studies provide a more precise insight into the human information processing cycle accomplished in different road users' tasks. From these methods we expect a better understanding of the effect of technologies already existing or planned, as well as recommendations as to how new technologies should be designed and incorporated in order to be compatible with the human information processing cycle.

Human behaviour aspects and traffic safety in the DRIVE I programme

These principal methods for human reliability analysis can also be found in the part of the DRIVE I programme which puts the emphasis on the study of safety problems. One category of projects is working particularly on human behaviour aspects and traffic safety. The projects falling in this category can be subdivided into the following subclasses:

1. Hazard and accident analysis and identification of the functions to be improved.
2. Human behaviour analysis and the impact of new technologies.
3. Driver surveillance and data recording.
4. System safety.

The first two groups are clearly linked to the two types of human reliability analysis; the third, although related to this approach, puts the emphasis on different aspects; and finally the fourth group concentrates on questions of technical reliability.

Hazard and accident analysis and identification of functions to be improved

In the framework of the system analysis approach this group of projects concentrates on the analysis of malfunctioning and failures in the system. Extracts from the mass of information and statistics on traffic accidents, the most common causes of accidents in the Community, have been examined by three projects (V1040, V1042, V1062). The data collected will determine the scenarios that most often lead to accidents, so that new road transport information systems can be designed and their effect evaluated. This data will also determine the needs which define the basis of requirements for new technologies to improve safety, such as automatic speed control in vicinities of intersections or the enforcement of speed limits. If in rear-end collisions drivers could react one second sooner, about 90 per cent of the accidents could be avoided.

Different road users are concerned by these requirements: not only motorists and lorry drivers, but also pedestrians and cyclists (vulnerable road users). DRIVE has made a special effort not to develop technologies solely for motor vehicles at the expense of pedestrians and cyclists. Two types of activity have been undertaken in corresponding projects to improve safety and comfort for such road users who are an essential part of the system.

1. The needs and habits of vulnerable road users have been taken into account in providing planning tools for traffic networks. A database will be used in conjunction with a vulnerable road user behaviour model, as help for the infrastructure owners to adapt passages or crossings to the needs of these users (project V1031).

2. One of the areas of direct interaction between two system actors, pedestrians and car traffic, has been concentrated particularly on pedestrian crossings. Important improvements are proposed using telecommunication technologies (V1061).

Human behaviour analysis and the impact of the new technologies

The driver can be considered as the crucial human factor in accident causation (Parkes and Ross, 1991), so the specific analysis of human reliability is essential for this type of operation and has been undertaken by the projects in this group.

In order to increase human reliability new technical systems will be made available which are intended to facilitate the driver's task. Effectively these new systems can supply information in higher quality and greater quantity, but the information presented will be helpful only when matched to the task which is to be executed by the driver: better performance by the driver in relation to his task is influenced essentially by the content, the mode of presentation of the information and the moment when it is supplied. So the information has to be really needed at the appropriate moment, it should be presented in the most perceptible way and it should be easily understandable. This means that the information has to be compatible with the information processing operations, the expectations and mental representations of the driver.

DRIVE I distinguished two driver populations: drivers in general and elderly drivers, as their number will grow considerably in the coming years. To meet the needs for the population of elderly drivers, new ATT systems have to be designed in a way that compensates for that part of their capabilities which are decreasing. The question of whether this type of design should constitute a basis for all ATT systems will have to be discussed in the light of complementary results (V1006). The research carried out on the specific human reliability analysis of the driver population in general concerns three topics within DRIVE.

One project (V1017) concentrates on the elaboration of a set of appropriate methods to analyse the driver's behaviour in different driving tasks. By the extension of two models (Michon, 1985; Rasmussen, 1987) these tasks have been divided into four categories. The most important behavioural characteristics shown in these categories of driver tasks can be represented by well defined data types. Various methods and tools have been used in a 'multi-level approach' in order to collect and to cross-reference these data types in a large variety of driving situations, in which ATT technologies were used by the driver. The final objective of this project is to show the 'usability' of already existing ATT technologies (Parkes, 1991).

In another project, concepts for new ATT systems have been developed on the basis of behavioural analysis in different driver tasks (V1041). In the now almost classical categories of driver tasks (navigation, manoeuvre and control tasks) (Michon, 1985) experimental studies have been conducted and have revealed the information processing operations, constraints (as information and work overload) and mental representations of experienced and inexperienced drivers. From these results the needs in term of information and control assistance have been deduced. A prototype of a driver assistance system, the so-called 'Generic Intelligent Driver Support', has been designed and will be tested in different traffic situations which will at first be of a limited type, but is nevertheless considered as representative ('Small World') (Michon and McLauglin, 1991). This project contributes, at the levels of both concept and realization to a new approach, to the specification and design of ATT systems closely adapted to the driver's needs.

The third project of this group points in the direction of generalizing the design process in such a way that ergonomics principles for Man-Machine Interfaces (MMI) will be applied in a coherent way all over Europe so that these interfaces will be consistent between vehicles (project V1037). Different levels of interaction between the driver's task and the in-car equipment have been looked at and relevant standards for these levels have been assembled. The three basic types of standard relevant to the MMI have been found, but only one, the performance standard, is based specifically on how the driver integrates the system into his task. The development of these standards will be pushed forward by this project.

Drivers are not only at risk from information and work overload, but also from underload in certain situations. Driver underload is a particular problem on long journeys. Long periods of relative inactivity can lead to reduced awareness, drowsiness or even to the driver falling asleep at the wheel. Such incidents are a source of great danger especially as they tend to occur on open roads and therefore at high speeds. Monitoring to detect reduced awareness before it reaches dangerous levels is an area that has been examined (project V1004). A system using various sensors coordinated by a neural network has been found to be feasible and further work has been proposed in the area. The action taken in response to a warning from the system could vary from a warning to the driver to, in the most extreme case, the system taking corrective action. Complementary research in this area is still required.

Driver surveillance and data recording

These projects are concerned with recording the progress of the vehicles in traffic both by the infrastructure and from inside the vehicle.

External recording of vehicle progress is being examined for automatic policing purposes (project V1033). All motorists regularly see cases of illegal or anti-social driving being rewarded by reduced delays for the offending driver. As long as the possibilities of being caught and punished are small then drivers will perceive benefits from this kind of behaviour. If on the other hand being caught is a likelihood, or even a certainty, then this type of driving will quickly disappear. A list of seven applications for such automatic policing systems has been drawn up. It is of course necessary to examine the accept-ability of such systems and how their use would be viewed by legal authorities (Rothengatter and Harper, 1991).

The concept of the 'Black-Box' fitted in all large passenger aircraft is familiar to most people. In DRIVE I, research has been carried out to specify a similar system for road vehicles (project V1050). This system consists of an in-vehicle recorder (recording the position, speed, equipment status, etc.) of the vehicle and a reader (which could be used by traffic designers, police, driving instructors etc.) to extract data from the recorder. The data could be used to determine the cause of an accident, but perhaps more important data can be gathered into a database concerning both accidents and 'near-misses' which could be an important source of information for safety researchers (Fincham *et al.*, 1991).

System safety

If the work described so far and in other parts of the DRIVE I programme is to have any effect then the issue of implementation and the impact of the work must be addressed. Just as manufacturing industry has embraced the concept of 'total quality', so the DRIVE community must adopt the concept of 'total safety' in all ATT systems. Safety is not an add-on at the end of the design process, but is an inherent part of that process. As every decision is made the question must be asked 'How will this affect the safety of the system'. This is particularly true of safety critical systems where failure could lead to damage, injury or loss of life. Research has been conducted in DRIVE I into methods of identifying safety critical systems and ensuring the concept of total safety in their design (project V1051). Standards for software and hardware specifications, design, validation and verification will be defined and put forward for certification by appropriate national and European bodies (Clutterbuck *et al.*, 1991).

Conclusions

The research realized in the DRIVE I programme on human behaviour aspects and traffic safety represent an important step forward in the incorporation of road user needs in the definition and adaptation of ATT systems linked to European road infrastructures, at least at two levels:

1. On the project level the results will afford to have a better insight into both the knowledge and the tools which are available in order:

   ● to define new systems which will improve individual and overall road safety

- to understand the effects of existing advanced transport telematic systems on various driver tasks and populations and to choose those which can help to realize particular driving tasks
- to gain experience on the specification and design of new systems based on a solid ergonomic analysis.

2. On an interproject level, the group on human behaviour aspects and traffic safety has made an important effort in order to assemble the knowledge gained and the methods used or developed in their projects. This common work has resulted in guidelines especially for system safety, man-machine interaction and traffic safety systems. This allowed a presentation of the state of the art in the DRIVE I programme in this field and to found the new DRIVE II programme on a solid basis in this area. This important common effort resulted from European research collaboration, reflecting an interesting dynamic emerging from the DRIVE programme, with the objective to improve safety on European roads. The resulting knowledge and expertise has now been carried forward into the new programme by a special Kernel project which addresses this topic in order to accompany specific safety-relevant pilot projects.

References

CEC, 1990, *Towards Trans-European Networks, Progress Report*, Brussels: Publications of the Commission of the European Communities.

Ceccini, 1988, *Recherche sur le Coût de la Non-Europe*, Luxembourg: Publications de l'Office des Publications des Communautés Europèennes.

Clutterbuck, D. *et al.*, 1991, Drive safely, in, *CEC Advanced Telematics in Road Transport, Proceedings of the DRIVE Conference*, Amsterdam: Elsevier.

Fadier, E., 1990, Fiabilité humaine: Méthodes d'analyse et domaine d'application, in, Leplat, J. and de Terssac, G. (Eds) *Les Facteurs Humaines de la Fiabilité*, Marseille: Editions Octares.

Fincham, B., Fowkes, M. and Willson, P., 1991, DRACO – A driving accident co-ordinating observer, in, *CEC Advanced Telematics in Road Transport, Proceedings of the DRIVE Conference*, Amsterdam: Elsevier.

Gerondeau, C. *et al.*, 1991, *Report of the High Level Expert Group for an European Policy for Road Safety*, Brussels: Publications of the Commission of the European Communities.

Leplat, J. and de Terssac, G., 1990, *Les Facteurs Humaines de la Fiabilité*, Marseille: Editions Octares.

Michon, J. A., 1985, A critical view of driver behaviour models, in, Evans L. and Schwing R. S. (Eds), *Human Behaviour and Traffic Safety*, New York: Plenum Press.

Michon, J.A. and McLauglin, H., 1991, The intelligence of GIDS, in, *CEC Advanced Telematics in Road Transport, Proceedings of the DRIVE Conference*, Amsterdam: Elsevier.

Müller, P., 1990, Verkehrs(un)sicherheit – eine Collage zu Anlass und Zielen der Veranstaltung, in, *Verkehrs(un)sicherheit*, Dortmund: ILS Taschenbücher.

Parkes, A., 1991, Data capture techniques for RTI usability evaluation, in, *CEC Advanced Telematics in Road Transport, Proceedings of the DRIVE Conference*, Amsterdam: Elsevier.

Parkes, A. and Ross, T., 1991, The need for performance based standards in future vehicle man machine interfaces, in, *CEC Advanced Telematics in Road Transport, Proceedings of the DRIVE Conference*, Amsterdam: Elsevier.

Rasmussen, J., 1987, Reasons, causes and human error, in, Rasmussen, J., Duncan, K. and Laplat, J. (Eds) *New Technology and Human Error*, Chichester: John Wiley.

Rothengatter, J. A. and Harper, J., 1991, The scope and design of automatic policing information systems with limited artificial intelligence, in, *CEC Advanced Telematics in Road Transport, Proceedings of the DRIVE Conference*, Amsterdam: Elsevier.

Thomas, D. B., 1991, The feasibility of monitoring driver status, in Queinnec, Y. and Daniellou, F. (Eds) *Designing for Everyone, Proceedings of the 11th Congress of the International Ergonomics Association*, London: Taylor & Francis.

Timms, P., and Carvalho S., 1991, Inclusion of pedestrians and cyclists in network planning models, in, *CEC Advanced Telematics in Road Transport, Proceedings of the DRIVE Conference*, Amsterdam: Elsevier.

Warnes, A. M., Frazer, D. and Rothengatter, T., 1991, 'Elderly drivers' reactions to new vehicle information devices, in, *CEC Advanced Telematics in Road Transport, Proceedings of the DRIVE Conference*, Amsterdam: Elsevier.

3

The contribution of PROMETHEUS to European traffic safety

M. Emberger

Introduction

Mobility, a characteristic of our society and a vital need for our economic system, has become a symbol for our quality of life. Transport volume, and especially road transport, keeps growing with the increasing individual freedom and expanding economic activities, and we have reached a situation where limitations of mobility become critical. The automobile industry in Europe takes on the challenge to contribute to socially acceptable solutions for mobility. With the research programme PROMETHEUS, founded in 1986, the automobile industry complements the traditional research on vehicles with research on transportation and traffic. The primary concern of PROMETHEUS is safer, more efficient and cleaner road transport. Technologies for communication and information exchange in the vehicle and its environment will improve the cooperation between driver and vehicle, between individual road users, and between road users and traffic management.

The situation

The post-war economic growth in Western Europe, made possible by an essentially free trade system, has substantially increased the demand for transportation. In fact, the provision of increased transport capacity proved to be a major factor for economic growth, with—for a number of reasons—road transport taking the major share of the increase. This higher transport capacity has been achieved mainly by a massive increase in the number of vehicles, while the expansion of the road network and its related infrastructure has not kept pace. The consequences of this situation are felt by everybody: road congestion, accidents, pollution of the environment, noise, stress, unreliability of projected transport and travel schedules, and other discomfort.

And road transport is bound to grow further: A 40 per cent increase of vehicle mileage is predicted for 2010. This development reflects the inherent advantages of the automobile—autonomy, flexibility, coverage and reliability —and is further fuelled by the implementation of the single market in Western Europe and the changes in Eastern Europe. The projected extension and improvement of the road network will have only marginal effects on transport capacity. Thus the gap between the demand for road transport of people and goods and its predictable and dependable execution—its service level—will widen further. Rapidly increasing congestion will cause higher transport costs as well as mounting environmental problems.

A patchwork of isolated measures to improve the road network, vehicle technology or traffic control cannot provide a lasting solution for our traffic problems. Instead, a systematic approach looking beyond road and automobile is needed:

- How to balance supply and demand within the entire transport system, i.e. how can the different modes of transportation be merged to form the 'proverbial' integrated transport system?
- How can we optimize each individual mode of transportation as well as the 'modal split'?

The new approach

When launching the pre-competitive research programme PROMETHEUS in 1986—an acronym for 'Programme for European Transport and Traffic with Highest Efficiency and Unprecedented Safety'—Europe's automotive industry confirmed its traditional role as a leader in the drive for safer, more efficient and cleaner road transport.

Road transport continues to be automotive industry's prime concern, but PROMETHEUS breaks with the traditional patterns of thinking. Its aim is the improvement of the European transport infrastructure as an entity. Europe is rediscovering that an important key to a healthy economy is well-functioning transportation for goods and people, and learns to accept that the increasing interdependence of national economies is an irreversible and in the long run positive development. It is now widely realized that transport problems are not limited to road traffic and neither do they end at national borders. This, of course, is not a new fact, but it has been increasingly overlooked in the years of destructive competition between the different transport systems. Transport policy as an honest effort to obtain the maximum benefits for society was non-existent on a national, let alone on the European level. Admittedly, this is not an easy task, and the troublesome political times our continent experienced did not help either. However, there never and nowhere seems to have been a consistent strategic transport policy. PROMETHEUS was born into a progressive, fast changing Europe, although the sweeping changes in the

East and the subsequent need of the former West to re-orient itself could not reasonably be foreseen. The situation in the different parts of Europe is quite diverse, yet it all boils down to one common problem: transport demand and transport infrastructure are not in harmony. Regional or national activities may provide short-term relief for hot spots, but are no solution medium and long term.

As a recognized pioneer of truly universal thinking in its field, how does PROMETHEUS try to achieve its objectives? To improve transportation in Europe PROMETHEUS concentrates on the issues of travel and transport management, harmonized traffic flow and safe driving. PROMETHEUS predicts that the most modern information technology will allow the introduction of real-time data exchange and real-time control as new functions in road traffic, and to integrate vehicle and traffic control. On the basis of a 40 per cent increase of miles driven by 2010, PROMETHEUS systems shall improve road transport safety by 30 per cent and its efficiency by 20 per cent.

What will be the tangible results of PROMETHEUS? In concert with other European RTI research projects and backed by the promised intensified political support, PROMETHEUS will specify the interfaces between vehicle and infrastructure, vehicle and control equipment, and components of the vehicle control system and demonstrate the benefits to the users.

This goal requires an appropriate transport policy framework and complementary actions of authorities, for example those who collect, maintain and distribute traffic data. Future joint research of automotive industry, the communications, electronics and supplier industry, academic research institutions and transport and traffic experts will continue, and concentrate on those issues and applications that improve safety and reliability of information technology in traffic. Industry provides two-thirds of the annual PROMETHEUS budget of 90 million ECU, the remainder coming from the national ministries of research and technology or industry.

PROMETHEUS has found partners

In the autumn of 1991 PROMETHEUS, at the Fiat La Mandria test track near Torino, successfully demonstrated what had been achieved in three years of common research work. At this occasion PROMETHEUS presented to authorities and politicians an explicit offer and, at the same time, a request for active cooperation. Follow-up activities included a presentation to the European Conference of the Ministers of Transport in Paris in November 1991. Mr Raymond Levy, then chairman of Renault and president of the European automotive industry association, made statements on products that can be developed on the basis of PROMETHEUS results and implemented short term. Mr Levy detailed specific requirements for action from politicians and administration. While this list is too specific to be presented here, let me quote from a more general statement of Mr Levy: 'It [PROMETHEUS] offers technical components as tools for improving traffic by cooperation between travellers, transportation means and traffic managers. There is a need to foster close cooperation between national, regional and local government departments, operators and industry to develop a European strategy for the introduction and implementation of Road Transport Informatics. Indeed the future success of these programmes and perhaps the future mobility of European Transport depends on obtaining European-wide agreements not only on technical standards but on an effective consensus-based coordination and planning of the implementation of these systems.' The conference agreed on the foundation of ERTICO, the European Road Transport Telematics Implementation Coordination Organization, an important step towards practical application of PROMETHEUS and related activities, like DRIVE and its spin-offs.

With PROMETHEUS the automotive industry contributes to the development of a transportation system where information on the availability of transportation shall manage demand, and where autonomy, flexibility and coverage of road transport shall be preserved. First in this field with a comprehensive new approach, PROMETHEUS had aboard the automotive experts with their associated research groups on traffic engineering, telecommunication, artificial intelligence and micro-electronics. PROMETHEUS was actively involved in shaping the DRIVE programme initiated by the European Community. DRIVE and PROMETHEUS are complementary programmes for the definition and development of the vehicular and infrastructure systems of Road Transport Informatics, and a fruitful work sharing could be established. With ERTICO this development has reached a further step: The political powers and the providers and operators of transportation infrastructure joined and the work now can continue with a new quality. Starting out by tackling the immense self-imposed task all on its own, PROMETHEUS has succeeded in incorporating all relevant parties in its efforts, and to grow from an imaginative scheme on paper to a realistic operation.

Common European demonstrations

The activities of PROMETHEUS can be grouped according to the three tasks of the user/driver. Each of these tasks is organized in three subtasks, represented by the common European demonstrations.

1. Travel and transport management

Selection of the means of transportation, of the route and of the travel time.

Subtasks:
● Travel information services
  Informs the user on the ability of the whole transportation system— including all modes of conveyance—to meet their needs.

- Commercial fleet management
  A specialization of the PROMETHEUS approach for the transportation of freight.
- Dual mode route guidance
  Combines on-board systems and infrastructure support for optimized performance and early introduction.

Today's transportation systems did not develop independently of alternative modes, yet it is becoming increasingly visible that there is insufficient communication and team work. Customers as well as operators lack sufficient information on alternative transportation. This situation causes inefficient use of transport capacity, and higher costs due to waste of time and energy.

Travel information services

This service is aimed at the traveller and will offer a convenient choice of transportation means, routes, departure times, etc. Knowledge on transportation alternatives, actual traffic situation, available parking, etc. will support a well-balanced use of all modes for long-distance and urban travel, e.g. the desired diversion to public transport. Studies predict that fully developed travel information services may reduce motorized individual transport in cities by 15 to 20 per cent. These high potential benefits make it an attractive task to ensure availability and harmonization of the data bases of all the authorities and operators involved, e.g. traffic data, public transport schedules, etc.

Commercial fleet management

A pan-European data communication network for freight and fleet management to continuously monitor and control the flow of goods and vehicle utilization, is expected to improve goods transport efficiency by 30 per cent. On-line dispatching shall match transport demand and actual resources to avoid inefficiencies and waste. Actual and projected traffic data shall improve tour planning. Response to emergencies, vehicle maintenance and processing of transport documents are improved. The large potential benefits for the fleet operator, as well as for the society, justify the substantial effort to install the necessary prerequisites for a fleet management system. This requires common European standards for data transfer and traffic data bases as well as standards for the necessary communication links.

Route guidance

Should the traveller decide to use personal transportation, for at least part of the journey, a route guidance system can help to reach the destination by determining optimal routes in real-time, taking into consideration actual traffic conditions. Different information channels are under development for this purpose (e.g. RDS, GSM, beacons . . .). Naturally, route guidance is also an important aspect for commercial traffic. To preserve the flexibility and coverage of autonomous guidance systems, and to offer the driver in addition route recommendations reflecting the actual traffic situation, PROMETHEUS is developing a dual mode architecture with an autonomous and an infrastructure guided operation mode. This system is expected to improve efficiency by 20 per cent. The automotive and the electronics and communication industries are developing the necessary standards for route guidance equipment. For the required collection, processing and distribution of traffic data and the installation of the necessary infrastructure, industry has to rely on the cooperation of the authorities.

2. Harmonization of traffic flow

Cooperation with other road users and the traffic management centres.

Subtasks:
- Cooperative driving
  A general approach using information from the infrastructure or other vehicles, which may inform, warn or interact.
- Autonomous intelligent cruise control
  As a first step towards cooperative driving, this assistance system maintains adequate distance between vehicles in the same lane by speed control.
- Emergency systems
  Emergency call and rescue services, localization in the GSM system.

The driver's behaviour, his choice of speed and headway, and his manoeuvres, have a large effect on overall traffic efficiency and safety. As traffic density increases the sensitivity to disturbances grows. Studies of multiple vehicle accidents indicate that distance keeping to the vehicle ahead, overtaking and crossing are driver performance problems which call for improvement of his or her perception, planning and reaction capabilities.

Cooperation between vehicles, based on local mobile communication networks providing short range communication between vehicles and bi-directional connections to fixed roadside stations should significantly harmonize traffic flow and improve safety and efficiency of road traffic. Strategies to solve traffic problems locally by real-time advice or support, are a powerful complement to the global and macroscopic measures applied by traffic management. Cooperative driving is thus integrating vehicle and traffic control for the benefit of the individual drivers and the overall traffic. Cooperative driving addresses all objectives of PROMETHEUS. The most significant impact will be on safety: for 2010 a reduction of accidents with injuries of 9 per cent is forecast, if all vehicles and the whole infrastructure are equipped and the cooperative driving equipment is in the advisory mode. In the automatic intervention mode benefits would increase to 20 per cent, on motorways to 35 per cent.

The basic applications are:
- Intelligent manoeuvring and control
  - safeguarding and optimizing lane changing and overtaking by coordinated manoeuvres.
  - recommending proper behaviour to the driver or the vehicle control system.

- Intelligent cruise control
  - harmonizing speed and distance between vehicles in uni-directional traffic, first applications will be restricted to highways.

- Intelligent intersection control
  - supports interdependent longitudinal control of vehicles at intersections.

- Medium range pre-information
  - informs driver or vehicle control system of relevant static and dynamic data on safety and traffic related occurrences beyond the range of vision.
  - emergency warning via short-range communication to roadside beacons.

- Emergency call
  - emergency call via vehicle-roadside communication and GSM/mobile phone.

First applications of cooperative driving features depend on the existence of a short-range, bi-directional communication system between vehicles and beacons, with the potential of being upgraded by progressively adding higher level functionalities such as communication capabilities between vehicles. PROMETHEUS cannot expect to create a purpose-built communication system from scratch. As short-range communication is being introduced for some non-PROMETHEUS services, like automatic debiting, route guidance and traffic information, PROMETHEUS cooperates in the standardization work initiated for these other purposes to get a tool for practical low-cost implementation of some cooperative driving applications. When these applications have proved themselves in real life, PROMETHEUS will be in a position to motivate for the introduction of a more specialized short-range communication system. In addition, basic systems of some applications will need mandatory minimum equipment, e.g. low-price transponders for safe distance keeping.

3. Safe driving

Keeping control over the vehicle.

Subtasks:
- Vision enhancement
  Seeks to reduce accidents in low-visibility conditions with, e.g., video cameras or UV headlights.

- Proper vehicle operation
  Driving with adequate safety margin by improving driver actions and vehicle reactions by on-board information and warning systems, advanced vehicle control systems and support functions.
- Driving with reduced collision risk
  Investigates driving strategies for collision avoidance by identifying developing critical situations.

Safe driving means optimal control of the vehicle with respect to its environment and its driver's capabilities. Late perception of obstacles, misjudgement of safety margins in driving, improper behaviour or reaction in unexpected traffic situations—the cause of many accidents—are normally classified as human failures, like lack of awareness or vigilance. Modern sensors and data processors together with communication services now offer opportunities for driver information and support in critical situations. New technologies can provide important information on the environment:

- the state of the road (obstacles, surface, visibility, geometry) can be determined by on-board sensors or transmitted into the vehicle by communication links.
- local traffic parameters like the distance to the preceding vehicle, or the own speed or position can be measured autonomously, more global parameters can be obtained from traffic control centres via communication links.
- monitoring the state of the vehicle indicates the limits of safe driving and manoeuvring and holds a potential for driver support beyond alert or advice.

To support the driver effectively in decisions or actions, his or her needs and capabilities for interaction with the function of the vehicle must be prime parameters for the design of the future vehicle control system and its (wo)man/machine interaction. Indicators and control elements form the interface between driver and vehicle. They must be designed to preclude unsafe use and have to facilitate:

- information either in the form of advice for proper action or by alerting for safety relevant events;
- assistance in decisions or support in actions to reduce the mental and physical load;
- automatic intervention in emergency situations or driver mistakes.

The third mode should be restricted to very critical situations, to avoid an accident or to reduce its severity. PROMETHEUS does not advocate an auto-pilot to replace the driver. The driver is considered to be a responsible person with natural limits of control capabilities. Delegating uncritical tasks to technical devices enables him or her to pay more attention to the strategic aspects of driving and to always stay in the control loop.

Automotive industry devotes much effort to increase traffic safety by improving progressively passenger protection, convenience and comfort in vehicles. Administrations improve the infrastructure by complementary measures. Both efforts and increased awareness and experience of drivers have led to a decrease of the accident rate, the number of accidents per vehicle-kilometre. Nevertheless, a significant reduction of the absolute numbers is a strong public goal. Safe driving is the basic prerequisite for road transport and has direct consequences for traffic flow harmonization, and more indirect consequences for travel and transport management. The present situation in traffic safety can be characterized by a few figures: In Western Europe in 1989 there were 1.3 million road accidents with injuries, and 56 700 people were killed. The main benefit for PROMETHEUS safe driving vehicle control systems is in increased active safety. Better information and driver support are primarily meant to reduce accidents and dangerous situations. Improved convenience and comfort contribute indirectly to the same goal by reducing the stress for the driver.

The potential benefits of safe driving systems could be eroded or even turned into the contrary by risk compensation through the driver, that is by using the new technology to reduce the safety margins. To counteract their misuse, PROMETHEUS will embed the new devices into a coherent system where their functions are related to specified control strategies and cannot be used for other than the intended purposes.

Implementation of PROMETHEUS results

What does PROMETHEUS mean when it marks something as 'available for implementation'? This label signifies that this system or component does not require further joint research work. Based on the results achieved in PROMETHEUS the involved industrial companies can decide what products and services should be developed and marketed.

There are two categories of system developed by PROMETHEUS: autonomous and infrastructure-based. For short-term implementation these are mainly systems not requiring a dedicated infrastructure, potential acceptance and the availability also play a role. Some of these are autonomous systems, others piggy-back on certain 'foreign' systems already existing or being currently installed, e.g. for automatic debiting or for collecting traffic data. The design of these basic systems allows for future extension to a future fully integrated system.

Systems considered to be ready for short-term—before 1993—implementation are:

Basic systems for
- autonomous distance warning
- fleet management

- dual-mode route guidance
- travel information services

Full systems for
- UV headlights
- emergency call

Systems considered to be ready for medium-term—1993 to 1997— implementation are:

Basic systems for
- visual range monitoring
- friction monitoring/safety margin monitoring
- medium range preinformation

Full systems for
- autonomous intelligent cruise control
- fleet management
- dual-mode route guidance
- travel information services

Open system architecture

Many components and subsystems can be envisaged and developed by industry to solve or alleviate problems of safe driving. Simply adding them to the dashboard will not give the sum of their benefits, especially if they interfere with the driving task by distracting the driver's attention. Effectiveness requires a coherent structure of the whole control system. This applies to the software, i.e. the decision and control strategies implemented as computer programmes, as well as the hardware, i.e. the electric network for data exchange within the vehicle and with its environment. In accordance with the current trend in the computer industry, PROMETHEUS supports the concept of the Open System Architecture (OSA) for vehicle control as well as for traffic control. Standards will be based on the present feasibility studies.

Conclusion

How will the driver experience PROMETHEUS? Will it support him, or will it patronize him, will it take the fun out of driving? PROMETHEUS wants to support the driver. It will give him additional information and keep a watchful eye for him on the surroundings to let him know should he or she overlook something. It will not take something away from the driver, neither fun nor responsibility. He or she is still in full control of the vehicle and still is fully responsible. Cooperation is an offer, taking it up is a question of good will. PROMETHEUS will make driving safer and more enjoyable. On a higher level PROMETHEUS wants to pave the way to progress from control of the vehicle to the control of travel and transport, to allow mobile individuals move in a mobile society. Actually, you can already start preparing for PROMETHEUS. You do not have to wait for the computers to do all the thinking for you. Think PROMETHEUS now! Try to apply the PROMETHEUS spirit: for example even on your daily drive to work you can practise cooperative driving. You may be surprised how well it works.

4

MMI scenarios for the future road service informatics

O. Svidén

Introduction

This scenario is based upon ideas and preliminary R&D results both from the DRIVE programme and a number of EUREKA transport programmes like PROMETHEUS, CARMINAT and EUROPOLIS. The idea to sell new Road Service Informatics (RSI) as services rather than as products is coming from the telecommunication side (e.g. Minitel in France). The scenario synthesis is a personal construct. It is not necessarily a forecast of what will happen. The purpose is to show that a number of interesting possibilities to improve road traffic will be available in a few years time, and the scenario provides the reader with a vision of how these possibilities may be used.

New technology will not end up as an innovation that reaches the market unless it is socially accepted, and the driver acceptance is, to a dominant part, dependent on the man-machine interaction. If it is pleasing to use the new informatics services, they will be used. This is the question with this scenario: 'Will the future road transport informatics be sufficiently inviting to use?'

The following implementation and usage scenarios refer to a tentative future year 2000. First a technical scenario defines some of the road transport informatics systems and functions available on the market by that time. The crucial questions about man-machine interaction are described in the next section. Then follows a usage scenario that presents the dynamic interplay between the driver and the informatics services. At the end there is an outlook for the year 2010, and a short scenario describing the effects on traffic when mature road service informatics have reached a stage of market penetration.

Implementation of RSI technology—a technical scenario for the year 2000

The 1990s are marked by the early implementation of a number of carrier technologies enabling road vehicles to communicate data between themselves and the roadside. The first consumers of this new technology were the commercial and public transport fleets. The pioneer consumers of RSI were found in organizations for trucking, dispatching, fleet management of buses, taxis, limousines, demand controlled handicap vehicles etc. They were the competent buyers of technology before the market spread to the private cars.

Let us first see how far the RSI evolution has come by the turn of the century. Cars have now received an intelligent RSI visor, in front and above the driver. This RSI unit includes chips for route guidance, parking guidance and automatic debiting. It has a smart card reader for the personal TRAC (TRansport Access Card). It contains a transmitter and receiver for communication with roadside beacons. The standard RSI unit acts as an intelligent replacement of the dumb sun visor. The RSI unit has display optics at its edge and can be adjusted in position and brightness to fit the individual drivers. The driver can see one row of symbols and a line of text above the traffic scene in front of him. He normally 'experiences' the symbols with his peripheral, rather than his direct, vision. The principal and standardized information on the RSI unit is combined with more detailed head-down information on the integrated dashboard displays. He scrolls through presented information by using a joy-stick for his thumb on the steering wheel and enters a command with the other key on the wheel. When stationary, he can flip the RSI unit down to get access to a complete keyboard and screen for general inputs, programming of standard destination addresses etc. The RSI unit mentioned can be installed in a car within an hour at a cost of 100 ECU. (Many buy two units! The desktop unit is used for trip planning.) Purchase and installation can be done at any local RSI shop, the proper use of all RSI functions are taught by using the desktop PC driving simulators. The local RSI service shop also gives help, advice and driving instructions if needed. Teenagers use RSI simulation as a game before learning how to drive the car itself. The follow-up tutoring is then included as a function in the RSI unit.

Parking and route guidance are two function to assist car drivers in selecting the best parking options and thereafter efficiently guiding them to their choice. The parking fee is debited automatically from the personal TRAC smart card. The use of an anonymous prepaid card is an option to the driver. These smart cards can be 'filled up' with more money at any service station, or at the automatic cash dispenser directly from the owner's bank account. These smart cards open the doors to a range of road services available on a charge basis, beside parking and route guidance. A digital map is organized, updated and authorized. It is distributed in many forms via the computer network for trip planning, via beacons at roadside and by CD-discs for the in-vehicle use. The map includes not only the geometry of the road network, but all the traffic rules and regulations associated with it. Services, such as fuel, parking facilities, service stations, shops, hotels, restaurants are included in this digital mapping. It defines the entire service environment of our roads. Behind all these infrastructure offers stands a number of private and public service providers. En route, the driver enjoys full privacy. It is the infrastructure service that identifies itself to the driver! This is the crucial point. No identification of the driver or the vehicle is required. The only electronic check from the infrastructure side is to confirm that there is sufficient money available on the anonymous TRAC card. A will-to-pay is registered via the smart cards, but not the one providing the money for it or the user of it. On a business trip it is the company card that is used, keeping track of expenses and organizing data for the reimbursement.

The above measures for protecting the privacy and providing appropriate information on how to select and find 'road services' in a broad sense, makes the automatic debiting of service charges natural. When coming to a reserved parking place, the RSI unit informs the driver of the parking conditions. In fact the RSI unit now acts as an automatic parking meter in the vehicle. Parking can be seen as the starting point for the rational automatic payment of all RSI user charges.

Telecom networks and service development has been progressive for decades using the 'charge for use' principle. Automobiles and their value added features are traditionally sold as products. The RSI evolution has something of both—but no traditions. This was seen as an opportunity for the RSI industry for financing the early implementation.

One important RSI function is the built in park-and-ride options service. When entering a major city area, the driver can receive information on his RSI unit regarding a number of parking options. The computer estimates the time of arrival to the selected destination, whether it is reached by car or combined with a metro and/or bus journey (the last link by foot from the end of the car/public transport journey is included in the calculated time of arrival). Via the beacons the vehicle computer learns about the appropriate public transport timetables and displays the edited result of its analysis to the driver, together with the difference in costs for the alternatives routes. In the year 2000 these parking management and park-and-ride functions have proved themselves to be very effective for controlling traffic volumes in many cities. Demand management starts with better parking facilities. Cities with congestion problems invest in new ring road and parking facilities on top of the metro station close to the ring. The result is that many car users now prefer to park in the city periphery and take the lift down to the platform, rather than frustrate themselves in traditional city traffic. Those that need the van or car in the city, can experience driving in decreased traffic—at a price. This is the key to the mobility problem: by creating a true market, with efficient user charges system and the improvement of alternative transport modes, drivers are given two options. Some prefer to have a price reduction, others are willing to pay for improved city mobility!

The commercial vehicle fleets and public transport vehicles have been scrapping most of their outdated and redundant radio communication equipment in favour for the RSI units plus the mobile digital telephones. They also install and use extra RSI chips and software as add-ons. The functions tested are, dispatch planning, vehicle location, load and fleet management.

Commercial vehicles and executive automobiles used as mobile offices are also installing the chips and active controls needed for intelligent cruise control (with dynamic speed and distance keeping features). A lane keeping function is also tested. The later two functions are tested by some professional drivers in the cruise lane on certain motorways by the year 2000. An example of this 'spin off' from professional truck driving is the intelligent cruise control function with a dynamic speed and distance keeping function involving the vehicle in front. It is first of all seen as a convenience to driving at slow speeds in dense traffic. In the long run, with all vehicles equipped, ICC can act as a stabilizer to the traffic flow. The problem with shock waves propagating backwards from a disturbance in a row of vehicles, can be reduced with the ICC function. By now, additional roadside transponders at junctions also recommend appropriate speeds for entering the junction, depending on the traffic intensity. The ICC is now seen as an important first step to future safety functions.

In summary then, traffic has become slightly more efficient with the help of a number of road service informatics services by the year 2000. Vehicles are moving more smoothly than before, and by applying demand management and parking policies, that restrict city driving by pricing mechanisms, the city traffic has become less noisy and polluting. The smoothing of traffic flow and reduction of stress has decreased the number of rear end collisions and thus provided an improvement to traffic safety.

Man-machine interaction

The scenario above describes a number of RSI systems and functions and the expected impacts from the use of these services. There are potential benefits to traffic efficiency safety and the environment. But this favourable outcome is strongly dependent on social acceptance, market penetration, user behaviour and the change of driving behaviour with the help of RSI. This does not come automatically. The handling qualities of the equipment, the visual quality of the displays and the design of the RSI services must be made so attractive as to have a direct appeal to the users. The integration of the different functions must be made in such a manner that they do not disturb each other. The RSI should be supportive to the dynamics of the driving task. This means that information should appear as visual symbols, text, audible signals, or tactile feedback to hands and feet, in the very fraction of a second when they are needed. Information must be given priority, so that a 'commercial message' is downgraded or even silenced by an urgent warning or safety message from the traffic control centre. The RSI services should be inviting to use. The safety related information should be ever present, the different RSI services available at the tip of the finger, and be so discreet as not to disturb discussions in the vehicle. The on-board computer should edit the information, such as warnings, guidance, advice, traffic rules, vehicle status, in such a way that information comes naturally. No instruction manuals should be needed. The tutoring is on-line and self explanatory. To arrive at this ideal situation a number of MMI principles have to be adopted:

1. Support the driving task and give appropriate dynamic information in the two control channels available to the driver:

   - Speed: accelerate, keep speed, decelerate, stop.
   - Steering: keep course, change lane left or right, turn at intersection.

2. Edit the general information for decision making whilst driving, so that the RSI does not overload the driver with information. This means that information has to be screened and/or delayed according to the following tentative priority list:

   - Safety: warnings, advice, system status, tutoring.
   - Traffic: guidance, rules and information, tutoring.
   - Navigation: calculated arrival time at destination, street names and numbers.
   - Service Options: park-and-ride options, user charges, 'yellow pages' information and tourist information, names, addresses and services available.
   - Communication: business messages, private telephony, booking of services and commercial advertising.

There are many technical solutions on the above MMI specification. The problem of synthesizing and presenting the information is done by R&D groups in the PROMETHEUS, CARMINAT as well as in DRIVE programmes. The integration of different pieces of information into a general purpose RSI display unit is one of the most demanding problems in order to get user acceptance. By the year 2000 we will certainly have a number of alternatives available of first, second and third generation RSI units, and all the time it is the dynamic man-machine interaction that acts as the driving force to the integration and design of the RSI.

The development of a user friendly RSI unit can be related to the development of integrated displays and head-up displays for aircraft. During the last 30 years there has been a move away from the knobs and dials epoch. Modern aircraft display edited information on data screens, and by simple pointing at the screen different functions can be selected, as well as a number of options. In military aircraft the most dynamic and safety related information for the pilot is presented head-up, as an image seen superimposed on the environment seen at distance outside his canopy. By careful editing and presentation of a few symbols, such as velocity vector, a horizon line and graphic information on speed, altitude and heading, the pilot is helped to project the consequences of his split second manoeuvres, and get guidance for navigation at the same time.

A spin off from the aircraft head-up display technology can be useful for driving a car in traffic. If some relevant symbols and perhaps a line of text, at selected instances en route, are shown to the driver, and presented as an image superimposed to the traffic scene, it can be perceived by the user as a 'customized information on a gantry' at a constant distance in front of the vehicle. The information is there if and when the driver wants it, and it is edited and customized to the trip planned. The dynamic speed and steering guidance is related to the vehicle speed, lane chosen and position in traffic. Figure 4.1 gives you a hint of the principles.

The evolution of a good MMI design has to be compared to the evolution over the last decade of the personal computer. Most of the hardware has been almost the same, in principle, whilst the software has improved the user friendliness for word processing, diagram design and illustrations by new generation programs every second year or so. The vehicle RSI is a PC applied to the driving conditions. This means that to operate the equipment, select functions and services, and decide among options. For the desk top PC you use a 'mouse' to point with. In the vehicle application this has to be designed as a joy-stick or direction keys and an enter key to be manipulated by the thumbs or fingers when holding the steering wheel. Or it can be commanded by spoken words if a voice recognition system is part of the RSI unit. The smart card and its reader in the vehicle is assumed to act as a key to the car doors, engine lock, and act as a 'door opener' to all RSI services available through the system. It is assumed to be as small as a credit card, 3 mm thick and have a little transmitter in it, similar to the hand control for the home TV. It can have a little display window with LCD screen for text. With the above technical descriptions in mind we can now illustrate the use of these RSI equipment in an RTI usage scenario. Imagine yourself in the situation to make ...

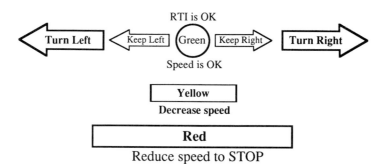

Figure 4.1 An example of integrated guidance. A minimum set of symbols are presented to the driver in a head-up display. The driver has two channels of vehicle control: 1. lateral control by steering and, 2. longitudinal control of speed by gas and brake pedals. Guidance information must be well edited and tuned to the dynamics of these two channels.

A trip from Cambridge to London in the year 2000

You have an important business meeting on Friday at 10.00 at an unfamiliar address in Docklands. The evening before was spent among friends in a pub. On the way home you make a trip plan for the next day's journey. Pulling the business card from the person you will meet through the vehicle card reader, the destination address is automatically entered: 'Murdoch Tower, floor 11, room 16'. The RSI computer suggests a departure at 8.15. This time information is automatically stored in your TRAC card.

At 7.30 in the morning the TRAC beeper alerts you. On the display you read 'Depart to London 8.15'. At 8.10 you get a new beep reminder. Some minutes later you walk up to the car on the driveway, direct the edge of the card towards the sensor at the upper end of the windshield and hear the door locks click open. Your card was identified as one of the four family cards authorized to access the vehicle interior. You put the card in the interior card reader, and as your TRAC is one of the three cards giving access to driving the vehicle, the engine starts and powers the systems and displays. A green symbol in the RSI visor, in the upper part of the field of your vision, indicates that RSI functions are checked and everything is in GO condition. You drive through the suburbs to the nearest artery out of Cambridge. Your speed is guided and you see the recommendations in the RSI. The green round symbol 'o' indicates the GO condition and that the speed is within the recommended speed range. A yellow minus ' – ' sign is a recommendation to slow down, a long red bar '—' recommends you to brake for a full stop ahead.

Beside the supporting symbols for your control of vehicle speed there are arrow symbols beside it for the navigation. A small arrow close to the speed symbols alerts you mentally to the fact that a turn to the left or right has to be made in a few seconds, and when the outer larger arrow also lights up it is time to do the manoeuvre. If it starts to blink it is high time, and a risk that you will miss the opportunity to turn. These arrow symbols also carry the colours of the speed symbol, so as to strengthen the speed recommendations and enable them to be seen when turning the head scanning the traffic before a turn manoeuvre. Both the speed and steering guidance in the RSI unit are meant to be 'directly experienced' via the peripheral vision, in parallel to your direct vision and attention to the traffic environment. Only when you want more detailed information do you have to look at the head-down displays.

As you are an experienced driver the tutoring system has tuned the RSI symbology to that level of 'strength and prediction' that fits your visual ability, personality and experience. This level of adaption is learned and remembered by your personal TRAC. When performing a routine drive through your own neighbourhood, the RSI and its information is seen only dimly as a confirmation. As the RSI is a learning system, it suppresses the information which it knows you to know by heart. The start of your journey from home is just of this kind. And the RSI is smart enough to know that any 'nagging' in this situation will only irritate the master, and its inherent risk of being turned down flat!

By 8.45 you are out of Cambridge on the motorway to London. You engage the ICC, and the intelligent cruise control keeps your vehicle coupled automatically to the speed limit or to a safe distance to the car in front of yours in file. The RSI gives you a short written message: 'Arrival time 09.50' With the ICC on and the confirmation of arriving to the very place of the meeting with 10 minutes to spare you can relax. You can drive with your 'minds off' as the pilots say. With the confirm key on the steering wheel, the text and numbers disappear. It will appear again only if something happens that changes the arrival time. Whilst in a relaxed drive you can prepare yourself mentally to the presentation at the meeting. The green 'o' is dimly on as a constant light monitoring that the ICC is doing its job within recommended speed range and safe distance. It stays green even when you feel a vehicle braking as a result of a car entering your lane in front of you. If you had been in manual mode, a yellow ' – ' would have recommended you to apply the brake in order to resume a safe distance in file. You remember the situation when the RSI was new to you, then it was also showing the yellow bars to you as an explanation to why the braking was applied. But now in the mature driving stage, when you rely on the ICC performance, you have authorized it to stay green as long as you do not have to interfere.

Some half an hour later the green 'o' starts to flash and a message comes up 'Drowsy? – Alert!'. It is a signal from the driver monitoring system, indirectly measuring jerks and delays in the steering movements, interpreting these as signals of decaying driver vigilance. It is a discreet warning to not overestimate the support services from the system, and a reminder that it is you as a human driver and decision maker that still have the full responsibility for the driving.

Some further minutes later the message 'Delay Arrival 10.15' comes up, and this certainly alerts you. The following information 'Fog and heavy rain N. London' confirms to you that something has to be done. You start to scroll through the option list and in the Park-&-Ride file you get an attractive offer 'Hemel Hempstead, Arrival 9.55'. Look deeper into the option and get particulars in the shorthand manner you have learnt to appreciate: Exit 19, guidance 3 km, parking place 34 booked, train leaves platform 4, 3 minutes after arrival, Liverpool Street, Circle Line, Tower Hill, Maglev to Murdoch Tower, arrival 9.55. With a press on the confirm key you go for the option.

Leaving the smooth motorway flow at exit 19 you are now in relatively unknown territory. The RSI alerts to a corresponding level, giving you ample recommendations to fork left, brake and stop in front of roundabouts, select suitable exits, turn left, right, change lane, brake, stop, wait accelerate in a long chain, until you see parking space no. 34 in front of you. Long ago you realized that map information would not have been of particular help in this situation. You have accepted the system and follow RSI recommendations. With a last look at the RSI display you read Wait 5 minutes, Platform 4, Arrival 9.55. You take away the TRAC from its reader, leave the car, hear it lock itself up after closing the door, and walk determinedly to platform 3. When your TRAC is pulled through the card reader at the platform entrance it beeps to you. Looking at it you are then reminded that it is platform 4 you should use. When passing the correct platform entrance, your card displays 'Wait 3 minutes, train 277, car 7, seat number 26'. On entering the train car you find that seat no 26 is in the non-smoking section, as was read automatically from the TRAC card when the place was booked as the train came into the platform. In similar fashion you are guided to your final destination, via train, underground Maglev and lifts up to the floor 11 room 16 for your meeting. You check the time on your TRAC to be 9.53. Just-in-time for a relaxed moment before entry!

On the way back in the afternoon, after some unplanned shopping, the TRAC advises you on train services, and most important of all, remembers the station where your car is parked! Once seated, the RSI gives a menu of destinations. You enter the first option: 'Home?' by a press on the confirmation key. On the way home in the old reliable, the next trips are planned and it is time for some RSI tutoring. As you are driving by yourself for some 30 minutes more, you can go through the lists of points and stars of your accumulated driving. First the negative list of points: as number 1 ranks my high driving speed entering curves and intersections. The RSI suggests an exercise where you engage the ICC for the driving in the familiar town of Cambridge and in the blocks around your home, just to show how the driving can be made more mature, efficient and pleasant to the environment. Reading the list of stars you find an approval of your response to demand management, by using park and ride options as a help to curb congestions and emissions in the city area. For this you have been immediately rewarded on-line by having lower costs for parking and the trips as a whole, rather than insisting to 'pay your way' through the sensitive London traffic. In a summary assessment the RSI puts you in bonus class B, close to the A level. This message motivates you to accept the suggested home town training. When coming back towards Cambridge, you leave the ICC on and the RSI unit in tutoring mode. You watch the vehicle behaviour, get some explanations in the RSI and note that in a funny way by driving as slowly as the ICC system prescribes you are moving more evenly through traffic, as carried by a green wave. Furthermore, as you move around the bend of Jesus Lane you see the traffic light at Bridge Street shift to green when you approach it. At first you are irritated by the slow speed, you miss the 'hunt to the next red light', but you can feel the decrease in stress level when reminded via the RSI unit. You decide to adapt to the tutoring and rather hunt for the A level, with the economic bonus going with it.

On Sunday morning you are taking your seventeen-year son out for a drive. With his new TRIC, Traffic Instructor Card, he is allowed to drive the car with strict speed and range restrictions until passing his level D.

Outlook for the year 2010

By the turn of the century the basic infrastructure for RSI is implemented over all Europe. Some 20 per cent of vehicles are equipped with equipment for RSI. A large number of local and national service providers are operating the systems and offer the services on a charge basis. The road service informatics is being developed gradually to fit growing user needs. The RSI technology has proved its performance, reliability and potential as an information enhancement between road users and the service oriented traffic control centres. The first years of the new century are marked by an RSI market penetration.

Ten years later the RSI evolution may have come this far: By the year 2010, standardized RSI systems are mandatory within the European Community for all new cars produced. The RSI impacts on traffic are visible. Traffic is dense, due to growing transport demands, but the flow is smooth and efficient. Safety levels are marginally better than a decade ago. A small positive effect on the environment from RSI has been screened by the growth in traffic volume. The cleaner air surrounding traffic today is due to the use of clean fuels and catalyst equipped engines. The operational driving task is increasingly regarded as a boring nuisance and the joy of driving has been lifted from a mechanical level to a tactical level. Road pricing lets the drivers experience directly the cost to society of their driving in congested traffic and within emission sensitive areas. The tutoring function represents a new game which will stimulate the driver and ultimately improve their standard of driving.

In the year 2010 road service informatics technology has proved its system performance and reliability as an information enhancement between road users and the service providers, and service oriented traffic control centres. At this time, drivers have shown to what extent they are willing to change their behaviour and improve their driving skills with the tutoring function. By now, it is time to plan the second step in implementing a number of radically new safety systems. To a limited extent some of these functions have been developed and tested by commercial vehicles. Now it is time to make an implementation strategy for the safe road traffic as a whole.

PART II
ROAD USER NEEDS

5

Road user needs

K. Rumar

Introduction

In this Chapter the needs, problems and limitations of various road users will be analysed and discussed. The reason is, of course, that the needs of the users should be one of the main starting points when we are trying to design RTI systems. For a long time road users have been almost synonymous with drivers. The present road traffic system is to a very large extent constructed for automobile traffic. One of the main effects of this approach is that the accident and injury risk for so called unprotected road users is much higher than for automobile drivers and passengers. So in this section the needs of unprotected road users will also be dealt with. Before we embark on the analysis of how RTI could meet the needs of various road users it could be useful to try to see the problem in a larger perspective. How has this role of the road users been changing during the development of the road transport system? How did we get into the present situation? Also, when discussing future technology, it is useful to have a historical perspective. History often has something to teach us about how to shape our future.

The changing role of the road user

The early species of mankind were continuously moving. They were nomads and moved from place to place. This picture did not change for some million years. Man developed into a creature adapted to living and moving within the natural environment, in day-light at slow speed, and with special sensory, central nervous and muscular properties for such conditions. This evolution, that was first postulated by Darwin, although containing some jumps, is a slow process of adaptation to the environmental conditions that all species have gone through.

The development of the road transport systems has followed much the same scheme as most man-machine systems, such as industrial production, weapon systems, home work, sea transport and many others (Rumar, 1985). In the first stages man was doing the work in direct contact with the environment. The work was mainly muscular, and the feedback on right and wrong predictions, decisions and actions were normally instant and perceived directly from the environment (see Figure 5.1). What has happened during the last hundred years or so is that the advancement of technology has given us equipment that has relieved man from most of the muscular tasks. The machines can carry out heavier tasks much quicker and for a longer time than the human muscles. We can now transport, produce, destroy, etc. large quantities, at long distances and high speed, with low costs, high precision and low technical failure rate. We have technically highly sophisticated vehicles, roads, information and regulation systems. This has, in turn, made us very dependent on transport.

Previously, the contact between man and environment was close and immediate. By putting one foot in front of the other we moved towards our

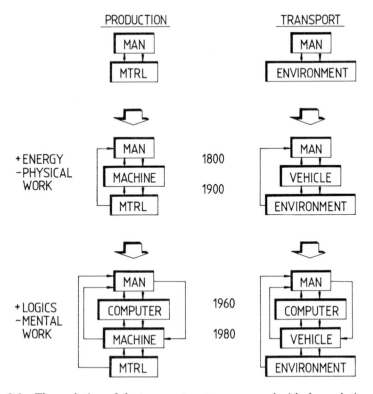

Figure 5.1   The evolution of the transport system compared with the evolution of the production system. The stages from walking/handicraft via external energy source to the introduction of logics can be identified. The human operator role has changed considerably; less feed-back, higher speeds, larger attentional demands.

goals. If we made a misjudgement or a misstep we were usually punished without delay. The signs and signals from nature and other animals were natural, and the decoding was often inherited. Technical evolution has been so quick that the adaption of man to the environment has been overrun. Furthermore, at the same time as man has been relieved of his muscular tasks, the mental demands, such as perception and attention, have increased. Consequently, we have today an out-dated human being with stone-age characteristics who is controlling a strong, fast, heavy machine in an environment packed with unnatural artificial signs and signals. The gaps and differences between various road user categories have become larger. Therefore the interaction between them has become more difficult. In the 'childhood' of road traffic the main problems were to create a reliable, economical, safe and comfortable vehicle that could stand the roads of those days. Man also had problems with building roads that could take the wear from the vehicles, that could stand various climatic conditions, and still remain fairly smooth. We now encounter new problems—to control the technical part of the system. We have gone from a situation with mainly technical problems to a situation in which the main problems are with the human being who is controlling the technical system, especially problems related to information and decisions.

During the last decade the electronic revolution (small powerful, low cost computers) has entered road transport. This happened earlier in the military, air, office, and industrial areas and later in the homes. The purpose of these computers is to relieve man of some of the mental tasks (see Figure 5.1). In traffic they were first used in traffic control systems (e.g. 'intelligent' traffic lights), but are now increasingly used in cars. The next step will be to start communication between the computers along the road and the computers in the car and between cars. This is required to develop information, warning and regulation systems to the same technical level as present cars and roads. This is the idea of Road Transport Informatics (RTI). Society (says that it) wants to use RTI to improve efficiency, safety, economy, environment and convenience of road transport. But which requirements should be specified for RTI from various users' points of view? What does he or she need? What does he/she reject?

Road user goals and limitations

Before we start discussing the needs of the road user we should try to establish the user motives of travelling. The motives and the goals are the basis for the needs. The primary goal for the road user is of course to reach the destination, but he does not accept that goal at any price. He requires a certain time, speed, safety, economy and comfort. These are the secondary goals of the user. In order to be able to analyse user tasks we have to transform these general goals into operational goals. There are many ways to describe this. A combination of Rumar (1986) and Brown (1986) gives the following operational goals (tasks):

1. strategic tasks (choice of transport mode, time of departure, localize targets, order of targets, routes),
2. navigation tasks (to follow the chosen or changed route in traffic),
3. road tasks (to choose position and course on the road),
4. speed control (choice of speed in and before every situation),
5. traffic tasks (to interact with other road users in such a way that mobility is maintained but collisions avoided),
6. rule compliance (following rules, signs, signals),
7. manoeuvring tasks (to handle the vehicle so that 3, 4, 5 and 6 are reached).

These tasks are relevant to drivers in general. The weight of the various tasks changes with driver category. A professional driver of a heavy goods vehicle puts different weights on these tasks than private automobile drivers. Drivers of motorcycles differ from bicyclists. A special category are the pedestrians. They do not have the vehicle manoeuvring task (7), but they have limited problems with the road (3) and the speed (4) tasks.

Information technology is intended to support or replace user behaviour and performance. It is therefore essential that we know more about this crucial link in the transport chain (see Figure 5.1). To begin with it is quite clear that, in some respects, human performance is far superior to technology as represented by sensors, processors, artificial intelligence and actuators. Take for instance the simple human task of steering a car along the road—the most simple of all driving tasks. Even in daylight, this is quite a complicated task to carry out by purely technical means. In adverse conditions, e.g. bad weather, it is close to impossible. Man has a wonderful quality to analyse and structure complicated visual stimuli in a fraction of a second. Naturally such areas are not the ones we should work with when we try to help the human operator in traffic. Man also has several limitations and it is primarily here that support from advanced information technology is wanted. The most obvious, but at the same time most overlooked reason for human errors, is that we are living biological organisms, not mechanical or electronic preprogrammed robots. This means that now and then we do not work or perform adequately—we all make errors now and then. The unreliability of human performance is one of the most reliable human characteristics!

Physiological limitations: Some of these limitations we all have since they are inherited from our ancestors, e.g bad visual performance in night traffic, inadequate speed perception, no respect for high speeds, bad peripheral vision, low capacity for time sharing. They are probably due to immaturity of parts of the organisms. Still other physiological limitations are found in the higher age-groups, e.g. impaired vision and hearing, slower reactions, bad memory. These are normally effects of age deteriorated organisms.

Psychological limitations: Here we find human limitations acquired during our life as well as limitations of a more basic psychological nature. One of the most critical characteristics of present road traffic is that it is not self learning. It does not immediately inform the driver that he is behaving rightly or wrongly. On the contrary, a driver may behave badly all his life and benefit from it, e.g. drive with short headways, make crazy overtakings, overestimate his visibility in night traffic, exceed speed limits, does not slow down for children at the kerbside. Of course such a situation creates bad driving, bad drivers, establishes errors. Due to lack of continuous feedback (or even negative feedback) road users improve very little just by participating in traffic. Inadequate expectations are created and established.

Another psychological limitation is our limited capacity to process complicated and extensive information in a short or limited time. This problem is most pronounced in city centre intersections with signs, signals, cars, cyclists and pedestrians. Here the human brain sometimes becomes overloaded. The way out is to concentrate on a few of these pieces of information. Sometimes this choice is wrong.

Sociological limitations: The car, its body and its speed seriously impairs normal human interaction and communication. This has many adverse effects on road user behaviour. Let us compare human behaviour on roads with human behaviour in shops and side walks. Queueing is quite common in both situations, but behaviour is quite different. Nobody would dream of breaking or pushing himself into a post office queue, but that is almost normal behaviour on streets. When walking, it is normal to allow the other person to enter a door first. When driving, similar behaviour is exceptional.

Sometimes we make mistakes on foot, but it is very simply solved by an apology. In driving almost every little mistake is taken by the other driver as a provocation, and the car has no signal for 'excuse me—that was not my intention', so the hypothesis that is was a provocation is not falsified. Besides this effect of lack of normal communication, the closed in situation of the driver creates an isolation, an anonymity, a feeling of 'everybody on his own' which releases behaviour that is normally blocked by social norms, social contacts. Drivers tend not to take their responsibility seriously: to be reckless, to break the traffic rules and laws are very low on the subjective criminology ranking of most people.

These are examples of three areas where adequate use of modern information technology could improve the situation considerably by compensating for human limitations.

The role and effects of information technology

Information technology has a very large potential to rationalize road transport:

- by advanced and adaptive trip planning and route planning and route finding (shorter driving, quicker delivery),
- by eliminating a lot of paper work (e.g. invoices, customs documents),
- by fleet management (goods and managements),
- by predictive traffic control (decreased congestion),
- by simplified debiting (e.g. taxes, parking),
- by smoother traffic (constant speeds and few stops),
- by distributing vehicles on existing roads and thereby improving capacity on existing network considerably.

Effects like the ones just mentioned will also have an indirect positive influence on the economy and environment (less petrol, less pollution). But what about safety? Again there is no doubt that this new technology can support or even replace the driver in many critical situations. To some extent such equipment already exists (e.g. in the form of anti-lock-brakes). Other examples are:

- intelligent vehicle characteristics (braking, stability, steering, suspension, shock absorbers, etc.),
- electronic vision (detection, recognition) in low visibility conditions (fog, night, rain, snow, etc.),
- headway control to prevent rear end accidents,
- lane change control,
- intersection control,
- anticollision systems (emergency braking),
- speed adaptation (to road and traffic conditions),
- overtaking assistance,
- continuous and individual driver education (immediate positive and negative feed back),
- decrease of driver stress due to adequate information at the right moment,
- checking of driver condition (e.g. fatigue, intoxication),
- interaction with unprotected road users.

Human reactions

To a very large extent it is man himself that will limit the safety effects of the new systems. When we continue our work we have to keep in mind that:

- computerization of road traffic must be acceptable to the public and to the users (compare the reactions to computerization in industry). Studies show that some important factors here are no threat to personal integrity, no threat to driver control of vehicle, filling the needs of the user, system reliability, user friendly, costs of investment and of usage.

- The new equipment must not introduce any new problems such as distraction (compare the mobile telephones of today), overload of driver information channel.

- Any measure introduced will be used by the driver to the best of his interest. That is to say every measure introduced to improve safety will be used by the driver to improve safety only to the extent that he considers himself to have a safety problem. Most drivers experience considerable mobility problems (congestion and repeated delays), but their safety problems are subjectively small.

- When the new advanced information systems have been introduced the driver situation and tasks will change radically. When we try to predict the future effects of these systems we base our predictions on present driver behaviour and mainly look at the new systems as something additive, not interactive. But driver behaviour will most certainly change as an effect of information available and requests possible. But how? We do not know that.

- If it becomes much more simple and effective to drive maybe each car will be used more and longer distances driven. That in itself would impair safety, since we know that the correlation between exposure and accidents is high.

- The life time of cars and drivers is steadily increasing. Consequently many vehicles will, for a long time, not be equipped with the new type of systems. How can safety be improved in traffic where a varying proportion of the vehicles are equipped?

- Half of the victims in road traffic are unprotected road users. If we really want to improve safety we have to include them into the system. But how can we equip pedestrians and cyclists with advanced information technology?

One big safety advantage of the new technology, that probably is not appreciated by everyone, is the ease with which violation of safety rules will be enforced. Another possibility offered by this technology, that may prove very important both for research and for the courts, is the black box which makes it easier to find out what really happened just before the accident.

Conclusions

The conclusion from the discussion is that by means of this technology we will most certainly improve mobility in future road traffic. The effects on economy and environment will also be positive. But what about safety? The effects on safety will not be positive unless we really build the systems around the road user and his needs and not just technically advanced. It is very important that we now analyze and consider the future safety effects from the introduction of information technology in road traffic—tomorrow it will be too late. We may have good motives and intentions with the introduction of this new equipment but the road user will still use any new tool for his own advantage. We must be prepared for this interactive reaction of the road user, and design the system to meet such reactions.

Information technology in road traffic should aim at decreasing present safety problems of the road users—not just a group of car drivers, but most road users. At the same time the technology should be acceptable from the user point of view—being user friendly, not threatening personal integrity and driver personal control over the vehicle. Furthermore, the new technology must not introduce any new safety problems such as distraction and information overload. Possibilities to improve safety are almost innumerable provided we can, in a reliable way, solve problems like the ones just mentioned, e.g. advance warning, speed reduction, electronic vision, collision avoidance. But the most challenging possibility is perhaps to use this technology in order to improve road user behaviour. For the first time we will now have the opportunity to give adequate and individual feedback in road traffic.

References

Brown, I. D., 1986, Functional requirements of driving. Paper presented at the Berzelius symposium Cars and Casualties, Stockholm, March 1986, 19 p.

Rumar, K., 1985, The role of perceptual and cognitive filters in observed behaviour, in, Evans, L. and Schwing, R. C. (Eds), *Human Behaviour and Traffic Safety*, London: Plenum, pp. 151–170.

Rumar, K., 1986, In vehicle information systems. Paper at 3rd IAVD Conference on Vehicle Design and Components, Geneva, March 1986, D33–D42.

6

What should the vehicle know about the driver?

R. Onken

Introduction

Developments in new techniques for knowledge processing have produced functions which can be exploited for process control in two ways:

- autonomous process control and
- autonomous or knowledge-based assistance for the human operator in process control.

At the first glance, looking at the system authority level, the latter application seems to be less demanding. However, looking at the functional complexity, knowledge-based assistance for the human operator implies not only the incorporation and processing of knowledge about the control task and the technical means to be employed but also about the way to achieve the task under all possible constraints and events. Comprehensive knowledge is also needed to be incorporated and processed about the human operator, the human factors and the normative and individual behaviour. The development of these functions is extremely challenging. Nevertheless, experience shows that knowledge-based assistance can also make sense and could be theoretically considered as part of the desirable system knowledge base. Therefore, it should be asked, what knowledge can and will be held ready in foreseeable assistant systems in order to make them a useful device? That is the question which will be dealt with in the following, focusing on the assistant system for the driver of road vehicles, also called electronic co-driver. Road vehicle management and control is the special kind of human process control we are concerned with in this paper. Today, vehicle control is still performed, almost exclusively, by the human driver. This might change considerably in the future.

The road traffic congestion, as well as the information stimuli proliferation to the driver from without and within the vehicle, is imposing a great deal of stress and workload on the driver. In the first place, the driver needs support in all levels of driving tasks for getting his alertness directed to the relevant information sources, in particular, to let him make best use of his excellence in skill-based and rule-based driving performance (Rasmussen, 1983). Secondly, the main purpose should be to support the driver in his knowledge-based performance with regard to problem solving and planning, i.e. for navigation and also for short term manoeuvre decisions, where his capabilities are most limited.

There are a great number of development programs underway for intelligent assistance for the driver devoted to the design goals of increased driving efficiency and safety. They all have to deal with the issue of knowledge about the driver. The way this question is tackled decides whether the driver will actually use the system. Without driver acceptance the assisting function cannot be effective. The level of system adaptability to the driving situation and to the driver has been found to be crucial in order to achieve driver acceptance (Smiley and Michon, 1989; Kopf and Onken, 1990). Many of the approaches for intelligent operator support in other fields, such as airplane cockpit systems for pilot support (Wiener, 1985), have failed because of unsatisfactory answers to fundamental requirements, mostly ergonomic, including the question of vehicle knowledge needed about the human operator. Some of the approaches were highly encouraging (Smiley and Michon, 1989; Dudek, 1990; Kopf and Onken, 1990; Onken 1990).

In order to illustrate the functional necessity of comprehensive knowledge about the driver and how it becomes effective for the assisting functions, the basic functional structure of a driver assistant system will be outlined. Based on this it can be derived systematically, what kind of knowledge base about the driver (driver model) has to be provided and how this can be achieved. An example will be given for an implemented experimental system which makes use of the most essential parts of the presented driver model.

Baseline functions of driver assistant

What are the necessary functions of a driver assistant system in order to provide the aforementioned support the driver is looking for? The structure, as depicted in Figure 6.1, shows the general functional scheme of the assistant system, labelled as *driver assistant*. It comprises the interfacing between the assistant system and its functional environment as well as the main internal functions. Basically, three interfaces between the driver assistant and its external world have to be considered, the interface to the:

- external information sources (e.g. radio, communication),
- vehicle information sources (e.g. sensors, data from machine perception, vehicle system outputs),
- driver.

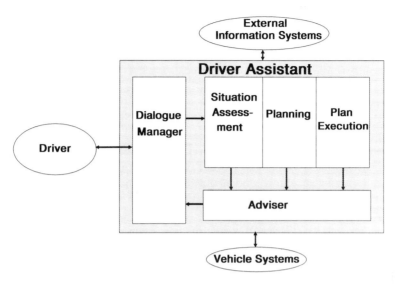

Figure 6.1 Basic functional structure of driver assistant.

All possible environmental impacts like weather, other vehicles and threats, which can be detected by vehicle sensors, perceptual devices or through communication, can be considered as covered by the above mentioned interfaces.

The interface with the driver is likely to be the most demanding one. The information media have to be chosen very carefully in order to ensure easy handling and command delivery, as well as easily perceivable information, to be provided by the driver assistant. This interfacing function is performed by the *dialogue manager*. In its main subtask it handles all messages to the driver, in order to present comprehensive and timely messages in the context of the current situation and the pertinent driver needs and priorities. For instance, advices containing navigational information, possibly on request of the driver, or system status information or situational hints and warnings have to be ranked, sequenced and formatted the way the driver wants to receive them. Knowledge about the current situation, as well as the driver's status and action sequences, is needed in order to achieve this with acceptable performance.

The main inputs into the dialogue manager are the advices and warnings from the *adviser* module of the driver assistant to be addressed to the driver. The adviser module, in connection with the dialogue manager, is serving the key purpose of the driver assistant, i.e. forming supportive advice for the driver. The adviser module and the dialogue manager, in turn, are drawing on information from the remaining driver assistant submodules for situation assessment, planning and plan execution.

These submodules can be considered as service modules for the adviser module and the dialogue manager. They autonomously keep track with all occurrences and aspects relevant for the accomplishment of the driving task through situation assessment, preparation of alternative plans, the selection of the plan to follow based on the situation assessment and they provide a reference for plan execution performance. As such, they are principally representing a machine capability of autonomous driving, i.e. of carrying out all of the functions the driver is trained to perform. There might be differences in performance quality compared to the driver, but the principal functionality should be the same.

The design criterion for these adviser service modules might be either to copy the driver performance as closely as possible or to look for the globally optimal function accomplishment, possibly even beyond the driver capabilities. It has to be carefully sorted out when one or the other should be adopted as a design criterion in order not to run into acceptance problems. For instance, for certain adviser functions such as the generation of plan advices, it can be desirable to look for the absolutely best solution, subject to certain information and criteria, which could not be generated by the driver because of mental restrictions like working memory limitations. Machine performance as close as possible to that of the driver might be important, for instance, for driver monitoring purposes and adapting messages to the driver's needs and priorities within the safety limits.

As mentioned earlier, the service modules intrinsically represent the capability of driving the vehicle autonomously. Here, we will only consider the use of this co-driver capability for driver assistance. We can also think of applications where the system might take over control, possibly at the driver's request. Active intervention by the driver assistant might be appropriate for certain cases on the basis of these capabilities for autonomous:

- situation assessment,
- planning and decision support, possibly decision making,
- action taking for plan and decision execution.

All of these three driver-like service functions of the driver assistant are, of course, subject to the global objectives of the driver which are assessed or fed into the system by one or the other means (Kopf and Onken, 1992).

The planning function solves the problems which are tracked down by the situation assessment function. The planning horizon can be long-term, as well as short-term, always taking the driver's strategies and intentions into account. Once the planning decision is made, the plan execution function comprises all services which can be made available for carrying out the plan agreed upon. Evidently, a great deal of knowledge is needed in order to have these functions work satisfyingly. In the first place, knowledge is needed about how to react in the face of a given situation. The situation is defined here by the set of determinants on the basis of which driving actions are effected (cf. Huguenin, 1988). There are driver external determinants as well as driver internal ones. The possible situations can be classified globally as those which can be considered quite familiar to the driver and those which are characterized by unfamiliar situations. For familiar situations, there exists a pertinent, well-defined and unambiguous reaction to be carried out. These situations are usually anticipated through expectations, and are the situations which are dealt with by the human operator through skill-based or rule-based behaviour (Rasmussen, 1983; Van der Molen and Bötticher, 1988). For unfamiliar situations, there are several alternatives to be considered for an appropriate reaction, i.e. a more or less complex selection process has to be carried out among these alternatives, what can be called a decision-making or planning process. These are the situations which could be dealt with by the human operator through knowledge-based behaviour, including the situation assessment and planning function (Rasmussen, 1983; Van der Molen and Bötticher, 1988). These should also distinguish between irregular and regular situations, depending on whether inadequate driver actions or events have occurred or not.

Therefore, symbolic or other kinds of knowledge representation have to be established for:

- classes and subclasses of possible situations and possible transitions from one class to another,
- (re-)actions associated with regular situations,
- action alternatives associated with unfamiliar situations or situations due to irregularities, derivation procedures for the relevant alternatives,
- guidelines for the selection process among relevant alternatives.

This representation includes the knowledge about the driver: his mental state, behavioural performance, intentions and possibly error contingencies. In order not to have to consider the full set of concepts every time, which are included in the knowledge representation, problem reduction can be employed, for instance, by breaking down the driving tasks and driving phases well-defined by transition conditions.

Incorporation of knowledge about the driver

The brief outline of the basic functions of the driver assistant shows clearly that understanding and evaluation of the driver's actions are needed. Therefore, most of the driver assistant functions are based on continuous information about the expected driver behaviour. Also actual information about the driver's state would be helpful, as well as the driver's intentions and error contingencies. Since the person actually driving the vehicle is chargeable to provide this information permanently, the driver assistant has to make use of other information sources or has to generate it on its own. This very crucial subfunction has to be readily available by means of a knowledge-based dynamic driver model (Figure 6.2), covering all of the above mentioned aspects and being incorporated within the driver assistant functions.

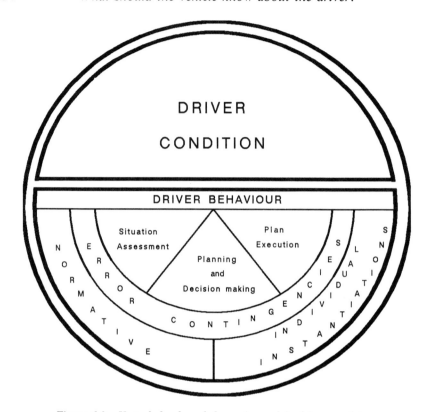

Figure 6.2   Knowledge-based dynamic model of human driver.

There is a great deal in the literature about human factors and human behavioural aspects. They cannot all be cited here, but see: Johannsen *et al.* (1977); Schank and Abelson (1977); Sparmann (1978); Rasmussen (1983); Godthelp (1984); Kraiss (1985); Boff *et al.* (1986); Rasmussen *et al.* (1987); Van der Molen and Bötticher (1988). Most of them do not deal with knowledge-based models. The pieces of information have to be put together from all sources available and have to be integrated into the structure of the knowledge representation. Lacking parameter values have to be established by additional experiments.

The kinds of behavioural functions to be considered in this model are again those already discussed for the adviser service modules i.e.

- situation assessment
- planning and
- plan execution

This model will primarily yield the driver's condition (status), intents and actions to be expected at every instant, possibly in the form of ranges. For the determination of these ranges, safety limits or thresholds of comfort are to be taken into account. Information about the driver actions is not necessarily confined to the control actions. Generally, all observable driver activities, including driver eye movements, are included. Both kinds of driver models can be used: that of a normal driver, and that of the individual performance of the actually driving person. The normal driver model describes the average driver, possibly as the average of a certain category of driver (professional, age categories, etc.). A distinction must be made between a normal model not considering driver errors, and the one including driver mismatching contingencies. The same is true for the model of the actual driver. It can be concluded that several partial dynamic driver models might work in parallel, possibly structured in the following:

- normative behavioural characteristics (with and without driver mismatching contingencies),
- individual driver characteristics, state and behaviour (with and without driver mismatching contingencies).

Knowledge about the individual driver model is of great interest with regard to aspects of adaptivity to the individual driver. This is of prime importance for the dialogue manager, and also for the adviser service functions, such as situation assessment, planning and plan execution support. If this could be achieved comprehensively, there would be no necessity for a normative model to be used for this purpose. However, a complete, perfect individual driver model is not available, so far. In the meantime, the normative model is used to provide a kind of substitute for the lacking knowledge.

For extrapolating the traffic situation a model is needed of the other drivers involved in that situation. A normative model is mandatory for this purpose. For ethical reasons, specific knowledge or data about the driving individuals in the surrounding vehicles has to remain confidential.

By studying the outcomes of these models, ranges of driver actions are determined, as well as estimates of their intents. This, in turn, can be used to detect discrepancies with respect to the a priori assumption of the driver's intent, and to evaluate these discrepancies with respect to possible error causes. These discrepancies are not always caused by driver errors. Certain deviations of the a priori assumption of the driver's intent could lead to sensible, albeit unexpected, actions. In order to avoid false alarms, it is of great importance that the discrepancies due to driver errors are discriminated from those caused by sensible changes of the driver's intent. Driver errors might be caused by mismatches due to certain cognitive deficiencies, or because of violations of limits of the principal human capabilities. This complex task of discrimination is facilitated through modelling of the individual driver state and behaviour, as well as mismatching contingencies. Frequently used methods for knowledge representation of rule-based transitions, task structures and conditioned actions, i.e. driver behaviour, are:

- and/or graphs, decision trees
- automata, petri nets and recently
- artificial neural nets (ANN).

One of the most favoured tools for knowledge representation is the petri net as a place-transition net, which is very powerful for the modelling of event-driven concurrent systems (Shank and Abelson, 1977). Higher level nets can activate lower level nets according to the abstraction level of driver tasks. These nets allow explicit modelling of resource availability. Net theory is used for system analysis, design, simulation and verification. Petri net mechanisms are usually transformed to real time software by code generation or complete reimplementation. Also real-time petri net interpreters, which process a special net description language, can be used. Other forms of transition net representations, such as decision trees or finite automata, are used in a similar way.

As will be shown in the next section, individual situation-dependent human operator nonlinear behaviour can be modelled with surprising consistency by multi layer perceptron ANN. Driver identification is also possible and individual error contingencies can be modelled in this way, as well as a global picture of the driver's condition, which is important for the driver assistant subfunctions, such as advice adaptation to the driver and driver intent recognition. Artificial neural nets seem to be a suitable means to represent the driver's operational expert knowledge, also under uncertainties. Another approach to deal with uncertainties is the application of fuzzy logic (Kandel, 1986).

The incorporation of the driver model as part of the knowledge structure can be carried out separately for each of the driver assistant submodules, tailored to the special needs for these submodules. Another approach can be that all modules are drawing the information, necessary for their particular function, from a single model implementation accessible for all of them. It might be worthwhile to note that all driver assistant modules are drawing heavily on the outcomes of the knowledge-based dynamic driver model. Therefore, the more comprehensive the driver model, the more efficient the performance of the driver assistant. The baseline structure of knowledge-based driver assistant systems with the dynamic driver model incorporated, as outlined, is more or less a functionally maximized one. In particular with regard to the dynamic model of the driver, only parts of it have been realized in the existing system developments.

Two developments might be mentioned, where adaptivity to the driver have already been implemented in experimental systems, the GIDS (Generic Intelligent Driver Support) (Smiley and Michon, 1989) and the DAISY (Driver AssIstant SYstem) development. The latter system will be described in more detail, as an example, with regard to the incorporation of the knowledge about the driver.

Incorporation of knowledge about the driver in DAISY

As outlined in the previous chapters, exploitation of both knowledge about the driving situation and the behavioural traits of the driver is of central concern with regard to successful design of a driver assistant system. This will be illustrated in more detail by looking specifically at the situation assessment function as being achieved in DAISY (Driver AssIsting SYstem), an experimental monitoring and warning system with regard to the primary driving tasks. This system has been designed within the PROMETHEUS project for driver support on the German motorways. It is implemented in a driving simulator at the UniBw München for testing and is going to be partially evaluated in an experimental vehicle in 1993. DAISY is to meet both design goals, increase in safety and driver acceptance.

The architecture of DAISY is shown in Figure 6.3 (Kopf and Onken, 1992). DAISY consists of four basic modules:

- the *reference driver* comprising the situation assessment function and a normative driver model,
- the *model of actual driver*,
- the *discrepancy interpretation* module representing the adviser function and
- the *warning device* as dialogue manager.

The structure is adopted, in principle, from a pilot assistant system which was being developed in the same institute. There is some resemblance with the structure of GIDS, which is developed within the DRIVE project. Driver intent recognition is implemented into DAISY to a certain extent, as far as the driving goals can be determined without navigational planning. From the architecture of DAISY it has become clear that knowledgeable situation assessment is the crucial function, in particular with regard to the use of knowledge about the driver. Therefore, we will go into more depth here to show how situation assessment is achieved for DAISY and how the knowledge about the driver becomes effective.

At this point a description of the function of situation assessment seems appropriate, thereby focusing first on the reference driver, where the main part of situation assessment takes place. The situation assessment is based on a structured knowledge representation of the possible elements (objects or concepts with pertaining attributes) of the current situation. The structure should allow for information condensation in terms of extraction of relevant situational elements with resulting situation determinants and situation classification. The situational elements can be considered as concepts which are described by the pertaining attributes or procedural sequences. This leads to a situation representation which also includes the incorporation of concepts like the driver's state, intention, actual behaviour or ranges of

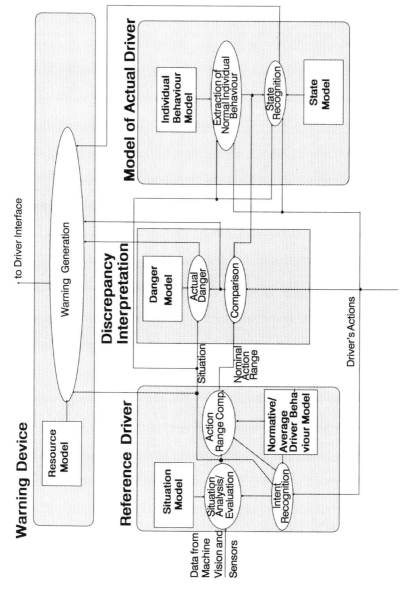

Figure 6.3   Architecture of DAISY.

behaviour and performance. The active function of situation assessment consists of:

- creation of actual values for situation elements and configuration of current situation on the basis of the relevant situation determinants (incl. classification),
- extrapolation from current situation,
- evaluation of the situation subject to the driver's goals (e.g. speed) and relevant constraints.

Examples for situational elements (concepts) are:

- the road with its attributes like number of lanes, geometrical parameters, exits or,
- surrounding vehicles, described by their relative position, speed (and acceleration), state of communication means (lights) and size (vehicle category) or,
- driving task related elements like the manoeuvre of approaching a curve or of lane change or, for instance, the manoeuvre to overtake a preceding vehicle.

The techniques used for knowledge representation are:

- multidimensional fields for lower-level declarative knowledge,
- decision trees for symbolic knowledge representation of actual situational facts,
- finite, deterministic criteria for symbolic knowledge representation of concepts like event or normative driver action sequences, known to be part of certain situation classes and therefore activated by them,
- analytical functions, describing behavioural aspects, for example.

Figure 6.4 shows an example for the criterion representation of the situational concept of lane selection and state of transition from lane keeping right to left and vice versa. Situation classification, whether implicit or explicit, is crucial for the situation assessment function. It is extremely difficult to achieve, if no knowledge about driver behaviour is utilized for this task. Therefore, obviously both situational and behavioural aspects have to be tied together within the same structure of situation representation. The concept 'state of lane change', for example, contains the geometric parameters of the road and the vehicle, as well as the driver's actions in terms of dynamic steering wheel activity. The integration of the behavioural model allows a decision to be made as early as possible as to whether the driver intends to initiate a lane change.

For situation classification, basically, traffic regulations have to be used. The next higher level is the use of normal driver behaviour, excluding driver mismatching contingencies. Above that, the most effective approach would be the use of knowledge about the individual behaviour of the driver actually driving the vehicle, including driver error contingencies. The latter requires a

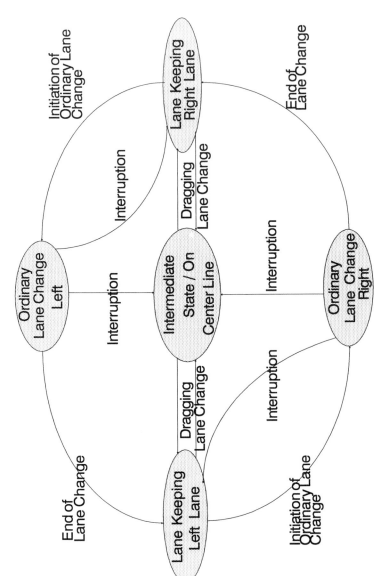

Figure 6.4 State transition diagram for 'lane change/lane keeping' situational concept.

great deal of learning capability on the side of the assistant system. It is available within DAISY to a certain extent. So far, for the purpose of situation assessment in the reference driver, DAISY only makes use of knowledge about normal driver behaviour without considering driver error contingencies.

Besides the situation classification and configuration, which is mainly a condensation process out of the set of all possible situations, the situation extrapolation depends heavily on what is known about the individual driver's intentions and behaviour.

Another aspect of requiring the same kind of knowledge about the driver concerns the adviser function itself, i.e. the warning function in DAISY. A warning has to be timely and it must not occur unless it is really needed. Depending on the situation and the driving style, the driver might allow only very small situation-specific time windows for warnings to occur. Therefore, if the situation is not assessed correctly and the knowledge of the driver is too sparse, the evaluation of the situation will not be sufficiently correct. This would produce nuisance alarms, or would possibly only bring about warnings when it is too late for the driver to react safely. In order to avoid this, the warning concept should take care to make use of the danger criteria used by the driver (e.g. time to line crossing for lateral control), and to provide gradual warning instead of a single alarm (Dudek, 1990; Kopf and Onken, 1990; Ruckdeschel and Dudek, 1992). This, in turn, requires a high degree of adaptivity to the driving situation and the individual driver. Both are implemented in DAISY.

Here, for the remainder of this chapter, the adaptivity of DAISY to the driver will be described in more detail. This leads us to the module of the model of the actual driver (Figure 6.3). Within this module, DAISY makes use of two levels of representation of the individual driver behaviour, both aiming at the driving style under normal conditions.

The first level of individual driver model is based on a statistical approach. The following example for the driving task of lateral guidance will illustrate the necessity and feasibility of warning adaptation by this approach. It is assumed that the driver's behaviour in terms of lateral time reserve depends on the following situational elements: velocity, road geometry, actual lane and occupation of adjacent lane(s). So, the situational parameter space has got five dimensions. For each of these dimensions, a criterion has to be defined for classification purposes, such that the actual situational cell in the parameter space can be determined anytime. Based upon this, during the learning phase, situation specific empirical probability distributions of the lateral time reserve can be determined and evaluated with regard to the extraction of a characteristic value of the time reserve. The driver-adaptive time reserve warning threshold is computed from this value (Kopf and Onken, 1992). Figure 6.5 shows the probability distributions of the lateral time reserve, left and right, as learned from an individual driver during three runs in a driving simulator associated to a specific situation. The driver-adaptive time reserve warning thresholds are easily derived from these distributions as mean one per cent percentile value.

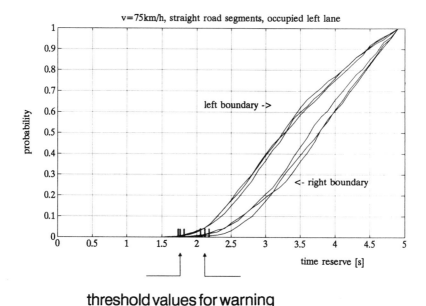

threshold values for warning

Figure 6.5   Probability distributions of driver-specific time reserve in lateral control (3 driving simulator runs, 1 subject).

The second level of individual driver model, not implemented yet, is based on the artificial neural net approach (Feraric *et al.*, 1992). If it is possible to get hold of all the pieces of situational information the driver uses, then these can be linked with the expected driver's actions through an artificial neural net. This constitutes a neural driver model. Similar approaches have been tried before by Shepanski and Macy (1987), Fix and Armstrong (1990), and Kraiss and Küttelwelsch (1991). These approaches, as well as our own results so far, are based on off-line training of the neural nets. The on-line training structure, as shown in Figure 6.6, would be feasible, if the nets could be implemented in analog hardware, or if they are prepared through a priori training. The benefit of this approach lies in the fact that the neural nets have got excellent interpolation properties, once they have been trained for the outmost situations in the situational parameter space. Continuous interpolative coverage of a large portion of the situational parameter space can be achieved that way by a single net.

Obviously, the style of driving does not remain the same under changes in the driver state. For a comprehensive individual driver model these time-varying effects should be considered. This leads to increasingly complex structures and tremendous training requirements. Therefore, for the particular application in DAISY, the driver behaviour under normal conditions with respect to the driver's state is aimed at. This seems to be sufficient, if there are reliable indicators for the driver's state being close to normal. Figure 6.6 shows

NET1: Training on mean of driver actions

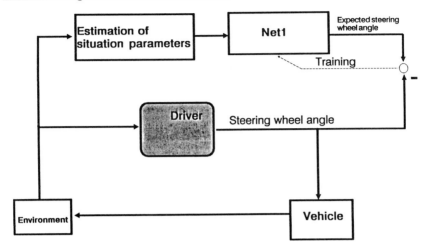

NET2: Training on deviations of driver actions from mean

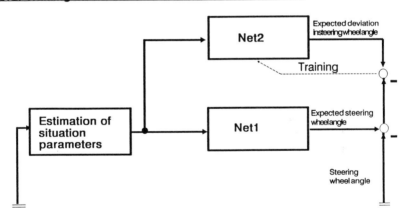

Figure 6.6   Learning of individual driver behaviour in lateral control through ANNs.

how the behavioural model of the individual driver can be achieved in DAISY for the task of lane keeping:

- ANN1 will be trained for the situation-relevant actions (here the steering angle) of the particular driver. All relevant situation-specific parameters are used as inputs for the net, which is taught with the corresponding effective steering action of the driver. After the training period, which might take some time, the output of ANN1 can be taken as a situation-specific mean value of the driver's steering actions to be expected.
- On the basis of the trained output of ANN1 and the driver action, ANN2 can be trained on the amount of expected deviations between effective

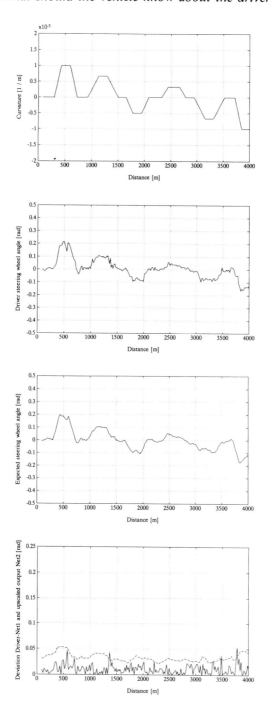

Figure 6.7 Driving simulator test results with ANN for wheel steering trained with data from a previous run.

driver actions and output of ANN1. This provides an index for the expected inconsistencies of the driver actions.

Thereby, ANN1 and ANN2 jointly describe the set of expected actions for the lane keeping task by the limiting hyperplane in the corresponding situational parameter space. This is what we call a driver model for the normal behaviour of the individual person actually driving.

This model can be used for adaptation of warning messages, which is the main purpose in DAISY. The situation-specific time reserve can be determined on the basis of the ANN outputs, which can be attributed to the particular driver. On the basis of this model of driver behaviour under normal conditions, driver identification can be accomplished to a certain extent and unusual changes from normal of the driver state can be detected. Figure 6.7 shows the modelling performance of an ANN, which was trained for a particular driver in a driving simulator and tested by different data, taken from another simulator run, with the same driver. The surprisingly low values for the inconsistencies are typical.

Conclusions

This paper deals with the effect of using knowledge about the human driver characteristics in driver assistant systems and which kind of knowledge is of prime importance for the assistant functions. This has been discussed in the framework of the functions involved to meet the situation-specific needs of the driver.

It has been shown that the knowledge of the driver's characteristics, his state and behaviour, has to be combined with knowledge of the driving situation in order to perform situation assessment with respect to the driver's needs. Here, not only normative driver characteristics, but also knowledge about the actual driver's traits are necessary. Validity of advice and warnings to the driver can only be ensured, if both adaptation to the individual driver and situation specificity can be achieved. As a sample for an implemented driver assistant system, DAISY has been looked at more closely from this perspective. The results, which have been achieved so far with this system, are very promising.

The following should be noted for the sake of avoiding misleading conclusions. Since it appears likely that the driver might like to use this kind of assistant system, he will trust the system more and more. This could lead, in the long run, to changes in the driver's cognitive style, which have to be taken care of in the design. However, this should be no reason to abandon the idea of knowledge-based driver assistant systems. Again, making use of knowledge about 'good' human co-driver behaviour can lead to solutions with respect to the mentioned long-term effect of 'overperfect' support to the driver.

Another issue is system reliability or system authority level. The higher the system reliability, the further one is closing into the capability of autonomous vehicle control, i.e. there are less and less safety barriers for the service modules for situation assessment, planning and plan execution to become autonomous. Nevertheless, the user of autonomous vehicles will be the human being as well. Therefore, knowledge about the user has to be available in this case, too.

The last issue which should be considered in the context of this paper, is that of protection of the driver's privacy. Through further technology advances, more and more information about the driver's state will be used in the system. The whole idea of knowledge-based driver support will collapse, if it will not be ensured that this kind of information is strictly inaccessible to anyone except the driver himself.

Acknowledgement

Sponsored by BMFT and Daimler-Benz AG under PROMETHEUS contract ITM 8900 A.

References

Boff, K. R., Kaufmann, L. and Thomas, J. P. (Eds), 1986, *Handbook of Perception and Human Performance*, Chichester: John Wiley.

Dudek, H.-L., 1990, Wissensbasierte Pilotenunterstützung im Ein-Mann-Cockpit bei Instrumentenflug, Dissertation UniBw München.

Feraric, J., Onken, R. and Kopf, M., 1992, Modellierung des individuellen Fahrernormalverhaltens mit Hilfe Neuronaler Netze und Interpretation der Abweichungen in einem Monitor- und Warnsystem, VDI-Tagung 'Menschliche Zuverlässigkeit'.

Fix, E. and Armstrong, H., 1990, *Modelling human performance with neural networks*, IJCNN.

Godthelp, H., 1984, *Studies on Human Vehicle Control*, Institute for Perception, TNO.

Huguenin, R. D., 1988, Fahrverhalten im Straßenverkehr, *Reihe Faktor Mensch im Verkehr* Nr. 37, Rot-Gelb-Grün-Verlag.

Kandel, A., 1986, *Fuzzy Mathematical Techniques with Applications*, Wokingham: Addison-Wesley.

Johannsen, G., Boller, H. E., Donges, E. and Stein, W., 1977, *Der Mensch im Regelkreis*, München: Oldenbourg.

Kopf, M. and Onken, R., 1990, A machine codriver for assisting drivers on the German autobahns, *9th European Annual Manual*, Ispra.

Kopf, M. and Onken, R., 1992, *DAISY, a knowledgeable monitoring and warning aid for driver on Germany motorways*, IFAC MMS, Den Haag.

Kraiss, K.-F., 1985, *Fahrzeug- und Prozeßführung*, Berlin: Springer.

Kraiss, K.-F. and Küttelwelsch, H., 1991, *Identification and application of neural operator models in a car driving situation*, IJCNN.

Onken, R., xxxx, Knowledge-based cockpit assistant for IFR operations, AGARD-CP 474.

Onken, R. and Kopf, M., 1991, *Monitoring and warning system for driver support on the German autobahn*, PROMETHEUS PRO-ART Workshop, Grenoble.

Rasmussen, J., 1983, Skills, rules and knowledge; signals, signs and symbols, and other distinctions in human performance models, *IEEE Transactions on Systems, Man, and Cybernetics*, Vol. SMC-13, No. 4.

Rasmussen, J., Duncan, K. and Leplat, J. (Eds), 1987, *New Technology and Human Error*, Chichester: John Wiley.

Ruckdeschel, W. and Dudek, H.-L., 1992, Modellbasierte Fehlererkennung in der Flugzeugführung, VDI-Tagung 'Menschliche Zuverlässigkeit'.

Schank, R. C. and Abelson, R. P., 1977, *Scripts, Plans, Goals and Understanding*, Hove, East Sussex: Lawrence Erlbaum.

Shepanski, J. F. and Macy, S. A., 1987, *Manual training techniques of autonomous systems based on artificial neural networks*, IJCNN.

Smiley, A. and Michon, J. A., 1989, *Conceptual framework for generic intelligent driver support*, Traffic Research Center, University of Groningen.

Sparmann, U., 1978, Spurwechselvorgänge auf zweispurigen BAB-Richtungsfahrbahnen, *Forschung Straßenbau u. Straßenverkehrstechnik*, Heft 263, BMFT.

Van der Molen, H. H. and Bötticher, A. M. T., 1988, A hierarchical risk model for traffic participants, *Ergonomics*, **31**, 4.

Wiener, E. L., 1985, Cockpit automation, in *Need of a Philosophy*, 4th Aerospace Behavioural Engineering Technology Conference, Long Beach, CA.

7

Determining information needs of the driver

Berthold Färber

Introduction

The question, what kind of information do drivers need, seems to be very simple, namely: all information which improves safe driving. But, which kind of information is it, in which modality and in what sensory channel should it be presented?

It is quite obvious that these questions cannot be answered in an all-embracing way. Different user groups, like professional drivers, novice drivers or truck drivers need different kinds of information. Even within the same group of drivers, information needs vary from situation to situation. If, for example a driver enters a foreign town, he requires totally different information compared to the situation he is in on his way to work. A possible way out of this problem is to present as much information as possible and to transfer the problem of information selection to the user. A look inside some modern hightech cars suggests that this was the idea of the designer and manufacturer. From an ergonomics and traffic safety point of view his 'solution' is not appropriate. On the other hand it is unrealistic to mourn the 'good old times', where cars were only equipped with speedometer, signal lights and headlight control. New and helpful systems, like antilock braking or route guidance systems, need an interface to the driver.

A top-down approach of driver information needs

This kind of cost-benefit analysis of driver information must distinguish two aspects of human information processing:

- human beings as purely rational information processing systems,
- drivers as complex 'systems' with rational and irrational decisions, changing motives, attitudes and behaviour.

69

Drivers as complex systems

To regard drivers as complex and partly irrational information processing systems is doubtless the most realistic view, but leads to a nearly insoluble problem. No measures exist to determine the actual motives and attitudes (especially irrational ones) of the driver. Even if they would be available in the near or far future, the predictions of actual behaviour are quite insecure for at least two reasons:

- behaviour is determined by multiple motives, and
- attitudes are not strongly correlated with actual behaviour (see Fishbein and Ajzen, 1975; Stapf, 1982).

A pragmatic objection could be that it is possible to measure the change of lateral and longitudinal acceleration, steering wheel angle or steering motions, and then predict the motivational status of the driver on the basis of those parameters. But, even if we are able to predict from a number of objective data the motivational status of a driver with sufficient reliability (e.g. aggressive or drowsy), what are the consequences? If we inform him: 'you are driving aggressively—slow down' he will feel his mind is made up for him. As a consequence he may reject the information system.

A second aspect to be considered is what Wilde (1974) calls risk homoeostasis. Even if the theory of Wilde is not fully testable, it highlights a problem which must be held in mind. To provide a driver with additional information (e.g. about an impending danger) changes his risk estimation. As a consequence he will—at least partly—compensate the benefit of the information by more risky behaviour. The main conclusion of the above is that the complex view of drivers with motives and irrationalities is realistic, but not very helpful for the determination of drivers' information needs. Instead, this view can serve as background for the determination of driver information needs. The actual determination must accept the restricted view of men as rational information processing systems.

Drivers as rational information processing systems

This restricted view of a driver considers men to be multichannel, multi-processing systems. Both aspects are equally important, but the multichannel aspect is often underestimated. Even if it is obvious that human beings gather information not only by ears and eyes, the proprioceptive-tactile cues for car driving are currently more or less neglected. The most convincing example for the influence of proprioceptive-tactile cues is the attempt to drive in a driving simulator. No existing simulator in the world is able to generate realistic lateral forces in sharp bends. As a consequence, curve driving is extremely difficult in driving simulators. Other examples are 'the seat of the pants' cues which are especially stressed by test or rally drivers. Lessons we can learn from this are twofold:

1. Not only visual cues like the optical flow field (Gibson, 1950; Lee, 1976), but also proprioceptive-tactile information is an important source for adequate behaviour adaptation.
2. The development of more comfortable cars, which it undoubtedly positive in relieving the driver's physical and mental workload, has a negative side effect. Proprioceptive-tactile cues are being filtered out. A good example is the improvement of truck cabins to avoid the degeneration of the spinal column of professional drivers. But, in specific situations these cues are urgently needed to guarantee appropriate behaviour at the right time. The advantages of this kind of information are faster processing times without loading the visual channel.

Assessment of experts

Within the PROMETHEUS project several new technologies for improved driver information and assistance are being developed. An evelution of some of these systems by experts of the PRO-GEN group comes to a pessimistic evaluation. The benefit of systems which inform the driver about impending danger or about other cars is estimated extremely low (Table 7.1). Automatic intervention seems to be the only effective way to improve safe driving. Additional information seems to be useless or at least ineffective. So, the question arises: does a driver really need new or other information in future cars, or is it sufficient to display what we are presenting now? Are all additional pieces of information only gimmicks? Two objections can be made against this evaluation. First, accident statistics, which are a main basis for the evaluation, are very often not sufficient to 'look behind' the mutual dependency of factors causing an accident. This holds also (more or less) for in-depth accident data. Accidents are a complex phenomenon which can be hardly cut into pieces. Second, the single systems and pieces of information are treated separately. Synergetic effects are not considered here.

Three levels of car driving and driver information needs

It is widely agreed (McRuer *et al.*, 1977) that car driving can be subdivided in three behaviour levels: navigation, manoeuvring and control. On the navigation level the driver chooses his route on a macro- and micro-level. On the macro-level he determines his destination, preferred departure and arrival time. On the micro-level he chooses between different routes, depending on the traffic situation and his needs. He will, for example, decide on different ways if he is on a business trip or a sightseeing tour.

On the macro-level he needs information about the best travelling time. A reliable offer for his departure time can optimize traffic flow and use of road

Table 7.1   *Maximum and expected accident reductions for different*
PROMETHEUS functions (from PRO-GEN report, 1991).

	Accidents with injured	
CED assessed	Maximum Reduction	Expected Reduction
Visual enhancement by UV-headlight (information)	1%	−1%
Visual enhancement by image processing (information)	7%	−1% to 2%
Proper vehicle operation (automatic intervention)	15%	6% to 9%
Collision avoidance, version 'stop at obstacle' (automated intervention)	30% to 40%	25%
Co-drive, version B and C (automatic intervention, automated driving)	27%	20%
Autonomous intelligent cruise control (automatic intervention)	3%	3%
Emergency warning systems (information, warning)	<1%	<1%
Dual mode route guidance (information)	<1%	<1%

capacity. This information should also include the closings of roads at night, a factor which is becoming more and more important especially for truck drivers.

On the micro-level the driver needs only three pieces of information:

- where am I?
- how far is it to the next decision point? (junction or intersection)
- what do I have to do there? (turn left/right or go on).

Generalizing the information needs on the navigation level of car driving leads to the following recommendation. Ask yourself, if the driver has an information deficit. Ask furthermore, if this deficit is relevant to perform his driving task on the level under discussion (navigation, manoeuvring, control). If this question is answered positively, ask, how he can get the information in the most easy, cheap and reliable way. Then present the information in this form.

On the manoeuvring level the driver has to react to other traffic participants like cars, cycles, pedestrians etc. and to obstacles. This kind of reaction (covering partly the control level) has to be adequate, performed right in time and should not disturb the traffic flow. What are the most important information deficits at this control level? To derive information needs at this level of control, it is useful to consider the phylogenetic development of the human race. The development and the survival of human beings was guaranteed by an ability to detect moving objects at the periphery of the eye (to see the enemy or the tiger coming). In contrast, it was not necessary for our ancestors to detect fast decelerations or accelerations in the area of central vision. For a 'running speed' of 40 km/h this ability was not essential. Consequently, evolution did not develop this feature. Deficits can therefore be observed when the driver has to estimate the deceleration rate of a sharp braking leading car. Visibility problems under poor sight conditions can also be explained by this phylogenetic approach. Human beings have been 'daylight hunters'. Therefore, our night-vision is restricted, leading to information deficits under poor visibility conditions.

What can we derive from the above? A good guideline to determine driver information needs in a top-down approach is to consider the phylogenetic development of humans, and the information acquisition and information processing capacities resulting from this development. Typical deficits can be identified on this background. However, it is not possible to evaluate the importance of these information needs. The evaluation problem can be better solved by a bottom-up approach.

A bottom-up approach for the determination of driver information needs

The bottom-up approach covers mainly three methods: questionnaires, simulations and the use of prototypes. These methods differ with respect to costs, reliability and ecological validity.

Questionnaires

The simplest idea to identify driver information needs is to ask potential users, comparable to marketing studies. This procedure seems to be a cheap and effective way to discover driver information needs. The problems of such studies are manyfold: (1) drivers can only comment on systems they have experienced; (2) they will quote and favour those information systems they have in their own cars. The reason for this behaviour can be predicted from the theory of cognitive dissonance (Festinger, 1957), which can be summarized as follows: after a customer has decided to buy a specific car, he finds many reasons, why this particular car and the implemented systems are optimal for him. This psychological mechanism helps him to cope with the problem of not being able to buy a luxury instead of a middle class car. This means that what we can get from questionnaires are ideas from drivers about their needs, which are based on currently existing systems and biased by the mechanism of cognitive dissonance. A look ahead is not possible.

A more reliable method which is related to questionnaires can be adopted from experimental cognitive psychology. This method is named 'free recall'. It is also restricted to the evaluation of existing systems, but has the advantage to be free from social desirability or cognitive dissonance.

Method

Ask subjects the following question: 'Imagine your car!' (alternatively 'Imagine an ideal car!'). 'What displays, controls, information systems are in that car?' The subjects have to report, by free recall, all systems they can think of. The idea of this method is that subjectively important systems will be named first, subjectively irrelevant ones will be named later on, or be ignored. From the mean position of each element the subjective value of an information system can be derived. The method was successfully applied in a study of Faerber and Faerber (1988) to evaluate displays and controls in cars.

Simulation

If a designer or an engineer has used the top-down approach coming to an idea about an information need, how can he test whether this idea is promising? The development of a technically perfect system and the test 'on the market' is not only an expensive enterprise, it is also unsatisfactory from an ergonomic and traffic safety point of view. So, before constructing new driver assistance systems like vision enhancement, distance warning etc., the development of simple fakes is useful.

An illustrative example. To define the requirements of a voice information system, a car was equipped with radio data communication and a simple voice output system. Subjects drove in a natural traffic environment where specific traffic scenes were initiated by the experimenters. In these situations voice warnings were presented. For example during an overtaking manoeuvre the warning was presented: 'your oil temperature is too high; let your car coast to a standstill and check the oil level!' (Faerber and Faerber, 1985). Of course some of the subjects doubted the trustworthiness of this message. They had the (right) hint that the system was a fake. To make the warning really credible for the subjects the oil-measuring stick was manipulated, pretending to the driver that his oil level was too low. After the oil check, subjects were really convinced of the value of a functioning voice information system implemented in a car.

What can we derive from this example? The development of simple, but realistic fakes is a very effective way to decide whether new information meets driver's needs, or if it is only an irrelevant gimmick. Simulations in a driving simulator (even in a simple one), or in modified cars require the full creativity of designers, engineers and psychologists. For an engineer, who is involved in highly sophisticated technical developments, it is sometimes hard to imagine simple and 'dirty' fakes. But for a first test it is only necessary to check the impact on the driver. The technology behind is arbitrary.

Prototypes

The development and test of prototypes to evaluate technical questions is an everyday business for car manufacturers. The use of prototypes for the detection and evaluation of driver information needs is doubtless the most realistic, but also the most expensive way. Therefore prototypes should only be used when simulations are impossible for technical reasons. For both methods, simulations and prototypes, a central question remains unsolved: what are the decision criteria for useful or useless information? Accident rates, objective or subjective workload, comfort or anything else?

Evaluation criteria for information needs

Interestingly no detailed list for the evaluation of driver information needs exists or is under development. So we have to restrict ourselves to more general questions and guidelines. On the behavioural side an evaluation can be based on the following questions:

● What is the desirable (ideal) behaviour pattern of the driver?
● In what situations and to what extent he varies from this behaviour?
● Is the improvement of this suboptimal behaviour by the new information system significant or only marginal?
● What negative side effects can be expected from the information under investigation on other relevant cues?

A helpful question which covers mainly the environmental information can be formulated as:

● How would a competent co-driver behave?

Rally drivers have always used the help of a human co-driver. In every day traffic co-drivers warn the driver of impending dangers or guide him through a foreign town. A competent co-driver will also find the compromise between useful and annoying information. That means, the behaviour of a co-driver can be a good guideline for the determination and evaluation of information needs.

References

Färber, B., 1986, Abstandswahrnehmung und Bremsverhalten von Kraftfahrern im fließenden Verkehr, *Zeitschrift für Verkehrssicherheit*, **32**, 9–13.

Färber, B., 1987, *Geteilte Aufmerksamkeit. Grundlagen und Anendung im motorisierten Straßenverkehr*, Köln: Verlag TüV Rheinland.

Färber, B. and Färber, B., 1985, *Design and evaluation of speech information systems in cars*, IMechE, 189–91.

Färber, B. and Färber, B., 1988, *Sicherheitsorientierte Bewertung von Anzeige-und Bedienelementen in Kraftfahrzeugen. Empirische Ergebnisse.* Frankfurt: FAT-Schriftenreihe Nr. 74.

Festinger, L., 1957, *A theory of cognitive dissonance*, Stanford: Stanford University Press.

Fishbein, M. and Ajzen, I., 1975, *Belief, attitude, intention and behavior: An introduction to theory and research.* Reading, Md., London, Sydney: Addison-Wesley.

Gibson, J. J., 1950, *The perception of the visual world*, Cambridge, MA.: Houghton Mifflin.

Lee, D. N., 1976, A theory of visual control of braking based on information about time-to-collision, *Perception*, **5**, 437–59.

McRuer, D. T., Allen, R. W., Weir, D. H. and Klein, R. H., 1977, New results in driver steering control models, *Human Factors*, **19**, 381–97.

Stapf, K., 1982, Einstellungsmessung und Verhaltsprognose. Kritische Erörterung einer aktuellen sozialwissenschaftlichen Thematik, in Stachowiak, H. (Hrsg.), *Bedürfnisse, Werte und Normen im Wandel. Methoden und Analysen*, Band II. Paderborn: Fink, Schöningh.

Wilde, G. J. S., 1974, Wirkung und Nutzen von Verkehrssicherheitskampagnen: Ergebnisse und Forderungen im Überblick. *Zeitschrift für Verkehrssicherheit*, **20**, 227–38.

8

Degrees of freedom and the limits on learning: support needs of inexperienced drivers

J. Groeger

Introduction

Much of what is normally considered 'learning to drive' proceeds from the point at which drivers begin to undergo formal training. What the driver is taught at this early stage of his or her career will be understood in terms of any relevant skills (e.g. cycling) or knowledge (e.g. travel as a passenger) the learner driver already possesses. Similarly, the learning which drivers are required to do later, when vehicles, traffic regulations or performance capabilities change, depends greatly on earlier experience as a driver. What and how drivers learn is therefore an important consideration, not only for those seeking to minimize the difficulties inexperienced drivers have, but also in understanding the problems faced by more experienced motorists in adapting to new-found circumstances.

Basic processes in learning

Fundamentally, 'learning' is the process by which our reactions to events, or anticipated events, become reliably different. Sometimes what is acquired is completely new, e.g. in the first few driving lessons being able to start up the engine and move off from the kerbside. Sometimes what is learned is an improvement in a behaviour already acquired, e.g. later in a course of driving lessons moving off smoothly from the kerbside, having checked for other traffic. The person-in-the-street's view of how such changes in behaviour are achieved is that 'it comes with practice', and that some 'instruction' is needed. This is accurate but, like much home spun wisdom, is incomplete.

77

Practice

Several points must be made about the role of 'practice'. Firstly, contrary to the time honoured phrase, practice does not make perfect: practice only makes predictable. Frequently performing the same activity promises only that it is likely to be done the same way in future, not that it will be done well. A standard for performance must be laid down and adhered to, for practice to yield a genuine improvement in performance. Another popular notion about practice is that after a certain amount of it, performance remains more or less the same. This is very well illustrated by work reported by Crossman (1959), who in studying an expert cigar roller over a number of years and millions of trials showed that performance does indeed appear to 'stabilize' after long periods of practice. However, it may appear to 'stabilize' in several different places, depending on the amount of practice and where the data collection stops. In fact, Crossman's data, like many similar findings, fit the power law of practice. This brings us to the third point to be made about practice, the amount needed to acquire a skill. Schneider (1985) in discussing the training of high-performance skills suggests that at the very least 100 hours is needed, but that this depends on the complexity of the skill to be acquired. Typing at 50 words per minute takes a training time of in the region of 200 hours, while the training of a fighter pilot typically requires 350 hours over a two year period. Even with this amount of training time satisfactory performance may never be achieved, not everybody will make fighter command in 350 hours! One obvious consequence of this is that different types of individual will take a longer or shorter amount of time to gain the rudiments of a skill. After that a consistently high standard of performance requires a great deal of practice, even when the trainee understands the nature of the task and can perform it accurately. During this period of extended practice, Schneider makes the point that: 'For the novice, decisions are slow and uncertain, and the trainee appears very overtaxed. In contrast, the expert makes decisions very quickly and with little effort, and can simultaneously perform other duties' (op. cit., p. 286).

Feedback

The process of 'instruction' generally involves being explicitly told how to carry out an activity, performing it under the guidance of an instructor, and having 'knowledge of results' or feedback. Obviously the quality of instruction plays a large role in skill acquisition, but given an understandable description of how the task is to be performed, 'feedback' is traditionally seen in the learning literature as the key to effective adaptation (e.g. Brunswik, 1952; Anderson, 1983). However, 'feedback' is not necessarily always as beneficial as one might casually expect. In situations ranging from motor skills learning (Swinnen *et al.*, 1990) to judgement under uncertainty (Hoch and Loewenstein, 1989), both of which are surely fundamental to driving, detrimental effects of feedback have been reported. Swinnen *et al.* (op. cit.), using timing tasks, have demonstrated that instantaneous knowledge of results degraded learning relative to delayed feedback. Swinnen *et al.* conclude that 'very short KR-delays interfere with learning, perhaps by degrading the acquisition of error-detection capabilities' (p. 706). It seems likely that such error-detection capabilities are in turn particularly important in facilitating generalization of learning to new contexts. Hoch and Loewenstein (op. cit.), show that feedback may serve to fuel unwarranted confidence in judgements, by virtue of a hindsight bias, and that feedback may serve to diminish 'surprise' at a particular outcome, thus also inhibiting learning. Brehmer (1980) is particularly pessimistic about amount of learning which feedback in complex situations promotes, as are Balzer *et al.* (1989).

Given the above, if we are to design systems which are truly capable of supporting the behaviour of inexperienced drivers we need answers to the following questions. Firstly, what do drivers learn when undergoing instruction and how much practice is devoted to it. Secondly, what do drivers not learn during training and what scope do drivers have later on for meeting this shortfall. Thirdly, how far do the conditions obtaining in normal driving serve to maintain or promote desirable driver behaviours. In the next section data we will consider some recent data relevant to these issues.

Learning to drive

The first point which must be made is that there is no official syllabus which must be followed by UK driving instructors when teaching people to drive, but the standards of those offering professional instruction are monitored by the Department of Transport. For the pupil, there is also a closely controlled state driving examination which must be passed. The content of this test is fixed, following particular routes and incorporating certain mandatory manoeuvres, a sight test and a verbal test of the candidate's knowledge of the highway code. The actual time spent driving during the driving test is generally of the order of 30 minutes, with the test beginning and ending in the same (usually) urban location for each driving test district. There is no requirement for individuals to undergo formal tuition prior to taking the state driving test. As a result, it is particularly difficult to know precisely how much training any individual driver has had, or what that training might have consisted of. These issues are, as pointed out above, particularly important in designing appropriate support systems which are designed to promote learning.

As part of our efforts within the GIDS consortium (DRIVE Project V1041), studies were carried out with the intention of characterizing the verbal support requirements of novice drivers, as a preliminary to developing an instructional support system. They also help to illustrate what we can, at best, expect the newly qualified driver to have been instructed in. [Full details of this study are reported in Groeger and Grande (1991)]. Twenty drivers who had no previous driving experience learned to drive with one of four driving instructors. Each lesson was video recorded and analysed in terms of the support given by the instructor, types of situation encountered and faults committed. This required detailed analysis of over six hundred hours of video tape and verbal commentary. Overall the pupils, aged between 17 and 24 years, required 607 lessons to reach driving test standard (Average = 30.3). For a subset of these pupils (N = 10) tapes were reanalysed to determine the number of times each driving manoeuvre was encountered, whether or not the instructor commented upon it. It is these data which particularly concern us here, since they give an indication of the exposure during training to each manoeuvre.

On average these ten pupils required 30.6 lessons to reach driving test standard. Their first time test pass rate (6/10) was similar to the overall group (13/20), both of which are higher than the national average (51 per cent, Forsyth and Kompfner, 1991). Table 8.1 shows the average number of times during training each of our ten drivers encountered a particular situation or driving manoeuvre. What seems likely to be the most frequent manoeuvre, simple driving straight ahead, is excluded because its frequency of occurrence is difficult to obtain.

There are several remarkable aspects of the data presented in Table 8.1, but I will confine myself to discussion of two manoeuvres, simply to illustrate the points which can be made on the basis of the data set we have available. Let us first consider lane-changing. The definition adopted for lane-changing, was rather more restrictive than the reader may expect, in that this category included only those lane changes which were not at junctions, but only those where the pupil was being explicitly instructed in lane-changing. Obviously lane-changing is an important component of other manoeuvres, e.g. right turns at large junctions, but such changes of positioning were counted as part of the junction manoeuvre rather than separately as a change of lane. Given

Table 8.1 Average numbers of manoeuvres, supported and unsupported, encountered during training, excluding simple straight ahead (N = 10).

	Total	Supported	Unsupported
Moving off	152.3	87.3	65.0
Ahead near controls	131.3	35.2	96.1
Ahead near others	164.0	150.0	14.0
Overtaking	50.0	41.2	8.8
Lane-changing	20.5	17.5	3.0
Roundabouts	116.2	97.8	18.4
Bends	132.4	78.7	53.7
T-junctions	511.5	332.0	179.5
Crossroads	304.5	198.7	105.8
Slowing	35.9	27.6	8.3
Parking	109.2	65.8	43.4
Reversing	70.7	56.9	13.8

our definition, on average a pupil learning to drive 'from scratch', will lane-change about 20 times during lessons. Since virtually no practice was carried out between lessons, this is the sum of their lane-changing experience. In 85 per cent of cases the instructor considers that some correction or advice is required. Furthermore, other analyses have shown that the need for support does not decrease over the course of lessons and that the difficulties learner drivers have are largely connected with observation of other traffic, signalling and judgement. These problems tend to be more pronounced on lane changes to the right hand lane (i.e. towards the centre of the road). The other manoeuvre worthy of particular comment is overtaking. The definition adopted for overtaking was that any moving obstacle, which required the driver to move out of lane and subsequently move back into lane in order to pass, was counted as an overtaking manoeuvre. On average the driver carries out 50 manoeuvres of this sort during training. However, the vast majority of these are overtakings of slow moving road users (e.g. cycles, delivery vans, other learner drivers). As a result it is worth bearing in mind that learners have virtually no opportunity to practise normal or high speed overtaking under the supervision of an instructor. Even for the overtaking of slow moving road users, support is given by the instructor on over 82 per cent of occasions. As with lane-changing, the need for support remains constant across the course of lessons, and observation and judgement again account for many of the comments made. In the case of overtaking, speed choice (learners tend not to overtake quickly enough) and road positioning are also particularly difficult for the learner driver.

These analyses show that motorists who learn to drive with professional instructors have very little opportunity to practise certain manoeuvres. Furthermore, where they do practise these manoeuvres, they appear not to improve greatly by the end of their training course (they still require the same degree of support at the end as at the beginning). It may be that this way of analysing our data understates the amount of practice pupils actually get on the components of tasks. For example, it might be considered that the 'lane-changing' drivers do, as a component of right turns at junctions, actually (somehow) promote the learning of 'lane-changing' as a distinct manoeuvre. In the psychological literature such a phenomenon would be termed 'implicit learning' or 'transfer of training' (see Berry and Broadbent, 1988, for a recent discussion). Such solid empirical data as are available on implicit learning suggest that it is limited to situations where the learner's attention is focused on physically similar stimuli (Berry and Broadbent, op. cit.). It does not seem unreasonable to suggest that attentional strategies at junctions and away from junctions are different, and that minimal transfer would therefore take place. Recent data reported by Duncan *et al.* (in press) show considerable independence between components of the driving task, even within the same individual. This would clearly be expected to minimize transfer between situations. Furthermore, our own data suggest that whatever learning takes place at junctions (and it should be noted that the need for support decreases at all types of junctions as lessons progress), pupil's ability to change lanes when not near a junction does not improve. One must therefore be hesitant about accepting arguments about the benefits of 'practice' from circumstances which approximate those of 'implicit learning'.

We therefore contend, that even with completely professional instruction, pupils do not have the opportunity to adequately learn how to carry out certain manoeuvres. For those who do not learn to drive solely under the supervision of a professional instructor, while their driving test pass rate is higher (see Brown *et al.* 1987), the scope for adequate learning seems likely to be still less. Data currently being collected by the UK Transport and Road Research Laboratory are highly relevant on this issue.

While, as pointed out above, there is no requirement for individuals to undergo formal tuition prior to taking the state driving test recent data suggest that most do so. Forsyth and Kompfner (1991), report data based on a sample of 9344 respondents, which indicate that 98 per cent of all drivers who had recently taken their driving test had had at least some professional instruction. However, from the data reported there is no way of telling how many of the sample only had professional instruction, or what the total number of hours driving was prior to passing the driving test. Disaggregated figures are presented, which indicate that about 80 per cent of their sample had had less than 50 hours professional instruction (7 per cent had less than 11 hours), and 80 per cent had had less than 40 hours practice with friends (45 per cent had less than 11 hours). Forsyth reports that the amount of professional instruction and amount of practice are unrelated, and thus we cannot be sure how much pre-test driving experience candidates have (except to say that for 80 per cent of the sample the number of hours experience must obviously be substantially less than 90 hours). Forsyth and Kompfner report that about 80 per cent of their sample had estimated the number of miles driven while learning to drive was less than 1000 (12 per cent less than 200 miles). Whatever the amount of experience gained, passing the driving test does not indicate that a consistently safe level of driving performance can be maintained. Forsyth reports preliminary details of a follow-up study of these motorists, which indicate that within two years of passing their driving test 70 per cent of respondents 'felt they had only just avoided at least one serious accident, while 19 per cent were actually involved in one accident as a driver' (p. F9). A test of drivers, similar to that they had passed in order to gain their license, yielded a pass rate of only 52.7 per cent. With an extended test, i.e. twice the length of the original, the pass rate dropped to 44.6 per cent. Since 'test failure' occurs only because of dangerous or potentially dangerous driving, these data appear to suggest that over half the motorists who pass the state driving test were incapable of driving safely on the occasion they were retested.

These data strongly indicate that drivers cease to be able to meet the demands placed upon them by the road system, or have been inadequately prepared to do so during training. Our own data suggest that the latter may well be the case. Current developments within the DRIVE framework appear to have the potential to improve the training drivers receive, and to extend formal training into the period after drivers have passed their driving test, i.e. when feedback on performance is at its least and at its most ambiguous (see Brown *et al.* op. cit). In the next section I will consider important issues in designing driver support systems, and show by describing one such system, PSALM, how such systems might be realized.

Performance support

Issues in supporting drivers

Up to this point in this paper we have tended to discuss the novice driver as a member of a particular group. It is important to realize, however, that the variation in performance within any such 'group' will be extremely large. Recent research (see Kuiken and Groeger, 1991) has shown that motorists, when grouped on the basis of driving experience (time licensed and annual distance driven), do not differ on a variety of performance-based measures. However, the fact that there is very substantial variation between individuals demonstrates very clearly that, taken from the stand-point of the individual driver, support needs differ greatly across the driving population. It is also important to realize that even if motorists do not differ in performance terms, it may be more difficult for some drivers, than for others, to maintain this level of performance. It may also be that the performance level achieved is inadequate, or even dangerous, for some drivers in some circumstances. Because of this, and because of the difficulty of establishing what the needs of different groups actually are, we need to consider supporting individual drivers, rather than groups of motorists. Irrespective of the type of support needed, two issues must be addressed by designers of support systems: the way in which support should be delivered and the timing of support.

Within the GIDS prototype, it is envisaged that visual, auditory and tactile modes of communicating with the driver will be available. Underlying the GIDS design philosophy is the requirement that overloading the driver and communication channels should be avoided. Given the already extensive demands on the driver's visual capacities we assume that performance related support should not be presented visually, unless that driver is not required to monitor the external scene. Communication with the driver, during driving, for performance support purposes must therefore be largely confined to the use of the auditory or tactile modes (e.g. intelligent steering wheel or intelligent gas pedal). Which of these two modes should be used depends in part on the timing considered appropriate, the behaviour for which the support is intended, and whether the designer actually wishes to encourage learning.

There are obviously three options with regard to the timing of support messages: before, during or after the manoeuvre in question. Let us consider these options in turn. Messages given before an activity are, in essence, 'warnings'. The success of a 'warning', in terms of changing behaviour, will obviously depend on whether it is delivered 'in time' and 'contingent on performance'. Warnings unrelated to current behaviour would clearly be both irritating (reducing the acceptability of the system) and of limited educational use. Unless the driver knows what has provoked a warning, it will be very difficult to learn how to drive in such a way as to avoid a warning. This seems likely to lead to cases where drivers explore the boundaries of acceptable performance in order to determine what 'correction' is appropriate. Warnings must therefore be contingent on current performance which is unacceptable, or on a recent 'history' of unacceptable performance under similar conditions. If warnings are given which are perceived to be 'earlier' or 'later' than the driver deems appropriate, it seems very likely that the driver will consider them not to be based on current performance. Warnings which are considered too early are likely to require further, virtually 'on-line', warnings. Warnings which are 'too late' may compromise safety. Establishing the appropriate timing of warnings is therefore essential, both to maximize their efficacy and to diminish any side-effects which reduce the overall acceptability of the system.

Support given during a manoeuvre is prone to other difficulties. In the first place, real-time support is the most technologically demanding and may not be feasible except under very limited and well defined circumstances. Secondly, while on-line support is completely contingent on behaviour it may cause the driver to focus attention on the error detected, but away from more serious ongoing or upcoming difficulties. The technological demands and potentially dangerous distracting effects of on-line presentation are serious drawbacks to performance support during manoeuvres.

The third option is to deliver performance support after a manoeuvre has been completed, i.e. 'feedback'. This would allow the support given to be made contingent on performance, but would not necessarily prevent the error being committed again—though increasingly 'forceful' corrections might guard against this possibility. Feedback could be given immediately after performance (when the level of demand on the driver has diminished), or might be substantially delayed. It is usually assumed in the literature on human and animal learning that contiguity is an important aspect of learning. However, in many of the empirical studies carried out, the learning environment of the animal or human is simple, and the cause of the feedback is very often apparent, especially by comparison with the complexity of the driving environment. Complexity, in this case, refers to the number of potential stimuli present when learning is taking place, and the potential variations which might occur in each stimulus.

One reason why contiguity is important is that it helps to establish a clear causal relationship between an act and the feedback provided. The temporal proximity between the two events effectively establishes a 'meaningful' relationship between the act and the feedback obtained. That is, an overt 'semantic' connection between the two events may be helpful, but unnecessary.

It is conceivable that where the temporal proximity between events decreases their semantic proximity must increase for effective learning to occur. The author is unaware of any explicit tests of this prediction. A further difficulty, which also relates to the issue of the complexity of the performer's environment, is that it is not clear whether learning will take place when the same feedback is generated in response to what appears to the performer to be differing circumstances. Finally, there is some evidence that immediate feedback actually impedes learning of error-detection strategies and perhaps 'generalization', as discussed above. The basis of this effect is that where correction is provided immediately after an event the performer tends to use this to guide performance, rather than to develop an understanding of what behaviour actually led to the feedback in question. Thus, one would predict that the user would become increasingly system dependent and less likely to generalize performance successfully to new, but arguably similar, situations.

This raises a fundamental question about what we seek to do with performance support. Do we wish to teach people to become reliant on the feedback system, or to encourage the development of driving practices which would effectively reduce the need for the support system (because the individual consistently behaved in an acceptable fashion)? Given the current state of the art of support systems the latter would seem the more sensible and safe objective.

Before the next section let us briefly summarize. Support for driving performance which is tailored to the needs of groups of motorists may not be feasible and seems unlikely to promote effective learning among individual drivers. Instead the needs of individual drivers need to be addressed, at least as an intermediate goal on the way towards a system which meets the needs of different groups of drivers. The way in which drivers should be supported seems likely to depend on what behaviour is to be supported and what requirements the system designer has in terms of longer term learning. The precise scheduling of support requires substantial empirical research in order to realize the full 'tutoring' potential of support systems. In order to carry out this research, a prototype support system must first become available. It is essential that this support system has sufficient flexibility to allow research into the issues raised above. It is also essential that the system can be realized with current technology, albeit with some assumptions made about availability and reductions in cost in future technology. In the final section of this paper a support system which we believe meets these demanding criteria is described. It is currently being built as part of the GIDS prototype.

Personalized support and learning module (PSALM)

It is envisaged that the GIDS prototype will include a means of storing the performance profiles of individual drivers, both in terms of the frequency with which they have encountered particular situations and their history of 'abnormal' performance in each situation, built up from data provided by the

GIDS sensors and prototyped GIDS applications. Abnormal performance will be regarded as a deviation from a normal standard in terms of the functioning of each sensor/application, e.g. sudden harsh braking, attempts to make rapid and unusually large steering corrections, excessive accelerations and adoption of highly variable headways. It is envisaged that the data routinely collected by the GIDS sensors and other applications will be processed by PSALM, whether or not it has occasioned a 'warning' or other message by some other part of the GIDS system. The driver will be informed when a criterion of 'abnormal' performance has been reached, or when he/she requests support from PSALM.

Initially, estimated performance criteria, in terms of acceptable numbers of incidences of such abnormal performance will be stored along with the performance profile for each situation. Exceeding this criterion will result in a request being passed from PSALM to the dialogue controller (which schedules all communications to the driver) for a message to be presented to the driver, indicating that over a period of time there has been a tendency for (specified) deviant performance to occur in a particular situation, e.g. a tendency for that driver to require excessive steering corrections while taking the third exit from a four-arm roundabout, together with advice about how to correct the error.

In the short term it is envisaged that feedback on performance will be based on aggregated performance in a particular situation. Before such a database is built up, support during a manoeuvre will not be given. Support during a manoeuvre is, as pointed out above, perhaps the least attractive of the options available with regard to the timing of support delivery. Instead 'warnings' will be used 'on-line', or feedback will be provided when the driver has left the critical situation, or 'off-line' when the driver has come to a halt. As we acquire an understanding of the effects on performance of interactions between components of the GIDS system, and as knowledge grows of the appropriate time course of the messages required to support performance, it is envisaged that support for individual aspects of driver performance will become increasingly 'on-line'. Ideally this development of the way in which the prototype can potentially aid drivers, i.e. from off-line to on-line support, will be guided by the empirical evaluations of the first GIDS prototype planned for later this year. It is also conceivable in the longer term that PSALM could, once a criterion is exceeded, inform the driver that performance has not been adequate, and suggest a local route which requires performance of the relevant 'problem' activities. This route could then be driven for practice purposes, with PSALM giving support while a manoeuvre is actually being performed, so that the errors in performance are identified and corrected under PSALM supervision.

As previously mentioned, most of the information in drivers' performance profiles will initially be based on estimated normative performance. As data become available these estimated parameters of performance will gradually be replaced by empirically established performance indices. This will help to make PSALM less tentative and, in turn, help to identify the potential that GIDS-type systems have for improving driver performance. Speculating beyond the first prototype, it is considered that from a successfully implemented PSALM-type system, benefits would accrue not simply to an individual's driving performance, but also to information management within his/her vehicle and the collection of accident related information.

The application we have briefly described here attempts to take advantage of the fact that much of the information, which will be computed and collected by the GIDS system, will not actually be used if our current understanding of, for example, the collision avoidance system is accurate. It may be worth noting here that PSALM is perfectly analogous with the existing vehicle systems which sense and store information on physical indices of, for example, engine performance and running hours, which are used to indicate servicing requirements. The essential difference being that it is the driver's performance, rather than that of the vehicle, on which a 'diagnosis' or 'prognosis' is based. We plan to carry out very basic evaluations of PSALM in middle to late 1991, and hopefully extend PSALM to a wider range of road systems and develop a simulation based tutorial system, based on PSALM, in the course of the next few years.

Conclusion

The intention of this paper has been to show that drivers need support in order to attain and maintain reasonable levels of performance. Current driver training practices do not, and perhaps cannot, achieve this. Instead, it was suggested, we need to consider the ways in which current ATT/RTI developments can be harnessed in order to enhance drivers learning and, as a result, their ability to drive safely in a wide range of conditions.

References

Adams, J.A., 1987, Historical review and appraisal of research on the learning, retention and transfer of human motor skills, *Psychological Bulletin*, **101**, 1, 41–74.

Anderson, J. R., 1983, *The architecture of cognition*. Cambridge, MA: Harvard University Press.

Balzer, W. K., Doherty, M. E. and O'Connor, R., 1989 Effects of cognitive feedback on performance, *Psychological Bulletin*, **106**, 3, 410–33.

Berry, J. and Broadbent, D. E., 1988, The combination of explicit and implicit learning processes in task control, *Psychological Research*, **49**, 7–15.

Brehmer, B., 1980, In one word: Not from experience, *Acta Psychologica*, **45**, 223–41.

Brown, I. D., Groeger, J. A. and Biehl, B., 1987, Is driver training contributing enough to road safety? in, Rothengatter, J. A. and de Bruin, R. A. (Eds) *Road Users and Traffic Safety*, Assen/Maastricht, The Netherlands: Van Gorcum, pp 135–56.

Brunswik, E., 1952, The conceptual framework of psychology, in, *International Encyclopedia of Unified Science* (Vol. 1 No. 10), Chicago: University of Chicago Press.

Duncan, J., Williams, P., Nimmo-Smith, I. and Brown, I, in press, The control of skilled behaviour: Learning, intelligence and distraction, in, Meyer, D. and Kornblum, S. (Eds) *Attention and Performance* XIV, Hillsdale, NJ: Erlbaum.

Forsyth, E. and Kompfner, P., 1991, Driver training and testing, *Proceedings of Safety '91*, 1–2 May 1991, *Transport and Road Research Laboratory*, Crowthorne, UK.

Groeger, J. A., 1991, Supporting training drivers and the prospects for later learning. *Advanced Telematics in road transport*, Vol. 1, pp 314–330, Amsterdam: Elsevier.

Groeger, J. A. and Grande, G. E., 1991, Support received during drivers' training, in, Kuiken, M. J. and Groeger, J. A. (Eds) *Report on feedback requirements and performance differences of drivers.* Deliverable report DRIVE 1041/GIDS ADA 02. Haren, The Netherlands: Traffic Research Centre, University of Groningen.

Hoch, S. J. and Loewenstein, G. F., 1989, Outcome feedback: Hindsight and Information. *Journal of Experimental Psychology: Learning, Memory and Cognition*, **15**, 4, 605–19.

Kuiken, M. J. and Groeger, J. A. (Eds), 1991, *Report on feedback requirements and performance differences of drivers.* Deliverable report DRIVE 1041/GIDS ADA 02. Haren, The Netherlands: Traffic Research Centre, University of Groningen.

Swinnen, S. P., Schmidt, R. A., Nicholson, D. E. and Shapiro, D. C., 1990, Information feedback for skill acquisition: Instantaneous knowledge of results degrades learning, *Journal of Experimental Psychology: Learning, Memory and Cognition*, **16**, 4, 706–16.

9

Special information needs among professional drivers

G. Berge

Introduction

Information technology for use in cars and for general use in the traffic is under rapid development. This is reflected in many research and development programmes concerning road traffic informatics (RTI). Today, it is not a problem to buy RTI-devices. In Japan, where they have most experience with RTI, many bus companies are using a localization system developed from the Japanese CACS-project. Most of the Japanese car manufacturers also offer navigation systems as extra equipment in their cars. In USA they have for some years sold navigation systems with digital maps presented on a TV-screen in the car. This system has been used by several transport companies in California. Even mobile radio systems, which can give the position of the car, navigational help and normal communication, are marketed in USA. In Europe it is possible to buy the navigation system TRAVEL PILOT in Germany and the information system TRAFFICMASTER in Britain.

RTI involves control and information systems. The possible success of these systems lies in the importance of having a realistic view of drivers' information requirements. This is of value both for manufacturers and future users of the technology. An analysis of information requirements based on hypotheses about future technology, however, is difficult to make. For instance, one problem is to distinguish between needs, future demand, wishes and desires.

This paper presents results from a study concerning the special information needs of professional drivers. The study is part of the work in DRIVE I project V1024 driver information systems. The results are mainly based on qualitative semi-structured, in-depth interviews with representatives of certain driver groups. The purpose of the interviews was to get hold of the information various groups of professional drivers use to accomplish their daily work—*independent* of technological solutions. The interviews were conducted with Oslo and Athenian taxi, delivery van, long-distance and bus drivers. In DRIVE

I there were two projects dealing with information requirements specially related to systems for professional drivers (V1027 EUROFRET and V1044 FLEET). These two projects are mainly limited to information requirements related to fleet management systems. Other RTI systems which might be useful for professional drivers, however, are route guidance systems, traffic information systems and various telecommunication systems. Much work related to the information requirements of professional drivers is limited either to one type of system or to one driver group. The approach in this paper is more comprehensive. The main purpose here is to focus on the wide range of information requirements the professional drivers have, and to structure this information in an understandable conceptual framework.

The drivers needs are structured in relation to four information categories related to the main tasks of driving, and to six information levels. The paper gives a fairly good view of the variety of information requirements, thus giving a review of which types of information professional drivers need in various situations. But it does not show how representative the requirements are among all drivers. To give a basis for measuring future demand for new technology, further work is needed.

Typology of information and drivers

To be able to structure the professional drivers diverse information requirements it might be useful to introduce a typology of information. In the DRIVE project V1024 a typology based on the combination of four categories and six levels of information has been introduced. (Berge *et al.*, 1990).

Categories of information

Driving can be seen as an activity consisting of numerous tasks/decisions all based on information. The information drivers need to perform these various tasks/decisions can be divided into four main categories. The four categories are as follows. (1) *Basic information* which is vital for traffic safety and connected to the routine tasks of driving. (2) *Regulatory information* concerning traffic regulations given by the road authorities to regulate the traffic in general and to guide or direct the individual driver's behaviour. (3) *Additional information* which is useful and desirable for the driver because it may contribute towards making the traffic and the individual driving more flexible, efficient and comfortable. (4) *Service information* which in principle is not directly connected to the driving task or general traffic conditions, but useful for the driver related to the purpose of driving. The service information is group specific and can be valuable for making a successful trip, whatever the purpose might be. Table 9.1 shows the most important elements of the task of driving (OECD, 1988) and the information objectives connected to the different information categories.

Table 9.1

Type of information	Basic information	Regulatory information	Additional information	Service information
Driver tasks	● Road ● Traffic ● Speed control ● Manoeuvring	● Rule compliance	● Strategic (planning/ decision making) ● Navigation	● Driving purpose ● Work
Information objectives	Manoeuvring the vehicle in the traffic	Legal driving	Efficient and comfortable journey	Group specific

Levels of information

The term 'information level' is used in relation to the origin and source of basic information (Rumar, 1985). The term can also indicate the content of the information. In addition to the three usual vehicle, road and traffic levels, the typology of information developed in DRIVE project 1024, introduces three more significant levels: the trip, infrastructure and cultural levels.

The information the drivers get at the six different levels is as follows: At *vehicle level* the drivers get information about the vehicle, e.g. the vehicle's technical specifications and information about the goods or passengers in the vehicle, including information about documents and licences about the goods and vehicle required for international transport. At *trip level* they get information about the trip; duration, destinations, routes etc. At *road level* the drivers get information from and about the road and the immediate physical surroundings like location of bus stops, special lines for public transport and also road network information. At *traffic level* they get information on traffic density, average traffic speed, congestion warnings etc. At the *infrastructural level* they get information related to the physical environment and the infrastructure like available parking space, location of petrol stations, information about tourist attractions, hotels etc. And finally, at the *cultural level* the drivers get information about customs, language and behavioural aspects in the area concerned.

The typology

The different information categories and the various information levels can be seen in combination. This combination and the information categories are the basis for a typology of information.

Table 9.2

Levels	Categories			
	Basic	Regulatory	Additional	Service
Vehicle	●	●	●	●
Trip		●	●	●
Road	●	●	●	●
Traffic	●	●	●	
Infrastructure			●	●
Cultural				●

The useful combinations are marked with ●

Driver categories

There are several groups of road users whose information requirements may vary considerably. On the basis of different types of driving purposes or activities, one can distinguish between three main categories of drivers: (1) private car drivers using the car for private reasons; (2) mobile professionals using the vehicle in their work, but whose main task is not transport of goods or persons (maintenance, repair, emergency etc.); and (3) professional drivers whose main work is to transport persons or goods. All these groups need the same basic information. They also need more or less the same regulatory information. The special information needs of professional drivers, however, are mainly related to additional and service information.

Requirements for additional information

The professional drivers' need for additional information is mainly dependent on how well the driver knows the vehicle and how experienced he or she is, on whether the driver usually drives a certain route, whether the choice of route is freely made, or whether he or she is familiar with the surroundings and destinations for the trip.

Vehicle level information

The general need for vehicle level information among professional drivers contains the following elements: vehicle technical characteristics such as passenger capacity and dimensions; engine characteristics such as horsepower and fuel consumption rates; and routines for emergency situations. These are static information about the vehicle. Useful dynamic and semi-dynamic information on vehicle level is information about the current condition of the vehicle: state before and while driving; warnings if certain limits are exceeded;

car diagnostics and assistance with technical problems; car maintenance recommendations; and the vehicle history, such as previous accidents, last date of replacement of tyres, brakes and oil and major engine overhauls and problems.

Road level information

Road level information seems to be very important for professional drivers not driving fixed routes. This knowledge is a main part of their profession. The information they need and use in their daily work is information about the road network such as: general map information; one way streets and direction of traffic; road names and house numbering system; information and warnings about tunnels and bridges along the road, their dimensions and entry points; and road surface conditions on roads in the road network. It is also important for most professional drivers to know about special attention points such as closed roads and mountain passes, duration and reasons for closing, dangerous turns, steep descents/ascents and permanent or temporary obstacles.

Traffic level information

General traffic level information useful for most professional drivers is information about traffic density on given streets, average traffic speed in a given area, traffic congestion warnings and forecasts, reasons for such incidents, and approaching emergency vehicles. Useful traffic level information also includes forecasts about incidents and special events or arrangements that may cause congestion or prevent the smooth flow of traffic, such as: strikes, demonstrations, riots, public holidays and expected mass exits of citizens, malfunctioning traffic lights, and sports, concerts and cultural events. Professional drivers are generally more interested in, and put more effort into getting, traffic level information than other drivers. This is because strategic driving is the main part of their work. For taxi and delivery van drivers, listening to traffic radio whilst driving is considered part of their job.

It is meaningful to stress that the need for traffic information depends on the general traffic conditions in the area. Differences between Oslo and Athens can illustrate this point. In Oslo where the traffic conditions varies through the day and throughout the city, traffic information is regarded as useful and sometimes necessary to carry through the professional driving task. This is reflected in the information given to, for instance, taxi drivers through their internal in-vehicle information print-out system, and statements from the Oslo interviews. In Athens this is different. The centre of Athens is characterized by almost permanent traffic congestion. Information about the traffic is consequently not regarded as useful. Traffic problems and congestion can not be solved by information to the drivers alone.

Infrastructural level information

Important infrastructural information for professional drivers are; location of petrol/gas stations and opening times and fuel prices, distance to the nearest petrol stations when running out of fuel, location and opening times of car repair and service stations, services provided and prices normally charged, closest service station in cases of malfunctions or major technical problems, location of special parking places for heavy vehicles and location and prices charged at toll stations. Way of payment at all the forementioned places may also be useful. Location of the nearest telephone booth, cafeteria or rest area may also be useful and even in some cases important. For long distance lorry drivers it is vital to know about opening hours of customs, and routines and necessary documents to fill in and present to the border officers in order to clear car and cargo.

Weather information can be characterized as information on the infrastructural level. As long as the roads are open for traffic, professional drivers are not particularly interested in information about bad weather in order to find an alternative route. The drivers in question are experienced and trained to handle most situations and are more interested in reaching their destination in due time than enjoying the journey (Elvik, 1986:12). The professional drivers use the weather report to decide whether to put on tyre chains or not. A private driver like a tourist, on the contrary, would presumably be more interested in the weather forecast so that he or she—in case of bad weather— could change the route or even find another destination.

Trip level information

Trip level information concerns all information related to navigation, route guidance and trip planning. Concerning route guidance, the drivers interviewed in Oslo and Athens did not appreciate this kind of information when given as turn directions in or before crossings. They wanted more general route guidance. Taxi and delivery van drivers, for instance, were sometimes in need of information about a landmark near the destination address and a suggested route from this known place to the actual address.

Long distance lorry drivers often drive in unknown areas. One strategy used to find the right destination in unfamiliar cities is to hire a taxi to show the way. The need for route guidance information, however, is specially felt when roads are unexpectedly closed by an accident, roadworks, snow etc., and in countries where direction signs along the roads are local and do not indicate routes to follow to reach distant destinations. The long distance lorry drivers also expressed a need for ferry information, a benefit if reservation was possible.

Even if regular buses in cities drive along fixed routes, some bus drivers expressed the need for route guidance information. In case of an incident blocking the original route, they wanted to get information about alternative routes. Bus drivers also wanted to be informed in case of an unexpected congestion along the route, and to know the reason for this congestion.

Requirements for service information

In their work the various driver groups must cope with different situations which may require special information. Thus requirements for service information vary considerably between trades and types of organizations. Disagreements about information requirements also take place between personnel in the same transport organization. The disagreements may concern the information flow, who shall have what information etc. For instance it was registered that in order to confirm information about opening hours, delivery time etc. drivers of a long distance cargo company in Oslo wanted more direct contact with the customer. The head office did not approve of such direct contact because they were afraid of losing control.

Conflicts about information distribution often arise when new information technology is implemented in work organizations. Thus it is important to recognize that a driver's service information requirements are not always objectively given. The drivers must, however, have satisfactory conditions of work and be supplied with sufficient information to fulfil the objectives of the work tasks.

The main working tasks

For the professional drivers the main tasks of work include to load, or pick-up passengers, parcels, letters or goods at various places, and to drive to the right destination as safely and effectively as possible. It also includes contact with head office and/or the customers, the handling of goods or parcels according to regulations or instructions, managing the customs, doing the required paperwork, keeping of accounts and if required, to find suitable places for overnight stops. The following paragraphs will point out what service information professional drivers require in order to accomplish some of these tasks.

Loading and unloading

When loading and stowing goods, service information on vehicle level is of great importance to the driver. The driver must know all technical details about the loading capacity and regulations for his vehicle. The driver must further have information about the weight of the parcels and product speci-fications. He or she needs to know whether he or she is transporting dangerous goods, eventually how to handle the goods under transportation and what to do if an accident should occur. Drivers also need information about the actual work that has to be done at the customer's place. In the planning and performance of their work, it is advantageous to be informed about time to wait before unloading, expected duration of the unloading process, and which documents have to be signed and delivered. This is part of the service information at the infrastructural level.

New orders and new customers

The professional drivers will need trip information such as where to pick-up or deliver the passengers or the goods. Usually there is no problem to get the destination address, either from head office or from the customers themselves. The information channels most frequently used are public telephones (55 per cent) and face-to-face communication (about 25 per cent) (DRIVE project V1044 FLEET).

The adoption of the just-in-time concept demands drivers to be more precise than before. Expected delivery or pick-up time represents information that is both important and urgent to the driver. The just-in-time concept also influences the significance of getting traffic information and route guidance to help drivers to reach the destination at the agreed time of delivery.

Taxi drivers sometimes confront a special problem; customers who do not know where to go. They know the name of the person, firm or restaurant they want to go to, but do not know the address. Infrastructure level information about addresses of firms, restaurants and various institutions is required in these situations.

Address-linked information

Usually the drivers easily find the pick-up or delivery address. It is, however more difficult to find the right entrance or contact-person. Hence, drivers driving to several addresses sometimes express the need for what can be called address-linked information. This infrastructure level information is related to the destination address, i.e. access paths from the main road to the given address, access routes, entrances and floors depending on the firm, department or person to be reached, permanent obstacles on the premises, where to park the vehicle, location and type of loading and unloading ramp etc.

Use of new technology

As shown, the professional drivers use a lot of information in their work. To get the right amount of information at the right time to solve the different working tasks is important for the professional drivers' work environment. A new driver information system can help the driver in mainly three ways. (1) It can help the driver to make better decisions by giving the driver more information or information of higher quality than previously. (2) It can release the driver from taking decisions by taking over parts of or the whole decision making process. And (3) it can reinforce imperative information which the driver either is not attending to, or not able to attend to because of physical hindrances.

Although many new driver information systems are meant to facilitate the provision of information through an in-car information system, some types of information are less well suited for such presentation. Most of the basic information is of this type. If basic information should be presented in a new in-car driver information system, it should only be as a reinforcement of information as described above. Any information from controls within the car indicating malfunctions in the car lights and braking system, however, should be incorporated in a new information system. Any presentation of information inside the car, basic, regulatory, additional or sevice information, which duplicates the information the driver already can see in the area outside the car where attention is focused, is not to be recommended.

The information typology presented indicates the general priority of information to the driver. In a comprehensive driver information system where all categories of information are presented, basic information must be prioritized first, then regulatory information. The prioritization between additional and service information can be individually decided by the driver. The most important is that the presentation of additional and service information is not interfering with the driver's reception of basic and regulatory information. Therefore, the presentation of the different types of information in the typology may require different man-machine interface (MMI) solutions. (Midtland, 1990).

The information typology gives further a basis to decide on what kind of information systems should be open to all drivers, and which ones only to closed user-groups. It also indicates what kind of information the authorities should be responsible for providing to the driver, and what kind of information can be given on free enterprise. Further it indicates what kind of information can be given in autonomous systems and in infrastructure based systems.

The Athens and Oslo interviews do not give a basis to range the additional and service information according to importance, nor for measuring future demand for new technology. Nevertheless, it seems evident that many drivers could benefit if they could utilize the possibilities of new information technology in their work. The challenge of managing difficult situations with limited information would change, but because of high competition and pressure of delivery times, it seems after all that the drivers wish and need to get better additional and service information than they get today. The question, however, is whether they, in the future, will change to new technological solutions to get this information. The answer may lie in the two words; cost and simplicity.

References

Berge, G., Midtland, K., Kolias, V. and Papadoupolos, D., 1990, *Driver information requirements*, Oslo: DRIVE project V1024 Driver Information Systems, Interim Report I1-1.

DRIVE Project V1044—FLEET, 1989, *Field research activities with respect to freight and commercial vehicle management systems* (Progress Report 2/89).

Elvik, R., 1986, *Radio Haukeli*, TOI-Report 1986, Oslo: Institute of Transport Economics.

Midtland, K., 1990, *Drivers' requirements for reception of information*, Oslo: DRIVE project V1024 Driver Information Systems, Interim Report I1–5.

OECD Scientific Expert Group, 1988, *Route guidance and in-car communication systems*, Paris.

Rumar, K., 1985, *In-Vehicle Information Systems*, Sweden: Lindköping, Swedish Road and Traffic Research Institute (VTI).

10

Elderly drivers and new road transport technology

A. M. Warnes, D. A. Fraser, R. E. Hawken and V. Sievey

The problem

Safe and efficient roads with mass car ownership

The European general public, no less than traffic scientists, are acutely aware of several dimensions of today's traffic problems. The costs of congestion, delay and, to a lesser extent, pollution are leading complaints, and the negative consequences on public transport services are widely understood. These costs mar the much valued benefits of mass motorization. Fortunately traffic deaths and injuries directly affect only a small number, but underpin the consensus for vigorous road safety measures.

Some other implications of recent traffic trends are less widely understood. Contemporary roads and traffic present a greater diversity of traffic conditions and demands than thirty years ago. Today's driver needs to deal with high-speed motorways, more complex junctions and route choices, and more highly congested urban roads. At the same time, car driving has become more critical for satisfying mobility needs and less concentrated among middle-aged adults, men, the more affluent, the highly educated, and professional drivers. Recent decades have seen absolutely and relatively more adolescents and elderly people on the roads (Automobile Association, 1988). Different age groups have different physiological and cognitive capacities, differing durations of driving experience, and different attitudes to driving risks. Only a minority of elderly women drive (although their number is increasing rapidly), and more of them than men learnt to drive in middle age. The age at which people learn to drive, and the length of their driving experience, may be related to their age-specific driving competence (Warnes, 1991). Increasing dependence on the private car may be forcing people who are less confident or less able drivers to continue using their cars.

Research and policy objectives for the European road system

The objective of the European Commission DRIVE research programme is to make European roads safer, more efficient, and less environmentally damaging. This paper reports findings from Project V1006, 'Factors in Elderly People's Driving Abilities' (DRIVAGE). This has investigated the potential of 'road transport informatics' (RTI) for helping older drivers. The DRIVE hypothesis is that new forms of information technology can contribute to safety, efficiency and environmental objectives. The functions of RTI systems are potentially numerous, from automatic debiting associated with road pricing, through in-vehicle route planning and guidance systems, vehicle fault diagnosis and warnings, central systems of vehicle identification, emergency assistance and control, to interactive communication between drivers and traffic controllers. All functions, even unobtrusive systems of road charging, cause some change in the concerns and actions of drivers. These effects, which may or may not be beneficial, need investigation. Few functions are yet available as products or as prototype systems.

Problems associated with older drivers and possible solutions

RTI devices and systems if suitably designed may be particularly valuable for novice, disabled and older road users. Much is known about the limitations and difficulties of older drivers (Markovitz, 1971; Carp, 1972; Hillman *et al.*, 1976; Taira, 1989; Salthouse, 1990; Warnes *et al.*, 1991). One body of research has focused on the relationship between increasing age, and pertinent physiological and cognitive capacities. Decrements in visual function, reaction times, and mental processing capacity are most frequently documented, often on the basis of controlled laboratory trials. Accident data show the specific vulnerabilities of elderly drivers and in which traffic situations and driving tasks they are at most risk. The consensus is that complex situations make heavy demands on reactions, information processing and decision making (although there have been contradictory assertions). To date in Europe, elderly people have been most exposed to traffic hazards as pedestrians and car-passengers: risk factors for the former have been studied, but neglected for the latter.

Stereotypes of the hazardous elderly driver are less accurate as descriptions of the majority than of a least competent minority. Nonetheless some elderly (and non-elderly) drivers are handicapped by poor eyesight, slow reaction times, lack of muscular strength and dexterity or susceptibility to fatigue. If RTI devices could be developed that transferred necessary information from sight to hearing, i.e. from an over-loaded to an under-loaded channel, or which redistributed the timing of messages away from high frequency peaks, then support would be given to the driver's critical senses and psychomotor functions. One example may illustrate the general principle. Route finding decisions may contribute to a driver's difficulties at an urban road junction or motorway intersection. If the driver could be presented with information that enabled a clear path to be decided before the junction is negotiated, more attention would be available for tracking, braking and speed control.

The 'DRIVAGE' research on elderly drivers

Objectives and design of the research

DRIVAGE has developed new experimental methodology and conducted a questionnaire survey of elderly drivers. The three partners were the Traffic Research Centre (TRC) at the University of Groningen, Netherlands, and the Division of Engineering and the linked Age Concern Institute of Gerontology and Department of Geography at King's College, University of London (KCL). The partners have contributed respectively expertise in cognitive psychology, information science and social gerontology (Ponds *et al.*, 1988). A low cost interactive-video driving simulator has been developed at KCL and an adapted computer-based test of divided attention at TRC. These have enabled tests of performance on primary driving tasks with and without simulated RTI assistance. The objectives, participants and work programme of DRIVAGE have been fully described elsewhere, but are summarized in four goals (Warnes, 1989):

- to build a low-cost semi-realistic driving simulator using interactive video technology,
- to modify the software of a divided attention task experiment to investigate the impacts of simulated RTI devices,
- to carry out experiments on the implications of providing drivers with additional information about the vehicle or the road system on primary driving tasks (tracking and speed/separation distance holding),
- to investigate the specific impacts of RTI on older drivers.

The project studies selected aspects of the driving performance of elderly drivers (and control groups of younger, experienced drivers) with and without the assistance of RTI devices. The KCL experiment simulates following a 'lead' car and tests distance-keeping and speed-control abilities. The TRC experiment tests tracking performance. In both, the impact is tested of presenting RTI-based information with and without the presence of a distracting task (which represents the operation of ancillary vehicle controls as for heating, lights or wipers). In the Netherlands and the United Kingdom, a social survey of elderly drivers has collected information on car use and dependency, problems with driving and traffic conditions, driving tactics and strategies, and self-scaled ratings as drivers. The elderly driver subjects have completed both simulator runs and mail questionnaires. Six focused group discussions on elderly drivers' attitudes to RTI functions and devices have been held in London (Sixsmith *et al.*, 1990). A common database and a common programme of data analysis has increased the sample size for the specification of age-group differences and enabled comparisons between Dutch and British drivers.

The semi-realistic simulator

The semi-realistic driving simulator (SRS) has four components: a Renault body, a video disc player, a 660 mm screen, and an Archimedes 440 computer (Figure 10.1). The video disc holds images of real road scenes taken from a vehicle at 64 kph. These images are shown at a rate controlled by the 'driver' using the conventional brake and accelerator pedals connected to the computer. To minimize distraction of the driver's attention, the side windows and the field of vision beyond the monitor are masked with diffusing screens. The speedometer shows the simulated speed and, being driven by an electric motor, strengthens the illusion of driving by generating a noise level which is a function of speed.

Simulating the dynamics of a real car required iterative adjustments to the responsiveness of the accelerator and brake design. Different positions of the accelerator correspond to the speed demanded within the feasible range of the vehicle. A multi-ratio gear box was simulated by different constants in different speed ranges, but the drivers were not aware of this and were told that the car had an automatic gear box. A single parameter describes the accelerator's effect. A change in accelerator position implies a demanded speed change. Sixty-three per cent of the demanded change is accomplished within 10 seconds. Further details of the construction and control program of the simulator are available (Fraser *et al.*, 1990).

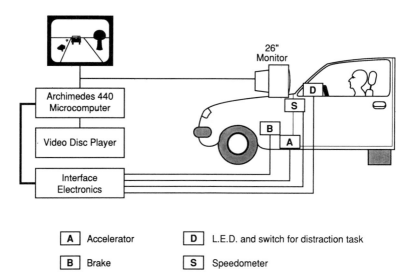

Figure 10.1   The components of the semi-realistic simulator.

The primary driving task simulated by the equipment was separation-distance keeping at variable speed. During a four-minute run, the lead car seen on the screen accelerated from a standing start to 70 kph and thereafter changed speed on three occasions. Subjects were instructed to 'follow the lead car at a safe and comfortable distance'. Data on the speed and separation distance of the following car, and on accelerator and brake positions, are stored every 16 frames of the display, corresponding to every 320 msec, i.e. about three times a second.

Alternative approaches to the representation of a RTI separation distance warning device were developed. The intention was to provide information that a potentially dangerous situation had arisen. Various warnings were devised to be activated when the two cars were 26 m apart or closer. A continuously presented head-up warning was first developed, comprising a variable height 'thermometer' column with a colour division at the recommended separation distance. This provided sufficient information to perform the car-following task. It could replace the normal visual stimuli. Some drivers were distracted totally from normal driving. The variable column was therefore replaced with a simpler head-up warning panel. This appears low-centre on the screen when the separation distance falls below 26 m and occupies 2 per cent of the road scene.

A second addition to the simulation was a small warning light mounted on the upper surface of the dashboard, just below the image of the road scene. This was intended as an unhelpful distraction. The light is cancelled by a spring-loaded switch mounted on the turning-indicator stalk which projects from the steering column: it is activated by the computer program five seconds after a cancellation. The number of cancellations during a test run of four minutes allows computation of the mean time to complete the task.

Five simulator runs were conducted with each subject. The first and last (R1 and R5) involved only the separation-distance keeping. The three central runs appeared in random order and involved: distance keeping with the distance warning panel (W); distance keeping with the distracting light (D); and distance keeping with both the warning panel and the distracting light (WD). Shortened versions of R1, WD and R1 formed a six minute practice session.

The principal hypotheses which have been tested and questions explored during the research are:

- that performance on the primary distance-keeping task is related to age,
- that there is a different effect of the warning panel on young and old drivers (tested by comparing the W and R1 runs and the WD and D runs),
- that the effect of the distracting light is related to age (tested by comparing the D and R1 runs and the WD and W runs),
- that driving experience (years of driving) is positively related to performance on the primary driving task,
- that actual driving distance during the last year is positively related to performance on the primary driving task in the tests,

- that learning improves performance within the experimental task (tested by comparing R1 and R5),
- that the sequence of the three central runs is unrelated to performance.

Results from DRIVAGE are being reported in a series of papers. The Groningen simulation is described by van Wolffelaar and Rothengatter (1990). The present paper concentrates on the drivers' performance on the primary distance-keeping task with and without RTI support and the distracting task. The results are analysed by age, sex and driving experience and in relation to the drivers' self-reported difficulties with driving and in traffic.

The experimental group of elderly drivers

The early 1990s are the years of fastest growth of the elderly driving population in the United Kingdom. Licence holding by 1989 had attained 50 per cent prevalence among those aged 60–69 years, but is still most common among males and above average income groups. It is less common at older ages. Fewer than one-fifth of women over 70 years of age and of those living alone have licences (Warnes, 1991). The age/sex groups which showed the greatest changes in licence holding between 1975/76 and 1988/89 were males aged 50 + and women aged 30–59 years (Table 10.1). Most of the European Community nations have experienced similar growth (OECD, 1985). This recent rapid spread means inevitably that today's elderly drivers are a biased sample of the elderly population. They tend to be younger, more affluent and more highly educated. Secondly, the present-day elderly driving population has characteristics specific to the first cohort in which a majority drive. High proportions have been professional drivers and learnt to drive late in life. A sample of elderly drivers today neither represents the contemporary elderly population nor future or past drivers.

The DRIVAGE study panel of elderly drivers completed a mail questionnaire and gave up half a day to use the simulator located in central London. Only those aged at least 55 years, resident in the London Metropolitan Area, and who had driven 1600+ km in the previous year were eligible. Ninety-six males and 58 females completed the mail survey and of these 121 completed the simulator test. The age range was up to 86 years, with 35 per cent in their sixties and 25 per cent aged at least 70 years. Seventy-seven per cent of the men, but only 54 per cent of the women were married. Among both sexes, 10 per cent were single and 13 per cent widowed (much under-representing the prevalence of widowhood in the population).

About half the sample lived within 15 km of central London and 86 per cent within the former Greater London County, i.e. within approximately 20 km from the centre. Three-quarters had lived in their present locality for at least ten years, and one-half for at least 30 years. The ages at which people left full time education were low, beginning at 12 years. Fifty-nine per cent had left school aged 16 years or less. Nevertheless, their educational attainment was

Table 10.1   *Car-driver licence holding: United Kingdom 1975/76–1988/89.*

Age Group (years)	Males				Females				Persons			
	1975 –76	1978 –79	1985 –86	1988 –89	1975 –76	1978 –79	1985 –86	1988 –89	1975 –76	1978 –79	1985 –86	1988 –89
					Percentages of the population							
50–59	75	72	81	85	26	26	41	48	51	48	60	66
60–69	59	59	72	78	15	16	24	28	36	36	46	50
70+	33	33	51	55	5	7	11	14	15	16	33	31
All ages	69	68	74	78	29	30	41	47	48	48	57	62

Source: UK Department of Transport, *Transport Statistics: Great Britain 1979–89*, HMSO, 1990, Table 2.15, p. 71.

high. Sixty-one per cent had passed school examinations (GCSE or equivalent) and 25 per cent had a university degree or equivalent higher education qualification. Many had professional and vocational qualifications. The volunteer panel was undoubtedly a high achiever group.

The reported incidence of illness over the last ten years was low. Fifty-nine per cent wore spectacles for driving, nine per cent were colour blind and 15 per cent had a hearing impairment. The highest ten-year incidence (30 per cent) of health problems was back pain, and a quarter reported arthritis or rheumatism. Other problems reported were hypertension (10 per cent), other muscular pain (10 per cent), bronchial illness (nine per cent), heart attack (six per cent), broken bones (six per cent) and diabetes (four per cent). The highest current prevalence rates were for arthritis or rheumatism (20 per cent), back pain (12 per cent) and hypertension (six per cent). Where an illness was mentioned, 64 per cent said that it had no effect on their driving. Forty-one per cent were taking long-term medication, however, but of these only seven per cent said that it currently affected their driving. Given that there are no comprehensive data on the characteristics of elderly British drivers, it is impossible to test rigorously the representativeness of our panel. Their concentration in the middle classes—91 per cent were owner occupiers—is probably replicated in the driving population. It is likely, however, that the panel under-represents those with the greatest health problems and driving difficulties. A control panel of 25 experienced drivers aged 30–44 years was also tested.

Results on simulated driving performance

Three summary measurements have been found to be the most useful ways of describing drivers' performance. These are: the mean separation distance between the lead car and the chase car; the standard deviation of speed difference between the two cars, and the mean delay in responding to speed changes. This section presents results in terms of each of these measures, and then considers the effect of the different tests that each subject completed.

Mean separation distance

This is a matter of the preference of the driver, rather than of ability or skill. One hundred and forty-six subjects completed the test runs and the associated attitudinal questionnaire. Of these, 121 were aged 55 + years, and 25 were aged 30–44 years. For both old and young drivers, the mean preferred separation distance was 33 metres and there was no significant difference between the age groups (Table 10.2). The standard deviations in both the samples were small and the difference between the elderly (+ 0.6m) and the control panel (+1.2m) mainly reflects the respective sample sizes. Those aged

Table 10.2   Mean separation distance on the SRS simulator (metres).

Test type	R1	R5	W	D	WD
Elderly subjects					
Mean	32.9	32.1	40.0	34.1	39.7
Standard error					
of mean	0.65	0.54	0.34	0.54	0.32
Standard deviation					
of mean	7.2	6.0	3.7	6.0	3.5
Coefficient of					
variation	21.9	18.7	9.3	17.6	8.8
Control subjects					
Mean	33.1	33.2	39.6	33.1	40.4
Standard error					
of mean	1.17	1.21	0.54	1.36	0.62
Standard deviation					
of mean	5.7	5.9	2.6	6.6	3.0
Coefficient of					
variation	17.2	17.8	6.6	19.9	7.4
p(H0)	0.90	0.44	0.52	0.52	0.37

Notes:
1. R1 and R5 are the first and last (fifth) runs of the unmodified car-following task. W are runs in which the 26 m or less separation distance Warning panel was operational. D are runs in which the Distracting dashboard light was operational. Note the minimal differences in performance between R1 and R5.
2. The coefficient of variation is the standard deviation expressed as a percentage of the mean. The probability p(H0) tests a null hypothesis and states the likelihood of the sample means for the old and control panels being drawn from the same population. A two value of p(H0) suggests that there is a difference between the populations.

70 + years were the only ten-year age-group with a significantly different mean separation distance (Figure 10.2).

With 121 elderly subjects, the results show clearly the considerable variation in car separation distances among both older and control groups. The standard deviations of the means from both R1 and R5 for the 121 elderly subjects were approximately 0.20 of the means and around 0.175 for the 25 subjects in the control panel. An inferential test of the difference between the R1 sample variances of the old and control panels indicates, however, no significant difference at the five per cent confidence level (calculated $F = 1.6$, critical value $F = 1.8$ for 24 and 120 degrees of freedom and $p = 0.05$). The comparable results for R5 produce a variance ratio of exactly 1.0. Over the five runs the separation distance variance of the elderly subjects reduced from 51.8 to 36.2 (-43 per cent), whereas that of the control panel increased from 32.7 to 35.4 ($+8.2$ per cent). Neither change represents a significant alteration in the variance.

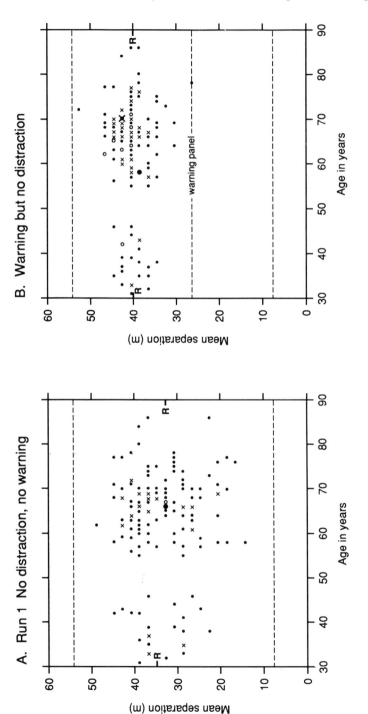

Figure 10.2   The separation between the lead car and the chase car, by driver's age.
A. Results for the first run given to each driver.
B. Data for tests where the driver was asked to respond to a warning panel (but without an additional distraction).

Two scattergrams (Figure 10.2) show the way that the warning panel altered both the drivers' driving style, and the mean distance between the cars. The horizontal dotted lines in Figure 10.2 show the limits enforced by the simulation. A driver whose mean distance is near to one of these lines must have been held on the limit for part of the test duration. The scattergrams show no apparent difference between the older and younger drivers. When the warning panel is active, the mean distance of the older and of the younger drivers increases by similar increments. Indeed the *mean* distance of every driver is more than the distance at which the warning panel was activated. The distribution of points in the scattergram suggests a normal distribution—means, standard deviations and standard errors are given in Table 10.2.

Standard deviation of speed difference

Two scattergrams (Figure 10.3) show the standard deviation of speed difference for each driver for R1 and for WD, where both a warning and a distracting task were presented to the driver. There appears to be a minimum of about 3 kph that drivers could not improve on, whereas the maximum value is not precisely defined. The distribution is not normal, and standard deviations, normal errors of means and significance of difference between means have to be viewed in this light (Table 10.3). Both groups of drivers did less well when asked to cope with the distraction and the warning, and on average the older drivers were more affected, as shown by the increase in the positive slope of the regression line which is located by the letter R on the two vertical axes. Notice, however, that a large fraction of the older drivers were better than the worst of the younger drivers. The two distributions overlap substantially. It would be difficult to identify a driver's age from his/her ability as measured by this test, and the converse applies.

The plots of standard deviation of speed difference show a few drivers with a performance much worse than that of the mass (Figure 10.3). This raises the question, 'Are there two groups of older drivers: "good" and "bad", with the good older drivers having performance indistinguishable from that of the younger drivers?'. To test this notion, the twelve worst drivers in each of the five runs were identified. If there are good and bad drivers, the bad drivers would be expected to turn up repeatedly, while if there is only a single population, then a driver occurring in one list would not be expected to occur in others. Table 10.5 (p. 114) shows the number of times a driver was found in the worst twelve in five runs 1,2,3,4 or 5 times, and the number of times this would be expected to happen by chance. Of the 60 cases (5×12), 8 drivers (7 per cent) account for 36 of the cases (60 per cent), so there is a strong suggestion that there are indeed a few 'bad' drivers in the older population. It should be noted that the terms 'good' and 'bad' here are derived from a single measure taken from a simulation of the driving task, and would be of greater value if independent measures confirmed the status of an individual.

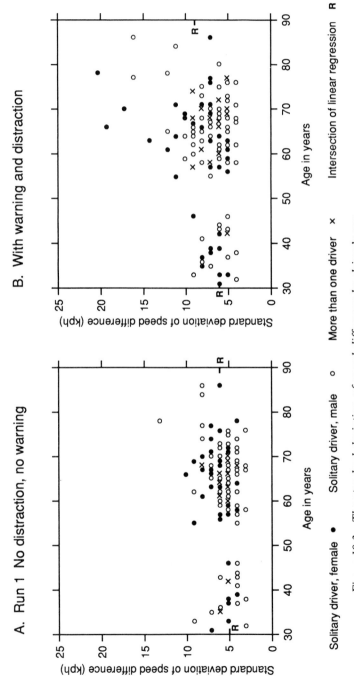

Figure 10.3   The standard deviation of speed difference by driver's age.
A. Observations for the first test taken by each driver.
B. Data for tests where the driver was asked to respond to both a distraction and warning panel.

Table 10.3   Standard deviation of speed difference (kph).

Test type	R1	R5	W	D	WD
Elderly subjects					
Mean	5.53	5.37	7.01	6.02	7.55
Standard error					
of mean	0.14	0.14	0.26	0.17	0.29
Standard deviation					
of mean	1.49	1.58	2.86	1.96	2.85
Coefficient of					
variation	26.9	29.4	40.8	32.6	37.7
Control subjects					
Mean	5.03	5.03	6.09	5.49	6.38
Standard error					
of mean	0.27	0.22	0.27	0.31	0.28
Standard deviation					
of mean	1.34	1.08	1.32	1.54	1.39
Coefficient of					
variation	26.6	21.5	21.7	28.1	21.7
p(H0)	0.11	0.37	0.016	0.15	0.04

These observations have implications for any policy or regulatory proposals concerning older drivers. It would not be appropriate to base decisions on average measurements, because these will underestimate the difference between the 'bad' older driver and the younger group, and will inappropriately label the 'good' older driver as different from the younger group, when many older drivers are no less competent than many of the younger drivers.

There was a difference between the standard deviations of the speed difference for the old (5.53 kph) and young (5.03 kph) drivers, although it was not significant at the 5 per cent level (Table 10.3). By run 5 the elderly drivers' mean performance had improved to 5.37 kph. This change was not significant at the 5 per cent level ($t = 0.81$, df 240). The control group mean was unchanged in run 5, but there was a marked though insignificant reduction in the variance from run 1 (from 1.80 to 1.17, -35%. $F = 1.60$). From run 5, the elderly and the control panels return variances which differ significantly, ($F = 2.07$). The theoretical value for df 120, 24 and $p = 0.05$ is 1.73.

Delay

This measure is found by time-shifting the records of the speed of the lead car relative to the chase car until the standard deviation of the speed difference is minimized. The measure is a reaction time for the system, including the driver's reaction time and the time for the car to change speed. It was expected that the delay would increase when extra tasks were to be performed by the driver, and this was found to be the case (Table 10.4).

Table 10.4 Delay to minimize standard deviation of speed difference (seconds).

Test type	R1	R5	W	D	WD
Elderly subjects					
Mean	2.45	2.36	2.60	2.95	2.61
Standard error					
of mean	0.11	0.93	0.09	0.10	0.10
Standard deviation					
of mean	1.21	1.04	0.96	1.13	1.12
Coefficient of					
variation	49.4	44.1	36.9	38.3	42.9
Control subjects					
Mean	2.33	2.39	2.69	2.93	3.80
Standard error					
of mean	0.20	0.21	0.18	0.26	0.18
Standard deviation					
of mean	0.98	1.04	0.87	1.26	0.87
Coefficient of					
variation	42.0	43.5	32.3	43.0	22.9
p(H0)	0.60	0.91	0.52	0.90	0.03

For the younger drivers, dealing with the distracting task or reacting to the warning panel increased the delay, and both tasks together appeared to impair the delay roughly additively. Note that the cognitive load for each of these tasks was set empirically at a level which was convenient for observing changes in performance. The elderly and the younger drivers returned insignificantly different mean system delay times of around 2.4 seconds. This measure showed the highest coefficient of variation of the three measures of primary performance, i.e. 42–49 per cent in runs 1 and 5 for both panels. Several results have confirmed that there is no significant learning or change in performance between R1 and R5 (Table 10.4).

For the older drivers each of the single extra tasks impaired performance, but the combination of warning panel and distraction task appeared to have been easier than the distraction task alone. This was, after all, what the warning panel was intended to do—to help the driver to drive well in difficult situations. Two interpretations can be put on this result. One is that the older drivers needed and used the warning panel, while for the younger drivers it was just a nuisance (though Figure 10.2 shows they responded to it). The other is that the younger drivers attended seriously to the distracting light, and cancelled it as a matter of high priority, while the older drivers were prepared to concentrate on driving well, and treated the cancellation of the distracting light as a mere distraction. This brings to prominence the need to consider attitudes as well as abilities when distinguishing groups of drivers.

The effect of the warning panel

The warning panel was displayed for only about 10 per cent of the duration of the simulator runs yet had noticeable effects on the subjects' performance on the primary distance-keeping task (but notice that while the panel was off when it might have been on, it still conveyed helpful information). It increased the separation distance between the lead and following car by around one-fifth or 7 m, with the older group of drivers showing a slightly greater effect than the control group (7.5 m or 23 per cent compared to 6.4 m or 19 per cent). For the elderly drivers, the change from run 1 is significant (Calculated t = 2.01. Theoretical t = 1.96 for df 240 and p = 0.025). Overall however no significant difference emerges between the older and younger drivers (Table 10.2). The warning panel also produces massive and significant reductions in the variance of the separation distance, by around 75 per cent for both panels (for elderly drivers F = 3.71, df 120 and 120, p = <0.01). As plots of individual subjects' simulator runs with and without the warning panel show, one of its clear effects was to reduce the duration of time at which drivers were close to the lead car. If replicated in actual traffic conditions, the maintenance of an increased minimum separation distance would be a safety benefit. It should also be noted that this increase in separation distance implies a similar one-fifth reduction in the simulated maximum lane capacity and an 'efficiency' disbenefit. The present study offers little guidance on the likelihood of the observed behaviour being replicated in actual traffic conditions.

When the warning panel is operational, older drivers return a standard deviation of the speed difference between the two cars of 7.01 kph, a 27 per cent increase, compared to 6.08 for the young drivers, a 22 per cent increase (Table 10.3). In other words, the younger panel maintained on the speed matching measure roughly a 10 per cent better performance without the warning panel and roughly 15 per cent better with the panel. The difference is significant with a probability of 0.016 under a null hypothesis. Another effect is substantially to increase the variance in the elderly drivers' perform-ance, although there is virtually no change in the variance among the control panel. It increases among the older drivers from 2.22 to 8.18 kph (F = 3.68, df 120, 120, p = <0.1). The warning panel had less effect on the mean delay or 'system response' time, with the older panel showing an increase of 6.6 per cent to 2.59 seconds, and the younger panel of 13.7 per cent, to 2.69 seconds (Table 10.4).

Change in performance with the distracting light

Operation of the distracting task of cancelling a light had relatively minor impact on the separation distance and the standard deviation of the speed difference, but a more substantial effect on the 'system response' time. Differences between the older and younger drivers were minor and never statistically significant. The older drivers' separation distance increased by around 5 per cent, but there was no change for the younger drivers (Table 10.2). The speed difference measure increased by roughly 10 per cent for both age groups, while the delay measure increased by 21 per cent for the older panel and 23 per cent for the younger (Table 10.5).

Interaction between the warning panel and the distracting light

When both the distance warning panel and the distracting light were operational, the behaviour of the older and younger drivers differed more markedly than in any of the simpler tasks. The older drivers' mean separation distance slightly decreased from the level of 40 m achieved with the warning panel alone, whereas the younger drivers showed a slight (and insignificant) increase in their separation distance (Table 10.2). Both older and younger drivers' ability to match the speed of the lead car was damaged by around 5–7 per cent, and a significant difference between the two groups was maintained with a probability of 4 per cent under a null hypothesis (Table 10.3). The most unusual results are for the 'system response' delay time, for the elderly drivers showed no change from the results obtained with only the warning panel operational, while the young panel returned a 14 per cent increase to 9.6 seconds. A significant difference at a level of 3 per cent emerges in the delay measure between the two age groups.

More than one explanation is possible for the divergence of the two age groups in this most complex task. It is clear that the younger drivers attended to the distracting light more rapidly, on average cancelling the light with a delay of 1.15 seconds during the run compared to a delay of 1.4 seconds for the older drivers. One hypothesis being examined through further analyses is that the older drivers attached less importance to the distracting light, and in practice chose more often to ignore it. In effect, the reason that there was little

Table 10.5   The number of times a driver was among the twelve worst drivers in a test.

Number of appearances	actual drivers	predicted drivers	actual cases	predicted cases
0	119	93.2	0	0
1	10	42.4	10	42.4
2	7	7.7	14	15.4
3	1	0.7	3	2.1
4	2	0.0	8	0.1
5	5	0.0	25	0.0
0–5	144	144.0	60	60.0

Predictions—a selection is made of twelve individuals at random from a group of 144, and this is repeated five times. The predicted frequencies are based on the number of times the same individual appears in the five selections. For instance, we predict that only one individual should turn up in three or more selections, whereas we find in practice that eight drivers are in the worst group three or more times.

change (or damage to) their performance on the primary task was because they concentrated upon it to a greater extent than younger drivers.

In summary the principal results and inferences from the experiments using the SRS are thus:

1. Drivers can and will increase their separation from the car in front if given information about this separation.
2. Many older drivers have performances in the range of younger drivers.
3. These is a small fraction of older drivers who consistently drive more erratically than any of the younger drivers.
4. Older drivers may have different priorities to younger drivers, and may be helped by suitable presentation of useful information.

Conclusions

When older drivers are asked to perform a single driving task, in this case the maintenance of a consistent separation distance from the vehicle in front, they perform only slightly less well than younger drivers as measured by the standard deviation of the speed difference between the two cars (Table 10.3). The difference is statistically insignificant and unimportant in relation to the variation in performance among the drivers of any—older or younger—age group.

As our colleagues in Groningen have found, however, when dual or multiple tasks are presented to drivers, the decreased abilities of older people become significant. The warning panel had a similar effect on old and young in terms of increasing the mean separation of the lead and following cars, but the older drivers became significantly less successful at matching the speed of the lead car. The presence of an obviously distracting task results in considerable differences in mean response by age, and the older drivers were far less attentive to the light-cancelling task.

The DRIVAGE results provide more comprehensive information on the relationship between age and aspects of (simulated) driving performance than have been available from many previous studies, and in particular have led to a distinctive interpretation of the characteristics of older and younger drivers. Our results indicate that the majority of older drivers perform indistinguishably from younger drivers on a range of simulated driving tasks. The principal contrast is that, particularly when faced with multiple simultaneous tasks, a minority of older drivers perform noticeably differently and worse. Correlates for this change in behaviour have not been identified: there is no evidence that sex, age (within the elderly age-group) or years of driving experience are responsible.

This intricate assessment of the relationship between age and driving performance arises from three characteristics of the project. Firstly, it has studied a relatively large number of elderly and non-elderly subjects, which allows examination of variation by age as well as group means. Secondly, the semi-realistic simulation enables measurement of both abstracted physiological and psychomotor capabilities essential to driving, such as the reaction times to the lead car's speed change and to the light, and aspects of driving performance which have more complex conditioning factors. Achieved separation distance, for example, reflects a complex balance between preferred distance and driving ability, and is dependent upon the speed of decision making and reactions as well as driving style, which in turn appears to be associated with the priorities attached to multiple simultaneous tasks. The third feature, which contributes previously rare analytical opportunities, is that the subjects who have experienced the simulator have provided, through questionnaires, a great range of personal, health and driving-history information. In other words, the results describe more of the complexity of driving than many previous laboratory trials. For this reason, our interpretations of the results are constrained by the provisional nature of existing descriptions of the driving task (Brown, 1989), and suggest that further research may lead to a deeper and broader understanding of driving.

Acknowledgements

Many have contributed to the DRIVAGE research programme at both King's College, London and the Traffic Research Centre, University of Groningen. The authors particularly wish to thank Andrew Tollyfield, who played a large part in the formulation of the project; Philip Neave and Andrew Davies, programmers; Barbara Rough and Judith Sixsmith who took the prime responsibility for the social survey of elderly drivers; and our assistants who carried out the interviewing, testing and data analysis. Most of all, however, we thank our respondents who took the time and trouble to provide the results of the study.

References

Automobile Association: Foundation for Road Safety Research, 1988, *Motoring and the Older Driver*, Basingstoke, Hampshire: Automobile Association.

Brown, I. D., 1989, *How can we train safe driving?* Verkeerskundig Studiecentrum Rijksuniversiteit Groningen, Groningen, Netherlands.

Carp, F. M., 1972, Transportation and retirement, *Proceedings of the American Society of Civil Engineers*, TE4, 787–98.

Fraser, D., Tolleyfield, A., Neaves, P. and Davies, A., 1990, *Construction and control of an interactive-video car simulator*, EC DRIVE Programme, Project V1006 'DRIVAGE', Deliverable 4, Division of Engineering, King's College, London.

Hillman, M., Henderson, I. and Whalley, A., 1976, I, Report 42, Political and Economic Planning, London.

Markovitz, J. K., 1971, Transportation needs of the elderly. *Traffic Quarterly*, **25**, 237–53.

Organization of Economic Cooperation and Development, 1985, *Traffic safety of elderly road users*, Paris: OECD.

Ponds, R. W. H. M., Brouwer, W. H. and van Wolffelaar, P. C., 1988, Age differences in divided attention in a simulated driving task. *Journal of Gerontology*, **43**, 151–6.

Salthouse, T. A., 1990, Cognitive competence and expertise in aging, in Birren, J. E. and Schaie, K. W. (Eds), *Handbook of the Psychology of Aging*, 3rd Edn, pp. 310–19, New York: Academic Press.

Sixsmith, J., Rough, B. and Warnes, A.M., 1990, *Elderly drivers attitudes to RTI devices: a report of group discussions*, EC DRIVE Programme, Project V1006 'DRIVAGE', Deliverable 3, Department of Geography, King's College, London.

Taira. E. D., 1989, *Assessing the Driving Ability of the Elderly: A Preliminary Investigation*, New York: Haworth.

Warnes, A. M., 1989, *Factors in elderly people's driving abilities*. EC DRIVE Programme, Project V 1006 'DRIVAGE', Deliverable 1, Department of Geography, King's College, London.

Warnes, A. M., 1992, Elderly people driving cars: issues and prospects, in, Morgan, K. (Ed.), *Responding to an Ageing Society*, pp. 14–32, London: Jessica Kingsley.

Warnes, A. M., Fraser, D. and Rothengatter, T., 1991, Elderly drivers' reactions to new vehicle information devices. In, Commission of the European Communities, *Advanced Telematics in Road Transport*, Volume 1, pp. 331–50, Amsterdam: Elsevier.

Wolffelaar, P. van and Rothengatter, T., 1990, *Divided attention in RTI-tasks for elderly drivers*, EC DRIVE Programme, Project V1006 'DRIVAGE', Deliverable 2, Traffic Research Centre, University of Groningen.

11

Cyclists problems: can RTI help?

M. Draskóczy and C. Hydén

The cyclist safety problem

Cyclists represent an important part of the traffic system—at least in some countries in Europe—especially in urban areas. In Europe, about 6 per cent of all road users killed in a traffic accident are cyclists, and the range is from 1 per cent in Greece up to 22 per cent in the Netherlands. As far as people injured are concerned about 7 per cent are cyclists, with a range from 1 per cent in Greece to 25 per cent in the Netherlands (UNO Statistics). It is a statistic which must be taken into consideration by any sort of project which intends to improve traffic safety. Although it is difficult to compare the exposure of cyclists with that of motorized traffic, some studies indicate that their risk of being injured in traffic is much higher than the risk of car drivers or passengers. Some results of an analysis carried out on Swedish accident data (Thulin, 1981) can be seen in Table 11.1.

The problem of cyclist safety is even more serious if the probability of not reporting some traffic accidents is taken into account. Statistical data presented in Table 11.1 are taken from the national traffic accident statistics, i.e. represent those accidents which had been reported to the police. More and more studies from different countries prove that police reported accidents are far from being a complete sample of traffic accidents (Thulin, 1987; Draskóczy, 1988; Stutts *et al.*, 1990). If traffic injuries reported to the police are compared with traffic injuries treated by hospitals, the two samples are usually of different magnitude, the hospital sample being larger. The samples are partly overlapping and both contain cases which are missing from the other. The under-reporting, i.e. the percentage of accidents not recorded by the police, is different for different accident types and for different road user categories. Thulin (1987) elaborated some multiplying factors for different accident types to estimate the true number of casualties using Swedish data (Table 11.2).

Table 11.1   The relative risk of being killed for cyclists (7–84 years) compared to car drivers. (Thulin, 1981)

	Car drivers	Cyclists
Risk of being killed		
per journey	1.0	1.3
per km travelled	1.0	8.3
per hour travelled	1.0	2.0

Table 11.2   Multiplying factors to convert the police recorded number of casualties into the true number of casualties, by accident type.

Accident type	Multiplying factor
car-car	1.37
car single	1.57
car-mc/moped	1.53
car-bicycle/pedestrian	**1.54**
mc/moped single	4.06
unprotected-unprotected	**4.55**
bicycle single	**24.30**

The proportions of the different types of accidents taken from the Swedish police records and the estimated 'true number of casualties', corrected by multiplying factors presented, can be seen in Table 11.3.

Analysing accident statistics, a new indicator called 'fatal equivalent' can be introduced to express joint value of the number and severity of all injuries of

Table 11.3   Police reported casualties by accident types and the corrected numbers of casualties by taking under-reporting into account. (Sweden, 1986)

	Police reported casualties		Corrected number of casualties	
	Number	%	Number	%
car-car	6684	44	9157	30
car single	3952	26	6205	21
car-mc/moped	554	4	848	3
car-pedestrian	1657	11	2552	8
car-bicycle	**1920**	**13**	**2957**	**10**
bicycle single	**352**	**2**	**8553**	**28**
		100		100

Table 11.4 Annual average number of injuries due to traffic accidents with their fatal equivalent and corresponding under-reporting ratio by road user category (Sweden, 1986–88) From: M-hosseini, 1991.

	Car	Mcycle	Moped	Bicycle	Pedestr.	Total
(1) Number of injuries	15,591	1,442	727	2,605	1,926	22,455
Fatal Equivalent	1,128	153	67	198	265	1,824
(2) Number of injuries	20,600	4,900	3,800	26,300	29,500	85,000
Fatal Equivalent	1,473	283	154	762	859	3,531
(3) Ratio of injuries	1.3	3.4	5.2	10.1	15.3	3.8
Ratio of FEs	1.3	1.9	2.3	3.9	3.2	1.9

(1) According to police reports
(2) According to police, insurance companies, hospital register altogether
(3) Underreporting ratio = number according to (2) divided by corresponding number according to (1)

a given category. Fatal equivalent is defined as the number of fatalities which could replace the mixture of injuries of all severities in the given accident category and would result in exactly the same cost. If fatal equivalent values of accidents of different road user categories, according to police reports and according to all sources, are taken into account, the situation is as follows (Table 11.4). One can see that if all the accidents are taken into consideration the number of injured cyclists is higher than that of car drivers and passengers. However, injuries in car accidents are in general more serious than cyclist accidents if single bicycle accidents, which are usually not reported, are also considered. Therefore, the cyclists' share in fatal equivalent value is lower than their share in number of injuries.

The main problem areas for cyclists and possible solutions based on road transport informatics (RTI)

The majority of bicycle accidents, as well as pedestrian ones, occur in urban areas, although the share of rural areas is somewhat higher for cyclists than for pedestrians. The rural cyclist accidents are much more serious than the urban ones, but even so the fatal equivalent value of the rural bicycle accidents is less than half of that of the urban ones (Draskóczy and Hydén, 1991). The main area of interest is therefore the urban traffic environment although inter-urban bicycle traffic, too, has to be taken into consideration when setting safety objectives. Two main areas can be distinguished when talking about the possibilities of RTI application. Part of the road system (especially motorways but the majority of the inter-urban road system, too) is assigned exclusively or primarily for motor vehicles that might travel at about the same speed and might be provided with similar equipment, etc. This part of the road system can grow to function more and more 'factory-like', and it is here that the more advanced technology will apply. The other part of the road system (first of all urban areas) is a multifunctional system, meaning not only that motor vehicles and other kinds of road users have to share it, but also that its function is not exclusively travel, and it is seen as 'living space' for people. This part of the road system is quite different, and the possible and necessary safety measures need, therefore, be different. Vulnerable road users use that latter part of the road system where the application of advanced technology is more difficult, and many different factors and diverging interests need to be taken into consideration when RTI measures are applied.

Analysis of different accident types might be a useful tool to define problem areas, but an international comparison of accident types is almost impossible, because of the different typology used at accident description in the different countries. The Swedish accident types for cyclists, the number of accidents by type and the corresponding fatal equivalent value can be seen in Table 11.5. One can only guess, using the Swedish typology, how many of the accidents occurred at intersections. Turning and crossing manoeuvres, that most probably take place at intersections, account for 64 per cent of all cyclist casualties, and 61 per cent of the fatal equivalent value. This is the minimum percentage of intersection accidents, but it is most probable that some of the rear-end, head-on or overtaking accidents were intersection accidents, too.

Table 11.5 The number of cyclists killed and injured, and the corresponding fatal equivalent by traffic environment and type of accident (Sweden, 1987).

	Urban areas		Rural areas	
	No.	Equiv.	No.	Equiv.
Motor vehicle—cycle	1486	99.4	210	35.7
Overtaking	69	3.0	26	4.1
Rear-end collision	73	3.4	23	8.7
Head-on collision	31	5.9	21	4.9
Turning/same direction	174	11.5	51	5.1
Turning/opposite direction	188	12.0	13	2.3
Crossing/non-turning	492	36.8	31	4.9
Crossing/turning	300	19.1	35	3.8
Crossing/unknown	43	2.2	3	0.2
Others	116	3.9	7	1.5
Cycle single	290	18.8	36	2.8
Cycle collision standing vehicle	66	3.2	3	0.5

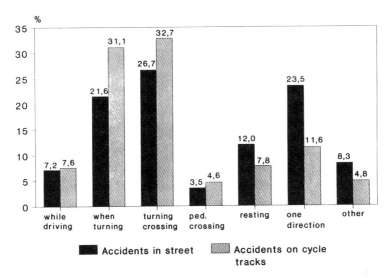

Figure 11.1   The distribution of accident types in the street and on the cycle tracks (from Ploss and Keller, 1989).

According to UK statistics 73 per cent of all bicycle accidents occur at intersections. The intersections seem to be of primary interest when defining main problem areas. The distribution of intersection accidents between signalized and non-signalized intersections is not presented in the national accident statistics, however a four-year sample in the city of Malmö in Sweden has shown that 32 per cent of the cyclist casualties occurred at signalized and 68 per cent at non-signalized intersections. This distribution might, however, be more dependent on the level of signalization at the place in question. A smaller scale bicycle- accident study was carried out in Germany (Ploss and Keller, 1989), by analysing, over a period of five years, bicycle accidents in some urban areas of Munich (946 accidents). Graphic representations of the results are presented in Figure 11.1. The main problem areas for cyclists are presented as examples of a situational/behavioural analysis based on a qualitative synthesis of different accident, traffic conflict and behavioural studies. This sort of problem analysis is a necessary introductory step when any sort of safety measure is considered in a given area.

Cyclists passing right through non-signalized urban intersections

As can be seen from the previous analysis this is the most frequent bicycle-accident situation. The conflicting partner might be a straight-crossing vehicle, as well as a right- or a left-turning one.

- *Conflict with a right-turning vehicle.* The conflicting partner for the cyclist in this situation is a vehicle coming on the same road from the same direction as the cyclist. The major problems in this kind of situation can be described as follows. The right lane or the right side of the right lane is the natural place for a cyclist riding straight on in an intersection. The same lane is assigned for the right-turning vehicles, too. As bicycles need minimal space, they often share the traffic lane with either a car overtaking the cycle in the same lane, or with a right-turning car that is caught up by the cyclist before the intersection. In any case the vehicle driver does not often see the cyclist, or forgets the overtaken cyclist, when his attention is needed for a busy intersection. Real interaction and mutual actions to avoid the accident are needed in this situation.
- *Conflict with a vehicle arriving from crossing direction.* The major problem in this case is partly perceptual (not seeing the bike because it is hidden by some bigger vehicle, i.e. a turning car or because its visible surface is relatively small), partly attitudinal (feeling that cyclists represent no threat for car drivers and expecting that they will take avoiding action). Conflicts occur most often when the motor vehicle just enters the intersection and the driver is overloaded with the task to watch out for all the necessary information.

Remedial measures

In a situation like this, when the main problem is a high probability of overlooking some important information, the best remedial measure is the locally lower speed for cars. It might function as a warning at a non-signalized intersection by advising a lower speed while crossing the intersection, or it might be an automatic speed control function activated by a transponder when the car arrives in the vicinity of the intersection.

Cyclists passing through urban intersections arriving on a bicycle track

The small scale accident analysis carried out in the city of Malmö has shown that about 25 per cent of all bicycle casualties took place at intersections where a two-directional bicycle track was running parallel to one of the roads, and the cyclists were arriving on the bicycle track to the intersection. Although no detailed accident data about one-directional bicycle tracks are available, the cyclist study in Munich (Ploss and Keller, 1989) has shown that cycling on a bicycle track is not safer than cycling on the road in Munich. Linderholm (1984) has carried out conflict studies, and compared the risks when the cyclists were passing right through signalized intersections, at intersections with and without separated bicycle tracks. He found the risks were approximately three times higher on a separated bicycle track than if the cyclists were travelling to the right on the road. In addition, Linderholm found that biking in the 'wrong' direction on a two-directional separated bicycle track at an intersection increased the risk considerably, compared with biking in the 'right' direction.

These results indicate that intersections with bicycle track are more dangerous for cyclists than other intersections without tracks, and that biking in the 'wrong' direction is particularly dangerous. The main problem in behavioural terms can be described as follows:

● From the car driver's point of view the conflict point with the cyclist, arriving on the cycle track, is just on the border of the usual intersection. This means that a driver who is arriving at the intersection is concentrating behind that point, to the intersection, and the driver who is just leaving the intersection is just relaxing. The situation is even more difficult when the bicycle track is located on one side of the road and is two-directional. As that kind of arrangement is relatively uncommon, it is not normal for the car drivers to watch out for cyclists, arriving from the opposite direction, compared with the arriving cars. Another sort of problem related to bicycle tracks is that if cars and bicycles share the same road, bicycles are more likely to be treated as similar vehicles, and their manoeuvres are taken into consideration more when car manoeuvres are planned. If the cycles arrive on a bicycle track there is a greater expectation for them not to 'disturb' cars, and to give way before entering the road. Right-turning cars have problems with parallel riding cyclists. Swedish and Danish studies (Karlshamnsstudien, 1989; Hydén and Almqvist, 1986) give more details about the interaction between right-turning vehicles and straight forward going cyclists arriving from the same direction on a separated bicycle track. The major interactional problems according to those studies could be described as follows: The right turning vehicle proceeds without stopping and seemingly unaware of the presence of an approaching cyclist. This was undoubtedly the most common case. When traffic is congested with both bicycles and cars, car drivers seem 'tempted' to try to use a 'too small gap'. This was the second most frequent case. A less frequent case was that of a driver who had started to cross the bicycle track with a big enough gap. When just crossing the bicycle track, the driver, however, had to stop because of a pedestrian. The car, therefore, unexpectedly blocked the bicycle track. The frequency of this case will, of course, be highly dependent on the geometrical layout, the signal timing and the frequency of pedestrians.

● From the cyclist's point of view there is, in general, a higher level of alertness when riding on the road among motorized traffic, than when riding on a separate cycle track. This means that entering an intersection from a cycle track needs a similar change of attitude and level of alertness as crossing the road for a pedestrian. The signals for such an attitudinal change are, however, often quite scarce.

Remedial measures

Warning car drivers arriving at an intersection with bicycle track might be a good measure to call their attention to the unusual situation. Some sort of warning might also be useful for cyclists, before they arrive at the point where they have to share the road with motorized traffic.

The expected effect of some RTI measures on cyclist safety

DRIVE SECFO (System Engineering and Consensus Formation Office) presented recently a short report, titled 'Areas, functions and sub-functions in advanced road transport telematics', in which the functions planned in DRIVE and PROMETHEUS are summarized. Traffic safety is only one of the aims of RTI application in road traffic, and vulnerable road users are not the main target group in this respect. However, they might be influenced by RTI application either directly, as target groups whose presence needs to be reported to car drivers or who are privileged or unprivileged by intelligent traffic signals, or indirectly, as RTI functions might influence users' attention or behaviour towards cyclists.

RTI measures planned in DRIVE and/or in PROMETHEUS with safety relevance for cyclists

The SECFO report on main areas of RTI application indicates clearly that RTI is going to improve car-car, car-traffic control, car-environment, etc. communication, but cyclists are more or less outside the field of its interest. They can not be equipped with high technology equipment and perhaps are not even ready to participate in the game. However, if improving traffic safety is one of the main goals of RTI application, this group of road users can not be forgotten because of their high share in traffic casualties. Cyclists, as part of the group of 'vulnerable road users', are mentioned in the report under the headline 'Traffic control', as traffic participants that can gain from RTI application as follows: 'Intersection state monitoring—Obtaining data on intersection traffic state including the monitoring of vulnerable road users (VRU).' Vulnerable road users are mentioned in another place, more as targets of information but in a way that might be very important from a traffic safety point of view: 'Conflict zone monitoring—Monitoring the total manoeuvring zone of their vehicle as well as other trafficants and their possible trajectories to detect possible conflicts. This includes detection of obstacles e.g. vulnerable road users and static objects.'

An overview of some RTI measures that might influence cyclist safety

RTI applications from the cyclists' point of view means primarily, application in urban areas. Three main areas can be distinguished:

- RTI application specifically for protecting or giving priority to cyclists (intelligent signals with cyclist detection, speed-reducing measures at specific places, etc.).
- RTI application by which motor vehicles are warned of the presence or possible presence of cyclists (conflict-zone monitoring, local warning, etc.).
- any other RTI applications which, although not having the aim to influence directly cyclist safety, provide the driver with information, warning, tutoring, etc. during driving and influences his behaviour (route guidance, navigation, traffic information, etc.). These applications have an indirect influence on cyclist safety.

Intelligent traffic signals with cyclist detection

Signal timing according to motor vehicle detection is a very common technique nowadays. Vulnerable road users are, however, very seldom detected although it is technically feasible. Riding against red for cyclists is much more common than driving against red for cars. The cause is not only that cyclists are less disciplined than car drivers, but also that cyclists, especially if they arrive on a bicycle track separated from car traffic, are almost never detected automatically. This means that even if they arrive at an intersection where there is no crossing traffic present, they have to stop and wait until a car activates the signal, or they have to push a button to get green. Their priority is thereby severely reduced compared with car traffic. As cyclists get a good overview of a traffic situation—by the use of both eyes and ears—the result of the low priority given to them is riding against red and non-usage of special bicycle facilities (riding among the cars gives higher priority). So riding against red could be reduced by intelligent, bicycle friendly traffic signals when the signal timing is based on the actual presence of cyclists and the cyclist traffic gets high enough priority when cyclists are actually present (by minimizing waiting times and providing crossing possibility for cyclists as soon as car flow allows it).

Speed reduction at intersections and at places where many cyclists are present

The measure is mentioned as a local speed enforcement in the SECFO document. The advantages of lower speeds at specifically dangerous places, especially if many vulnerable road users are present in the area, need not be detailed here as it was covered in the previous chapters. RTI technology can promote local speed enforcement in different ways. Drivers can be warned on a local speed limit, can be tutored if they do not keep speed limits, or can be enforced when it had been detected that they were speeding. Finally, speed limits can be ensured by an automatic speed-limiting function built into the cars. Studies of the effects of traditional measures to reduce speeds (humps, etc) indicate that there is one primary effect and that is the speed reduction itself. There seems to be little negative side effects, such as reduced attention, confusion, annoyance, etc. The net result is therefore, in the vast majority of cases, a very significant accident reduction, as well as a reduction of accident consequences. Speed reduction is very attractive for vulnerable road users, as it seems as if this group benefits more from a speed reduction than other groups. Besides it is one of the very few safety efficient solutions for vulnerable road users in areas where they are frequent and heavily mixed with car traffic. Injury reductions of up to 60–80 per cent are quite possible (Hydén, 1990). The reduction rate is strongly related to the relative and absolute speed reduction that is actually obtained. The very significant effects mentioned above are demonstrated locally at intersections and in residential areas. It is still to be tested and verified for larger parts of urban systems, and particularly for the main road network in urban areas where the vast majority of injuries to vulnerable road users occur. There are no reasons why speed reduction in new areas should work less efficiently, from a safety point of view, than in the areas already tested.

Summing up, there are strong arguments for the testing of different systems to reduce speeds at locations where the safety problems of vulnerable road users cannot be treated effectively in any other way. As positive safety effects are very likely, it is also important to evaluate clearly the 'tangible' and 'non-tangible' effects in other respects, e.g. delay for car drivers, stress to car drivers as well as vulnerable road users and others, energy consumption, noise, air pollution, 'social climate', etc. This is particularly important for two reasons:

- Speed reduction is not used to any greater extent, even though most attempts to estimate the costs and benefits of this kind of measure produce quite beneficial end results. The fact that in spite of this speed reduction is not used may indicate that the cost benefit analysis is not complete.
- Both PROMETHEUS and DRIVE give very high priority to safety. This means that actual safety benefits must also be obtained. As vulnerable road users in total represent 30–50 per cent of the injury problems—and as speed reduction seems to be a key issue in the attempts to increase safety for vulnerable road users—it seems 'unavoidable' to use speed reduction as one part of the strategy. It is therefore important that all the other effects that are caused by a speed reduction—positive and negative—are well known, so that the strategy can be designed in such a way that any negative effects are minimized.

Warning drivers of the presence or possible presence of cyclists

Warning drivers of danger is only useful if the driver has spare capacity to attend to the warning, realize the danger and decide on the optimal avoiding action, or if the situation is so unambiguous that one predefined avoiding action (e.g. braking) can be applied with high certainty. In general, urban traffic does not fit into this category and therefore the possibility to warn on actual danger is very limited. Apart from this hazard warning (if one can not ensure that every hazard is noticed by the system and the driver is warned of them) has a general problem of compensatory effect. If the driver has equipment that is said to warn him when there is any danger present, he need not concentrate so much on danger detection. Urban traffic is extremely varied and hazards, in the form of vulnerable road users, can appear at random, so no artificial intelligence to detect hazards, or forecast dangerous behaviour, can be regarded as totally reliable.

The driving task is, in general, less demanding on inter-urban roads, because vehicle movements are more predictable and less interference comes from outside the road. Warnings on the presence of cyclists might be useful, especially when visibility is restricted because of darkness, fog, etc. Bicycles are quite unexpected road users on these roads and their low speed, i.e. the high relative speed when a car keeps up with them, makes the meeting quite unexpected for the car driver. Conflict-zone monitoring might be a good safety measure in this case. In urban areas warnings on cyclists can be applied in a more general way; warning before the driver arrives in an area where cyclists, or eventually specific high-risk groups (e.g. schoolchildren), can be expected. There are traditional ways to warn drivers on the probable presence of vulnerable road users by traffic signs, but RTI technology could fit more to the actual situation, giving warning only when the presence of, e.g., school children is really probable. Warning on the actual presence of cyclists might be useful, e.g. at intersections where turning cars cross the parallel running bicycle path. Possible effects of the warning system are partly dependent on the way it is introduced. If the user expects that the system will give a warning every time some danger is present and, on the other hand, if the system does not warn there must be no danger, the behavioural modification and the compensatory effect might be so great that it counteracts the positive effects. The long term credibility and therefore efficiency of the system is also closely related to the frequency of unnecessary warnings. If the system, however, works reliably, it might modify attitudes and exceptions in general regarding those situations where the system usually warns.

Tutoring drivers while driving

Tutoring in general might be a very efficient way of influencing driver behaviour. Tutoring means in this respect that the driver is informed immediately and consequently about mistakes and violations he has made. This sort of tutoring might play different roles. It might inform the driver who is not aware of the mistake or violation, e.g. driving at speeds higher than the local speed limit; or at an unsafe speed in a given situation, running against red, etc.; or it might warn the driver who habitually behaves in a particular way that his behaviour is not safe. The value of the tutoring function is highly dependent on the intelligence of the system, i.e. if it is able to realize mistakes in a sophisticated way, and on the behavioural rules built in the system, what sort of behaviour is expected from the driver when he interacts with vulnerable road users and, what sort of behaviour is defined as mistake. One way of tutoring, with regard to the actual hazards drivers are involved with, is to use the conflict concept. Serious conflicts are found to be a good indirect measure of accident risks and assuming that hardware and software can be made sophisticated enough to actually record these serious conflicts, they can be used to tutor the driver. The driver should in addition be assisted—as far as possible regarding technical limitations as well as other limitations that exist with regard to what could be transferred to the driver while he is actually driving—so that he draws the right conclusions regarding his own behaviour and his own possibilities to reduce risk at a 'similar event' in the future.

The tutoring function and its technical base is a tool by which the climate of traffic can be influenced according to different value-based political decisions. Apart from the general usefulness of the tutoring function for inflencing driver behaviour there is a timing problem, especially in traffic situations where the full attention of the driver is needed. On the one hand tutoring is the most efficient if it is given as close to the bad behaviour as possible. On the other hand tutoring can not be given at any moment during driving because it disturbs the driving task itself. One of the difficult tasks when developing tutoring devices is to find the right way and timing of presenting the message to the driver. A possible solution might be to record the event, as and when it happens, and allow the driver to play it back when he has time to look at it.

Giving different, not safety relevant information to drivers in urban traffic

Driving in urban traffic, as it has been analysed above, is a situation that demands a high level of alertness and attention from the driver. Partners in urban traffic are quite often less visible and less predictable than partners on roads, where only, or mostly, motor vehicles are travelling. Therefore, it is of primary importance to determine how drivers get different sorts of information while driving in urban traffic, because to distract the driver might be highly dangerous. Any RTI functions, that provide drivers with information while driving in urban traffic, need to be tested from a traffic safety point of view.

Summary and conclusions

Cyclists constitute about 20–30 per cent of traffic accident casualties, if the under-reporting of accidents and the seriousness of the injuries are also taken into consideration. Application of RTI technology must not, therefore, neglect this group of road users, even if the main target group for road transport telematics is car drivers. Cyclists are present primarily in urban areas and the vast majority of their accidents occur there. The very presence of cyclists and pedestrians in the urban areas means that the urban traffic system is heterogeneous and interactions between road users are more complex and less predictable than, e.g., on motorways where the system is much more clearly organized and simplified. Sophisticated technology is much less likely to be applied in such an inhomogeneous system, partly because only some of the participants can be equipped, and partly because such a system always tends to work less 'factory-like', less according to the formal rules. Traffic conflict and behavioural studies have shown that conflict-free action in inhomogeneous traffic situations seems to be made possible not so much by mechanically applied rules and automatic control, but rather by the mutually flexible interaction of the partners.

In spite of the difficulties mentioned in the previous paragraph, there are possibilities to improve cyclist safety by RTI technology. It is also possible to take into consideration that RTI functions implemented for the sake of car traffic do not endanger cyclist safety.

References

Draskóczy, M., 1988, *A Bács-Kiskun megyei halálos balesetek okainak feltárása*, (Causes of traffic fatalities in the county of Bács-Kiskun—analysis of traffic injuries and fatalities of a county in Hungary recorded by the police and by the ambulance service), KTI Report, Budapest.

Draskóczy, M. and Hydén, C., 1991, Safety objectives for cyclists and pedestrians, DRIVE V1062. WP 4.3 Report, Department of Traffic Planning and Engineering, *University of Lund, Bulletin* **96**.

DRIVE SECFO, 1991, Areas, functions and sub-functions in advanced road transport telematics, Version 1.0, March 1991, Brussels.

Hydén, C. and Almqvist, S., 1986, *Konflikter mellan bilister och cyklister i signalreglerade korsningar* (Conflicts between car drivers and bicyclists at signalized intersections), Säktra, Bjärred, Sweden.

Hydén, C., 1990, *Hastighetsdämpande åtgärder på huvudgator* (Speed reducing measures on main roads) Department of Traffic Planning and Engineering, University of Lund.

Karlshamnsstudien, 1989, Cyklister sikkerhed i kryds, adfaerds- og konflikt-registrering, Notdisk Komité for Trafiksikkerhedsforskning.

M-Hosseini, A., 1991, *General estimate of main accident problems in different European countries*, DRIVE V1062. WP 5 Report, Department of Traffic Planning and Engineering, University of Lund.

Ploss, G. and Keller, H., 1989, Analysis of bicycle accidents as a basis for the design of urban cycling facilities, *Proceedings International Conference on New Ways and Means for Improved Safety*, Tel Aviv, Israel.

Stutts, J. C. *et al.*, 1990, Bicycle accidents and injuries: A pilot study comparing hospital- and police-reported data, *Accident Analysis and Prevention, 22*, 1, pp. 67–78.

Thulin, H., 1981, *Risker i trafiken för olika åldersgrupper och färdsätt* (Traffic risks for different age groups and modes of transport—based on information from traffic accidents reported to the police and from results of a study of travel patterns in Sweden made by National Central Bureau of Statistics in 1978), VTI Rapport 209, Linköping.

Thulin, H., 1987, *Trafikolyckor och trafikskadade enligt polis, sjukvård och försäkringsbolag* (Traffic accidents and traffic injuries according to the police, the hospitals and insurance companies), VTI Meddelande 547, Linköping.

12

Pedestrian problems: can RTI help?

O. M. J. Carsten

The pedestrian safety problem

In virtually every European country pedestrians account for between one-seventh and one-third of all road user fatalities. The proportion varies from 15 per cent in France and Belgium to 35 per cent in Great Britain (United Nations, 1990; Department of Transport, 1990). A large part of the explanation for the prominence of pedestrians in the casualty statistics is to be found in their vulnerability in an accident. British statistics for 1989 reveal that, in car-pedestrian collisions, pedestrians had a 50 times greater risk of injury than car occupants and a risk of fatality that was 1072 times as great (one car occupant died in such collisions, as compared to 1072 pedestrians) (Department of Transport, 1990).

A standard way of comparing the situation for different types of road user is to look at risk of an accident per unit of distance travelled. Pedestrian risk is significantly higher than that of the average road user. A recent analysis for DRIVE project V1031 compared the pedestrian shares of total travel and road casualties in three European countries (Tight and Carsten, 1989). The results are shown in Table 12.1, which clearly indicates the high risk of pedestrian travel. Yet the amount of time that pedestrians are exposed to risk can be considered to be much less than the time for vehicle occupants. Whereas vehicle occupants are exposed to accidents throughout their journey (even when stationary), pedestrians are for most of their journeys not at risk, since they are segregated from vehicular traffic (very few pedestrian accidents occur on pavements). Pedestrian risk normally only arises at road crossings. Risk per unit of time is thus very high for short intervals during a journey and negligible on other parts of the journey.

The general effect of RTI systems on pedestrians

RTI systems for improving the vehicular traffic situation are often aimed at

133

Table 12.1   Pedestrian share of total distance travelled and of road casualties.

	Share of travel	Share of casualties
Great Britain	6.0%	18.4%
The Netherlands	3.6%	8.3%
Sweden	2.1%	9.0%

increasing capacities and reducing travel times on urban roads. They will generally tend to increase vehicle speeds and vehicle flows on urban roads (the exception that proves the rule here is the application of RTI systems for demand management). Such effects will almost certainly cause safety problems for pedestrians, since pedestrian accidents are closely related both to speeds and to flows.

Vehicle speed

The severity of pedestrian injury in collisions between vehicles and pedestrians is directly related to the speed of the vehicle at impact. At 30 km/h the probability of pedestrian fatality is 5 per cent; at 50 km/h it is 37 per cent; and at 70 km/h it is 83 per cent (Hass-Klau, 1990). Given a speed of less than 25 km/h, pedestrian injuries are likely to be slight, while at speeds greater than 55 km/h there is greater than even chance of a fatality (Ashton, 1982). Thus any RTI measures that have as a by-product increased travel speeds for vehicles in urban areas will tend to increase the severity of pedestrian accidents. This would be the case, for example, with systems such as integrated route guidance and urban traffic control, which have as their aim increased network efficiency and reduced journey times. In Britain, 91 per cent of pedestrian casualties occur in situations where the pedestrian is not using a crossing facility (zebra or light-controlled crossing) (Department of Transport, 1990). Similarly, in Bradford and Groningen, two of the cities used for experimentation in DRIVE project V1031 'An Intelligent Traffic System for Vulnerable Road Users', two-thirds of pedestrian casualties occur away from junctions (Tight and Carsten, 1989). These accidents away from junctions and away from crossing facilities will be particularly susceptible to increased travel speeds of vehicles, although all pedestrian accidents could suffer. And small increases in speed will have quite large effects on the rates of serious injury and fatality.

Flow

In the past few years considerable effort has been devoted to developing empirical models of traffic safety. These models use observed relationships between traffic and site conditions on the one hand, and safety on the other, to build generalized predictive models for given situations. Some of this modelling activity has been devoted to pedestrian safety, and here results obtained in widely dispersed locations and in different situations have produced similar results.

The overall message from this research is that pedestrian safety is highly correlated with vehicle flows. Thus for British pelican crossings, the best fitting model to predict pedestrian casualties uses a cross-product of vehicle and pedestrian flows (Davies *et al.*, 1991). The model developed to predict pedestrian accidents at British signal-controlled urban four-armed junctions uses the following predictor variables: two-way vehicle flow on the arm, two-way pedestrian flow on the arm, London factor, land use (shopping or non-shopping), gradient of the opposite arm in the 50 metres approaching the junction, vehicle inflow per second of green time, and sight distance (Hall, 1986). A Canadian model to predict traffic conflicts between pedestrians and vehicles at signalized junctions, using data from Montreal, found that the total number of such conflicts at a junction was best explained by a power function of the cross-product flow of pedestrians and vehicles. The vehicle flow was defined as the total hourly flow of vehicles approaching the junction, and the pedestrian flow as the number of pedestrians crossing all arms of the junction per hour (Javid and Seneviratne, 1991). In all these models, an increase in vehicle flow will result in a commensurate increase in pedestrian accidents, casualties or conflicts. It follows that any new traffic system that aims to handle increased traffic flows is likely to result in an increase in pedestrian accidents, unless it is accompanied by extensive traffic calming. It also follows that measures aimed at reducing traffic flows, in particular demand management schemes, are likely to produce safety benefits for pedestrians.

RTI systems for pedestrian safety

While the effect of vehicle-centred RTI systems on pedestrians is likely to be negative, it is also possible to create RTI systems which are aimed directly at increasing pedestrian safety. Conceptually, one can imagine three types of such system:

1. An in-car device to detect pedestrians and perhaps automatically execute a manoeuvre to avoid an impending collision.
2. A device carried by a pedestrian which would detect approaching vehicles and advise the pedestrian on whether it was safe to cross.
3. Indirect systems which affect pedestrian-to-vehicle interaction.

Each of these approaches will be reviewed in turn.

In-car devices

The detection of pedestrians from a vehicle is no doubt technically feasible. The problem is what to do with the information once a pedestrian has been detected. A pedestrian walking at a speed of 1.5 m/sec takes 2.33 seconds to clear a 3.5 m wide lane. In that time a car travelling at 50 km/h will cover about 32 m. This means that a pedestrian warning device in a vehicle, travelling on a road with a 3.5 m lane width, would have to warn all approaching vehicles within 32 m when a pedestrian leaves the kerb. Yet in reality many pedestrians walk faster than this hypothetical speed and are often prepared to accept gaps that the system might regard as unsafe. Thus drivers would receive many false warnings and probably learn to ignore the signal and eventually to disconnect it. Devices that automatically applied the vehicle brakes would solve this problem, only to create another—that of emergency braking in unnecessary situations. Here again the temptation would be to disconnect the device. Problems also arise for vehicles which are closer than 32 m to a pedestrian who is still on the pavement. The issue then becomes whether that pedestrian intends to cross. A device to alert the driver about pedestrian presence on pavements would provide virtually continuous warnings in many urban situations, and would fail to distinguish emergency from ordinary situations. The alternative device that detects pedestrian intention demands capabilities that are not conceivable with current technology. A paper presented at a recent PROMETHEUS workshop concluded that no device could solve this problem (Fontaine and Malaterre, 1990).

Pedestrian-carried devices

A portable device to advise pedestrians on when it is safe to cross could well be technologically feasible in the not-too-distant future and, given advances in artificial intelligence, might even be realizable in the near future with an intensive development programme. Such a device has already been conceived in the literature (Armsby, 1988). The difficulties lie in the application of such devices. One problem is that, if they came into general use, vehicle drivers might learn to rely on them and yield to pedestrians even less than they do now. This could result in increased delay for pedestrians as well as in greater collision speeds and hence higher risk of serious or fatal injuries. A more severe drawback of such a device is that, outside some minority groups, few pedestrians would probably use it. No doubt the elderly and the sight-impaired could be persuaded to strap on the device and obey its warnings. But it is highly improbable that it would be used by some of the pedestrians who are at greatest risk, such as children at play or the alcohol-impaired. Such a device would therefore produce little safety benefit and might even have negative safety consequences.

Indirect measures

It therefore seems more fruitful to use RTI in a more indirect manner to enhance pedestrian safety and mobility. Two indirect approaches have promise. The first is the one that has been adopted by two DRIVE projects,

V1031 'An Intelligent Traffic System for Vulnerable Road Users' and V1061 'PUSSYCATS'. These projects have sought to improve the situation for pedestrians in the traffic system by the use of RTI detection devices linked to traffic signal controllers. The signal timings are made more responsive to pedestrian needs and thereby pedestrian safety and mobility is enhanced. Vehicular response to these devices is by means of the altered signal timings.

V1031 decided to examine the benefits to be obtained from installing pedestrian responsive systems on arterial roads that carry heavy traffic. Arterial roads ('A' and 'B' roads) account for over half of the pedestrian accidents in British urban areas and for about two-thirds of pedestrian fatal accidents (Department of Transport, 1990). Yet on such roads, traffic engineers are often unwilling to provide a pedestrian phase on signals out of concern that this would cause unacceptable vehicle congestion. Pedestrians can therefore incur considerable discomfort and risk. The trial system was installed at a four-way signalized intersection in Bradford, West Yorkshire, on a section of the Bradford northern ring road. This junction had no pedestrian facilities for the reasons stated above. It should be noted that in Britain, a pedestrian 'green man' at a traffic light is only provided when all relevant traffic movements are stopped. The junction had roughly five injury accidents a year, mostly involving pedestrians. Conflict studies in the 'pre' situation revealed that most of the problems involved interaction between turning vehicles and pedestrians. The number of conflicts observed was 28 in a three-day period, of which 23 were car-pedestrian.

For the experiment, microwave detectors were installed on almost all the signal heads and targeted at a zone of pavement which would be traversed by the pedestrian in approaching the junction. The detectors were set to ignore pedestrian presence (i.e. stationary pedestrians) and to disregard movements away from the signal head. With the detector-activated system, sensing of a pedestrian approaching the signal head produced a two-second extension in the inter-green time between the phases for the two roadways. Since this extension was triggered by pedestrian approach, it was not 'wasted' as it would have been with a fixed alteration to the signal cycle. Flow studies at the junction prior to installation of the system indicated which approaches to the junction generally resulted in a road crossing movement. It was thus possible to prevent wasted activations caused by pedestrians who did not intend to cross. Conflict observations in the 'post' situation indicated that a quite large benefit had been obtained: the total number of conflicts was down to 19 with car-pedestrian conflicts reduced to 14. There were no observable congestion effects.

Another approach, being used in PROMETHEUS, is the development of a device that would ensure compliance by vehicles with urban speed limits. Trials are at present taking place of a vehicle with a device to limit speed to the overall urban speed limit (Almquist *et al.*, 1990). It is envisaged that, in future, devices could be developed that would adapt the speed limit of a vehicle to appropriate limits for the location or even to prevailing road conditions. It is thus possible to envisage traffic calming where the only mechanism is communication between the roadside and the vehicle, without the need to install expensive and often annoying road humps, chicanes, etc. Speed limits could also be made 'intelligent', so that the police could, for example, institute and automatically enforce lower speed limits in bad weather conditions. Such systems have the potential to result in a significant reduction in the number and severity of pedestrian accidents and might do so at far lower cost than current techniques. In addition to lower initial costs, improvements and alterations would become a matter of trivial effort.

Conclusions

There is a grave danger that many of the high-technology systems for increasing the efficiency of the road network will have severe detrimental effects for pedestrians in terms of increased delay and risk. Furthermore, some of the systems that have been proposed to assist in pedestrian safety may themselves either be conceptually unsound, or may even aggravate existing problems. Direct approaches in the form of in-vehicle and pedestrian-carried systems do not offer much hope. The most promising approaches lie in indirect systems which affect the interaction of vehicles and pedestrians. Here promising results have been obtained with systems for registering pedestrian demand at traffic signals using automatic detection. Initial trials with a system for infrastructure-based adaptive speed limit have shown that this too may have considerable potential.

Disclaimer

This article is based on a review of current developments in Europe. As far as the author is aware, the situation in the United States is even more bleak, in that no pro-pedestrian RTI systems are being considered. Indeed, traffic safety in general does not appear to be a major consideration in the US IVHS programme. Developments in Japan have not been reviewed.

References

Almquist, S., Hydén, C. and Risser, R., 1990, *A speed limiter in the car for increased safety and better environment*, Department of Traffic Planning and Engineering, Lund Institute of Technology, University of Lund.

Armsby, P., 1988, An 'intelligent pedestrian device' (IPD) to assist in crossing the road: some initial thoughts, presented at Universities Transport Study Group annual conference.

Ashton, S. J., 1982, Vehicle design and pedestrian injuries, in, Chapman, A. J., Wade, F. M. and Foot, H.C. (Eds.) *Pedestrian Accidents*, Chichester: John Wiley.

Davies, H. E. H., Winnett, M. A., and Farmer, S. A., 1991, Pedestrian safety, *Safety 91*, Transport and Road Research Laboratory.

Department of Transport, 1990, *Road Accidents Great Britain 1989: The Casualty Report*. London: HMSO.

Fontaine, H. and Malaterre, G., 1990, Safety evaluation of PROMETHEUS functions, presented at PRO-GEN workshop, November 28–29.

Hall, R. D., 1986, *Accidents at four-arm single carriageway urban traffic signals*, Transport and Road Research Laboratory, Contractor Report 65.

Hass-Klau, C., 1990, *The theory and practice of traffic calming: can Britain learn from the German experience?* Brighton: Environment and Transport Planning.

Javid, M. and Seneviratne, P. N., 1991, Expected conflicts: a measure of pedestrian safety in CBDs, presented at Transportation Research Board 70th annual meeting.

Tight, M. R. and Carsten, O. M. J., 1989, *Problems for vulnerable road users in Great Britain, the Netherlands and Sweden*, WP 291, Institute for Transport Studies, University of Leeds.

United Nations, 1990, *Statistics of Road Traffic Accidents in Europe*, Vol. 25, 1988, New York.

PART III
PROBLEMS IN VEHICLE SYSTEMS

13

Problems in vehicle systems

G. Reichart

Future vehicles, equipped with RTI-systems developed in research programmes like PROMETHEUS and DRIVE are intended to have a positive effect on safety, efficiency, economy, ecology and comfort. To assess the potential impact of the new RTI-systems, already in their development phase, is of great importance for the decision-making on what kind of systems should be further pursued. Specific interest is on traffic safety, the topic of Part 3, since the social costs of road traffic accidents are still unnacceptably high. However, to assess the impact on road traffic safety is far from being an easy and well understood task. It starts with the problem of a commonly accepted concept of what is meant by road traffic safety, e.g. the prevention/reduction of accidents or at least relative protection from an exposure to hazards. But besides this definition problem some other methodological problems exist. Each prediction on systems which have not yet been introduced into the market, has to be based on assumptions on systems effectiveness, number of equipped vehicles, the kind of behavioural adaptation and so on. Even if we base our assessment on experimental studies some aspects remain in dispute. Have the experimental scenarios, the test persons and the system characteristics been reflecting the real world conditions sufficiently well? Are the experimental measures the right and comprehensive indicators of impacts on traffic safety? Are there any habituation effects or behavioural changes which are likely to occur? This list of questions is by no means exhaustive, but it shows clearly how difficult it is to predict the future and how speculative all our assessment has to be. Nevertheless all our current development activities are guided by two expectations:

- improved driver information,
- some forms of driver assistance and supervision will have a positive impact on road traffic safety.

These expectations are based on numerous studies in the field of human factors and a lot of traffic safety research performed over the past few years.

However, this does not relieve us from justifying the potential safety impact of the RTI-systems under development by:

- reasonable assumptions,
- carefully designed experiments and
- experience with comparable systems

The contributions in Part 3 are examples of the ways in which sensible reasons for our ongoing research towards better road traffic safety can be provided. Chapter 15 by Malaterre and Fontaine on 'Driver safety needs and the possibility of satisfying them using RTI-systems' deals with an attempt to identify drivers' needs for the avoidance of traffic accidents. Based on accident statistics and in-depth studies of a limited subset of these accidents some general needs of the interviewed drivers have been identified. These needs are compared to 10 selected functions to be delivered by PROMETHEUS-systems under development. It is argued that PROMETHEUS could contribute in the long run to a reduction of traffic accidents up to 50 per cent. This expectation is based on many assumptions as the authors state themselves and the real effect will depend on various aspects, e.g. the level of functionality of these systems, the number of equipped vehicles or possible changes of driving behaviour. Nevertheless, these PROMETHEUS-functions offer a potential for accident reduction and it will be a challenge to the developers to exploit it as well as possible.

The contribution by Rothengatter 'Violation detection and driver information' reports on the DRIVE-project AUTOPOLIS, which investigated the possibilities for automatic policing. It is expected that systems which automatically register violations of traffic rules and regulations lead to their better observance by the drivers. An accident reduction up to 25 per cent is assumed, if such a concept is realized. However a lot of legal aspects still seem to be unresolved, and it would also be interesting to have a much better understanding of why these violations occur. It might turn out that some of the rules and regulations are impractical, unjustified or at least somewhat vaguely defined.

Janssen and Nilsson's contribution deals with 'Behavioural effects of driver support': the case of collision avoidance. It is well known that some attempts to improve the active safety, which means the prevention of accidents, have in practice not shown the expected benefits. What has been observed in many cases up to now is some sort of behavioural change by the users of new supportive systems. However, it is rather difficult to predict in which particular way the behavioural change is to occur. The contribution based on a nicely designed experiment clearly shows the kind of behavioural effects collision warning devices are likely to create. Such experimental results are even an important help for the ergonomic lay-out of such supportive systems and yield also arguments, why different solutions to a problem have to be investigated in parallel to find a solution which is as close to the optimum as possible. This is what precompetitive research programmes like PROMETHEUS can offer.

Much concern exists that the introduction of the new RTI-systems will create overload for the driver and distract her/his attention from the primary driving tasks. Verwey points, in his chapter 'How can we prevent overload of the driver?', exactly to this problem and proposes an adaptive interface as a solution for this problem. The work reported in his paper was part of the DRIVE I project GIDS (Generic Intelligent Driver Support) on interfaces that adapt to drivers' workload. After a brief discussion on various forms of adaptation and a presentation of ideas on when such adaptation should take place, a study on determinants of driver workload is reported. The main result is that road situation is found to be the predominant factor for the visual load of a driver. It will be obvious from this contribution that there is still a long way to go before convincing concepts for adaptive interfaces can be introduced. There are still a number of open research questions which need to be addressed, such as: the degree of individual control on the adaptation; the classification of traffic situations from sensor data; the relation between traffic situation, individuality of drivers, duration of situation exposure and workload.

The other chapters of this section all deal with various aspects of route planning and navigation systems. Alm gives in his paper, on 'Route guidance-deciding driver information needs', a comprehensive view on the many aspects to consider in designing a route guidance system which meets the information needs of drivers. He emphasizes the important role of task analysis and empirical research for the development of suitable interfaces for navigation and route guidance. Lorenz shows in his contribution, based on two different experiments, how much objective data and subjective experience can differ. The presented results stress the need to design systems in which their behaviour is transparent to the user (e.g. 'Why am I guided to a seemingly incorrect direction?'), and their usefulness is not impaired by an incomplete consideration of user acceptance criteria. This seems to be in line with what has been called 'design for usability'. Van Winsum deals with route choice criteria. His work emphasizes the need to orientate the route selection methods on the driver preferences. It seems to be necessary to differentiate according to the purpose of driving (e.g. business-, holiday-driving), the area (e.g. city road, rural road) and individual preferences (e.g. persons might dislike motor-ways or narrow city streets etc.) To feed all these aspects into a route selection system and guidance system will require quite an extensive dialogue between user and systems. This will create quite a challenge to the design of such systems. Schraagen develops some design guidance in his contribution on 'Information presentations in car-navigation systems'. His and the work reported in other studies confirm the superiority of simple symbols versus map displays in terms of navigation error. However, as pointed out in the papers of Lorenz (Chapter 19) and Alm (Chapter 17) there might be other information needs or impacts on acceptance which will require to have map displays available. Whether these map-displays should be drastically simplified and show up only on demand, e.g. to achieve confirmation of being still on a route to the final destination, needs some further research.

The papers in this part touch important aspects of the driving of future vehicles. The picture is by no means complete. Even in partial areas, like navigation and route guidance systems, there are a lot of unanswered research questions requiring still further work. Technology itself will lead the way, but to maintain a safe and efficient driving system will require a thorough consideration of system safety and human factors, in the interaction between users and systems under development.

14

Behavioural effects of driver support

W. Janssen and L. Nilsson

When it comes to considering the effects of driver support systems it is clear that this is not an area in which the simplest of expectations are true. Rather than being a passive receiver of system messages and actions the user interacts with the system, i.e. he places himself 'in the loop' instead of staying outside. The following example will illustrate that, because some support system is available, this interaction or adaptation will in fact occur at the behavioural level. Suppose you, as a travelling salesman, have an electronic in-vehicle navigation system that is guaranteed to save you 10 per cent excess mileage that you suffer from the use of conventional maps and the vagaries of roadsigning, etc. Knowing this you would probably not leave it at that, but you would rather plan an extra trip a day because you know you can rely on the electronics. Thus, though navigational performance will indeed be improved considerably, the net result will probably not be the reduction in mileage (or its associated accident costs!) that is commonly expected. It is not necessarily the case, as this example shows, that behavioural adaptation will make a support system counterproductive in its entirety, but rather that the gain is something else than expected, or that it comes together with effects that are to be considered as negative. There is as yet no general theory of behaviour that is capable of predicting what behavioural adaptation will consist of and how it will work out in terms of safety on the road. Therefore, it is only by experimentation of sufficient breadth that an impression can be formed of how a support system performs, both in terms of its primary goal—the function that it was designed to support—and of its anticipated adaptation effects. What we have said here may be illustrated by reporting an experimental evaluation of some Collision Avoidance Systems (CAS) that was performed within the DRIVE-project 'GIDS' (Janssen and Nilsson, 1990).

Collision avoidance systems and behaviour

One of the essential tasks in driving consists in dealing with other vehicles on the road. In future generations of vehicles, faulty user perception or decision making in this task could be corrected by so-called collision avoidance systems (CAS). A feeling for what a CAS is basically aiming at may be obtained by noting that an average driver, at least in Western countries, has a collision with sustained damage once every four or five years. Ideally a CAS should signalize that case and only that one. A few more cases may be added if narrow escapes are included. Even then, however, it will be clear that superb discriminative power will be required from a CAS. Such power is probably unattainable, because it would demand complete knowledge of what distinguishes collision from non-collision configurations in traffic well before the collision happens. Thus the system would not only have to recognize at a sufficiently early stage that a collision will follow if no action is taken, but it would also have to know that the driver will, in fact, not take evasive action in precisely this type of configuration. Therefore, the design of any CAS requires knowledge about driver perception and decision making that is not yet available.

The study to be reported here deals with the behavioural aspects of longitudinal systems, i.e. that could function in car-following situations. It was performed as part of the 'GIDS'-programme, which is one of the DRIVE-projects, aiming at the development of intelligent co-driver systems. Similar projects are under way under the IVHS banner in America. Although the technology of detecting the presence of obstacles and estimating their parameters of movement is by no means perfect, this paper assumes that it will become so within a reasonable number of years. That is, it is assumed that it is possible to measure with sufficient accuracy:

1. the *distance* to each and every object in the vicinity of the CAS-vehicle;
2. the *heading* direction of the object relative to the line of movement of the CAS-vehicle (that is, the object's bearing angle);
3. the *relative* velocity of the CAS-vehicle with respect to the object in the bearing direction (from which can be deduced, if the CAS-vehicle's velocity is known, the *absolute* velocity of the object).

Before any CAS can be implemented satisfactory answers will have to be provided to the following questions:

1. What should be the *criterion* for system activation?
2. What *action* will subsequently have to be performed, and how is the responsibility for action to be divided between the system and the driver?

There are two simple and discrete criteria for system activation whose effects need to be tested empirically. One is the *time-to-collision* (TTC) criterion. This criterion would identify whether the collision at prevailing speeds and distances would follow within a certain time interval. In the car-following situation TTC amounts to the distance between vehicles divided by their relative speed. The other criterion is the so-called '*worst case*' criterion. This assumes that the vehicle being followed by the CAS-vehicle could brake at full braking power at any moment. The system would be activated in those cases in which a collision would then follow, based on driver reaction time and braking capacity of the CAS-vehicle (Panik, 1984). Conceivable system actions comprise the following:

1. A continuous display of some critical parameter, e.g. TTC, as a means of informing the driver.
2. A warning, in case the criterion has been met.
3. The suggestion of an action to be performed by the driver.
4. System action ('active controls') which the driver may overrule by an action of his own.
5. System action which cannot be overruled by the driver.

Intermediate forms exist: there is a dimension of amount of system take-over which ranges from 'none' to 'total', and the question is what is the best in terms of resultant CAS-functioning.

Experimentation with CAS

It is impossible to say from armchair considerations what the effect of a CAS—defined as a combination of a particular activation criterion and a particular form of action—on driver behaviour will be, let alone what its ultimate effect on safety will be. The present study compared a number of candidate CAS in terms of their behavioural effects in a car-following task performed in a simulator. The candidate CAS were rather elementary combinations of criteria for system activation and subsequent action.

Systems tested

The following systems were tested:

1. A TTC-criterion (TTC < 4 s), coupled to a red warning light on the dash-board ('TTC + light'). Van der Horst (1984) has presented evidence that TTC = 4 s separates cases in which the driver has unintentionally got himself in a situation that he would judge as dangerous, from those in which the driver feels that he is in control. The warning light in this CAS went on whenever and for as long as the criterion was met.
2. The TTC-criterion, coupled to a warning buzzer ('TTC + buzzer').
3. The TTC-criterion, coupled to a 'smart' gaspedal ('TTC + pedal'). In this system 'active control' was involved, i.e. the simulator's accelerator pedal was activated by a 25-N increase in pedal force whenever and for as long as the criterion was met.
4. A 'worst case' criterion + warning light ('worst case + light'). The criterion was any configuration in which a collision with more than a 10-km/h speed difference would follow if the leading vehicle were suddenly to brake with a deceleration of $7 \, \text{m/s}^2$, including a 1-s reaction time of the following

driver and a 7-m/s^2 deceleration of the following vehicle. This criterion was taken from Panik (1984).

5. The 'worst case' criterion + warning buzzer ('worst case + buzzer').
6. The 'worst case' criterion + smart gaspedal ('worst case + pedal').
7. A seventh system was added as a minimal CAS, i.e. as a continuous display of a critical parameter ('bar'). A brake-distance indication, displaying the distance it would take the following vehicle to come to a stop, was continuously given in the form of a bar moving over the road in front of the CAS-vehicle. A 1-s reaction time and a deceleration of 7 m/s^2 were assumed.

Design

The design of the experiment was to have subjects (Ss) perform a car-following task twice, once without and once with a CAS. The within-Ss comparison on relevant variables then constituted the basic test of the effects of a particular CAS. The experiment was run in the fixed-base simulator of the TNO Institute for Perception. This is a simulator in modular form which comprises a supervisor computer, a mathematical model of the vehicle, and a CGI image generation system. The perspective view of the outline of the road and of the leading vehicle was projected in front of the mock-up of a vehicle (Volvo 240) in which the driver was seated. The field of view was 50°. The simulator was programmed so that a subject driving the CAS-vehicle would approach a leading vehicle, driving at a specified speed, sufficiently often to assess the S's behaviour in dealing with that vehicle.

A standard two-lane winding road was displayed with four curves to the left and four to the right (radii 200, 250, 330 and 500 m; deflection angle = 20°). Lane width was 3.6 m. The leading vehicle would enter the road when the CAS-vehicle was exactly 7 s away. It then moved on the road at the speed prescribed for that particular trial. This speed was either 60, 70, 80 or 90 km/h. The leading vehicle continued at this speed for 40 s, leaving the road at that instant, unless it was overtaken by the CAS-vehicle at an earlier instant. In all the leading vehicle appeared 48 times, 12 times at each of its possible speeds, in a (constrained) random order of speeds. In order to prevent Ss from driving in the left lane all the time, simulated obstacles were put on the left lane at irregular distances. That is, the obstacles functioned as a simple simulation of oncoming traffic which could not always be neglected. The distribution of distances between obstacles was rectangular over the range from 100 to 400 m, with an average of 250 m. CAS were not triggered by the obstacles.

The Ss were informed about the procedure, and were instructed to drive as they would in everyday life. Test runs were made, including appearances of the leading vehicle and of obstacles on the left lane, to familiarize Ss with the simulator and with the task. No explicit instructions were given to Ss as how to deal with the leading vehicle. There were 8 groups of 7 Ss each. One was a control group that drove the entire route twice without a CAS. The Ss in the remaining groups first drove without a CAS, and then with the appropriate CAS. It took 2 Ss a full working day to complete the experimental series.

Results

The dependent variables of this experiment were:

1. The distribution of headways between leading vehicle and CAS-vehicle, in particular: the proportion of total driving time in which headway was below 1 s (Evans and Wasielewski, 1983).
2. The average driving speed of the CAS-vehicle, over the 48 runs.
3. The overall irregularity in maintaining speed, indexed in terms of the average of the absolute values of average acceleration and deceleration.
4. The amount of time spent in the left lane of the road.

Of these variables the proportion of headways below 1 s is the most direct measure of a CAS-effect on car-following proper. The other dependent variables evaluate the possibility of CAS-effects spreading out to overall driving behaviour. In Figures 14.1–14.4 data have been plotted so that higher values on the ordinate correspond to effects that must be considered to be unwanted from a safety point of view. Decreases and increases are in terms of performance on the second ride in the experiment compared to that on the first ride.

Headways below 1 s—Figure 14.1 shows that all systems except one reduced the occurrence of short headways, relative to control group results. The largest decrease was obtained for the 'TTC + pedal' CAS.

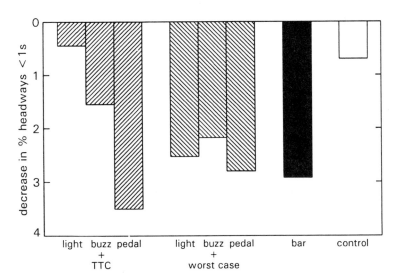

Figure 14.1   CAS-effects on close following.

Average speed—An increase in overall driving speed was associated with all systems except one (the 'TTC + pedal' CAS); see Figure 14.2.

Irregularity in speed—As Figure 14.3 shows all systems resulted in an increased variability in speed, relative to control, though to different degrees.

Time in left lane—One way of avoiding a collision with a leading vehicle is to move into the left lane. On a two-lane road, however, this may entail a risk by itself because of oncoming vehicles. Thus, drivers may in this way exchange one type of risk for another. Figure 14.4 presents changes in the overall percentage of time the CAS-vehicle was observed to be in the left lane under

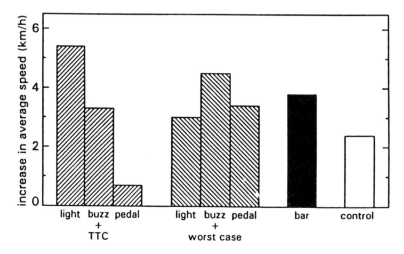

Figure 14.2   CAS-effects on average driving speed.

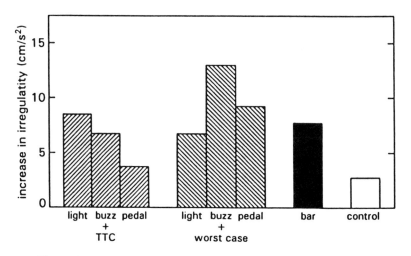

Figure 14.3   CAS-effects on irregularity in maintenance of speed.

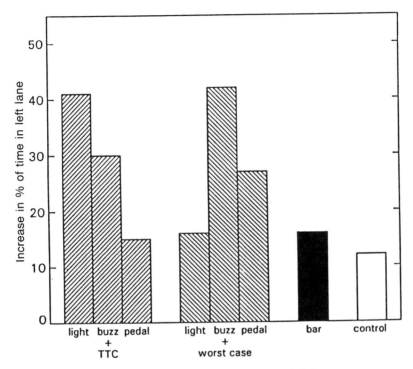

Figure 14.4   CAS-effects on driving in left lane.

the influence of the CAS. All systems showed an increase in this variable, though it is relatively minor for the 'TTC + pedal' and 'bar' (brake-distance indicator) systems.

The results of this experiment show that the use of the CAS studied here is accompanied by changes in the way in which the driving task is performed as a whole. The following effects seem to be associated with the availability of a CAS:

1. A change in the distribution of headways; in particular, a reduction in the occurrence of short headways.
2. An increase in overall driving speed.
3. An increase in acceleration/deceleration levels.
4. An increase in the time spent driving in the left lane of the road.

However, there are considerable differences among systems in the degree to which they are subject to these general effects. Figure 14.5 summarizes the results so that this becomes clearly apparent.

There appears to be one system which, while yielding a reduction in short headways, did not suffer from counter-productive effects in overall speed, speed irregularity, or driving in the left lane. This was the 'TTC + pedal'

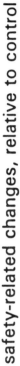

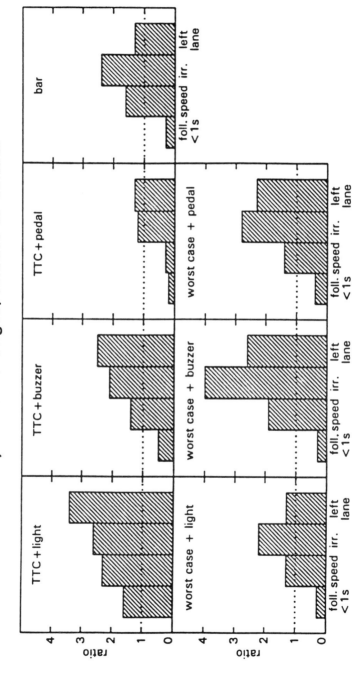

Figure 14.5 Summary of experimental results, in terms of change for CAS relative to control group. Ratios below 1 represent relative safety gains. Variables are those of Figures 14.1–14.4.

system. In all others the potential gain in safety obtained by the reduction in short headways was more or less offset by an increase in other, more risky, behaviour. Of these, the increase in average speed may reflect the trust CAS-users put in a system they know will warn when necessary. The increase in left-lane driving may be a form of anticipatory behaviour that avoids superfluous system triggering which would follow if the driver stayed behind the leading vehicle. This may in particular apply to the 'worst case + buzzer' CAS, which seemed to be a highly annoying combination to the Ss.

Conclusions

Experiments like the one described here form the basis for the design of a well-behaved CAS. Much more needs to be done before all the information needed will be available. For example, it needs to be checked whether experimental results can be replicated under realistic conditions in the field, and what form behavioural adaptation takes in the very long term.

What we hope to have made clear by presenting these CAS results is that the unexpected should be expected. Users of support systems do not feel the constraint of using a system only in the way intended by its designer and there is nothing bad in this. Providers of support systems, however, must be aware of the user's natural tendency to interact with the system in ways that serve whatever other purpose they have in mind.

References

Evans, L. and Wasielewski, P., 1983, Risky driving related to driver and vehicle characteristics. *Accident Analysis and Prevention*. **15**, 121–36.

Janssen, W. H. and Nilsson, L., 1990, *An experimental evaluation of in-vehicle collision avoidance systems*, DRIVE Deliverable GIDS-2.

Panik, F., 1984, Fahrzeugkybernetik, in, Papers XXth FISITA Congress, *Das Automobil in der Zukunft*, SAE-P 143, Paper 845101 (pp. 3273–84).

Van der Horst, A. R. A., 1984, *The ICTCT calibration study at Malmö: A quantitative analysis of video-recordings*, Report IZF 1984-37, TNO Institute for Perception, Soesterberg.

15

Drivers safety needs and the possibility of satisfying them using RTI systems

G. Malaterre and H. Fontaine

Introduction

Driving aid design is based on the view that what drivers need is to perform their task as safely and as comfortably as possible. Attempts are made to deal with the most frequently encountered difficulties, likely to lead to errors or malfunctions, and which could be the cause of an accident. This approach is to a large extent intuitive, and is not unrelated to what is generally thought about technological advances made possible by electronics and data transmission systems. It should, however, be backed up by the use of more objective methods such as analysis of the driving task, designed to pin-point the difficulties encountered by drivers, or accident analysis. This chapter will be devoted to a description of actual driver needs based on an analysis of report files representing accidents which occurred in France in 1989 (INRETS 1/50 accident report file, cf Fontaine *et al.*, 1990). In this instance, therefore, attention is focused only on safety. The different PROMETHEUS functions which could satisfy the identified needs, and thus make it possible to avoid the accidents in question, will also be examined.

Errors and needs

Drivers inevitably encounter difficulties when driving and may unfortunately make mistakes, some of which may lead to accidents. There are different ways of attempting to help drivers in this respect:

- by reducing difficulties, e.g. by modifying the infrastructure so as to separate or regulate traffic flows,
- by providing in their place, or conjointly with them, certain functions in the perception-decision-action loop common to all human operators.

It is therefore an essential requirement of any study to be fully aware of these functions and the errors which may ensue. Using the in-depth accident study (EDA) carried out at Salon-de-Provence (Ferrandez *et al.*, 1986) as a basis, it has been possible to analyse errors, by taking as a reference an operator model incorporating these functions (Malaterre, 1990). The Salon-de-Provence in-depth study provides a means of collecting detailed data in real time on the site of the accident, over an area covered by the SMUR (emergency medical service) in Salon. Approximately 70 accidents are subjected to in-depth analysis each year. Several studies have been carried out to describe the fact and factor sequences (termed accident mechanisms) in the minutes prior to the accident and which could explain the process.

The first of these (Malaterre *et al.*, 1986; Malaterre, 1990) made it possible to draw up accident mechanism typologies using an analysis of 80 accidents involving 115 drivers. Referring to work carried out by Kurucz *et al.* (1977) the perception, processing, decision, action sequence used constantly by the driver to regulate the system has been termed 'functional sequence'. It was assumed that one level of this sequence could be implicated in every accident, without this necessarily involving the responsibility of the driver in the legal sense of the term. For example, not seeing another user could be fully explained by obstacles impeding visibility. Nevertheless, it is the system 'data acquisition function' that is in question, or from an operator stand-point, his perceptive function in the widest sense of the term, i.e. this includes both the visual and information search functions. The aim was to identify which operator functions could account for the discontinuity, i.e. the accident situation. The rule adopted was to retrace as far back as possible the events preceding the accident. If, for example, a user tries to enter too small a gap, what is brought into question is not the manoeuvre itself, as it is the incorrect result of previous operations, the 'effects' as defined by Fell (1976), but the function corresponding to his information search if he has not looked in the right direction at the right time. A certain number of broad categories were therefore listed (see, evaluate, decide ... etc), together with the corresponding errors.

In another study (van Elslande and Malaterre, 1987), efforts were made to correlate accident mechanism categories with information or aid needs. This correlation is not always easy to make, as several mechanisms can result in the same need. Furthermore, when the complexity of the behaviour sequence is studied in greater depth, a diversification of needs can be seen. For instance, perception problems consistently call upon detection needs. On the other hand, problems caused by interpreting or predicting the behaviour of others could be connected to a series of diverse needs. An hypothesis will be put forward stating that each error corresponds to a non-satisfied need within the perception-decision-action loop. Error analysis using EDA backed up by an analysis of a representative sample of accident reports (Fontaine *et al.*, 1989) has made it possible to identify up to 17 basic needs. *For each user, the accident could have been avoided if the need had been satisfied.* A list of these needs is given below, with, in brackets, their distribution in the file analysed, i.e. 3179 accidents involving 6049 users (Malaterre *et al.*, 1991). It should be noted, that for nearly 20 per cent of these users, no need was identified. This corresponds to users who are generally passive during an accident, i.e. who are run into and who are not in a position to attempt evasive action. These needs are as follows:

N1.1. Driver status (8.4%)
This need requires the driver to have an average level of attention and skill. This is not the case when he is overtired and under the influence of alcohol, drugs or certain medicaments. This is a dual need: awareness (diagnostic) and action (not driving). These are not associated when the driver continues to drive, even though fully aware of his condition.

N1.2 Vehicle status (1.1%)
This need consists of driving a vehicle which is more or less in conformity with manufacturing standards. Here again, there are two levels which can be disassociated, in cases when a driver continues to use his vehicle, although fully aware that there are serious mechanical defects.

N2.1. Detecting a road-related difficulty (5.0%)
This need consists of the timely detection of a difficulty linked to road alignment, surface condition or general layout. This applies particularly to junctions or bends which are barely perceptible, poor road surfaces, ice etc.

N2.2. Obstacle detection (4.4%)
This need consists of the timely detection of any fixed or mobile obstacle, so as to be able to take evasive action or modify vehicle course. If the obstacle appears on the carriageway too suddenly for the collision to be avoided, it ceases to be a detection need and becomes a prediction need.

N2.3. Detecting oncoming users (7.1%)
This need consists of the timely detection of any oncoming user who is obscured due to alignment, vegetation, atmospheric conditions, or is concealed by other users.

N2.4. Transversal detection (19.1%)
This need consists of detecting the approach of another user on an intersecting lane at a junction, or similarly, for a pedestrian, detecting the approach of a vehicle on the roadway he is preparing to cross.

N2.5. Detecting a user outside the frontal field of vision (4.0%)
This need consists of detecting users hidden in vehicle blind spots (rear and lateral vision).

N2.6. Detecting a pedestrian (5.8%)
This need is a specific example of previous needs in that it involves pedestrians, whether visible or hidden by obstacles impeding visibility.

N3.1. Assessing speeds in relation to road conditions (3.7%)
This need consists of correctly assessing speeds in relation to road difficulties, in particular alignment. It differs from need N2.1 insomuch as a driver may have perceived a difficulty, e.g. a bend, but incorrectly judged the speed at

which it can be taken. Errors of assessment may be caused by both under-estimating one's own speed, and underestimating the difficulty in hand.

N3.2. Catching up on a slower road user (3.9%)

This need consists of a timely assessment of considerable differences in speed between the vehicle in front and ones own vehicle. This is found essentially in two different situations:

- When catching up on a very slow or stationary vehicle, e.g. as a result of traffic congestion or an accident, and particularly on fast-moving infra-structures where slow moving traffic is not frequent.
- In the event of a sudden slow-down in a line of traffic, when an inattentive driver is too late in perceiving this slow-down, in relation to the distance separating him from the vehicle in front.

N3.3. Estimating a collision course with another user (0.9%)

This need consists of correctly estimating one's relative movement in relation to another previously detected user, who is approaching a road junction from a different direction.

N3.4. Assessing gaps when overtaking or changing lane (0.6%)

This need consists of correctly assessing the time and distance required to overtake or change lane.

N3.5. Assessing gaps when joining or cutting across a traffic flow (0.6%)

This need consists of correctly assessing the time and distance required to join a more substantial or faster-moving stream of traffic, usually with right of way. This also applies when cutting across a traffic flow. This situation is usually encountered when moving off from a stop sign or joining traffic after turning and considerably reducing speed.

N4.1. Predicting that another user will move off or fail to stop (4.7%)

This need consists of predicting that a user, who generally does not have right of way, will cross a junction without stopping, or will move off without yielding right of way.

N4.2. Predicting the manoeuvre of another user (7.3%)

This need consists of predicting the manoeuvre of another user (changing lane, changing direction, overtaking, suddenly stopping), with the exception of the cases given above.

N4.3. Predicting pedestrian behaviour (1.9%)

This need consists mainly of predicting that a pedestrian, who has already been detected, will cross the road.

N5.1 Vehicle control (1.8%)

This need consists of being able to control one's vehicle in the absence of any apparent distraction, or incorrect perception of a difficulty (this applies particularly to novice drivers).

All these needs are not distributed equally between road user categories and types of road. Seventy-seven per cent apply to four-wheeled vehicles (which would seem, a priori, the only category that could be fitted with intelligent driving aids), and 29% occur outside built-up areas. This already provides an insight into what is at stake when it is known that certain driving aids will only be made available to large saloon cars. Figure 15.1 gives the breakdown of these needs, in and outside built-up areas. To give a clearer picture, needs have been grouped together into functional categories.

In built-up areas the overriding need is that of detection (mainly at junctions), followed by prediction (predicting the manoeuvre of a user who has already been detected). Outside built-up areas, detection needs, whilst fewer, are still in a majority, but status-related diagnostic needs (mainly driver-related: problems concerning alcohol or fatigue), together with space-time assessment needs, take precedence over prediction needs. Driver status-related problems deserve particular attention, and can be broken down as shown in Table 15.1. Users of 4-wheeled vehicles for whom the 'driver status' need was coded, can be characterized as follows:

- 55% involve loss of control. With regard to fatigue, this figure rises to 82%, the remaining accidents involve essentially rear-end or head-on collisions.

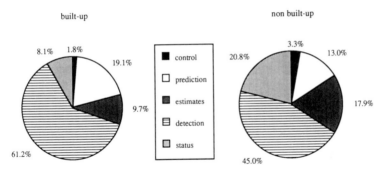

Figure 15.1   Breakdown of need categories in built-up areas (left) and outside built-up areas (right) for all types of user. This is based on 6049 users for whom 4549 needs were identified, 3214 of which were in built-up areas. These data are taken from the INRETS 1|50 file, 1989.

Table 15.1   Breakdown of 'driver status' need categories in INRETS 1|50 file (1989 data, 4-wheeler users only). The '% accidents' column includes all those for which a driver status need was noted for at least one of the drivers involved.

	% accidents	% drivers
Driver status	13.5	9.0
Alcohol	9.3	6.2
Fatigue	2.2	1.5
Inattention	2.0	1.3

- 53% occur outside built-up areas (80% involve fatigue).
- Alcohol-related accidents usually occur late in the evening, and fatigue-related accidents in the early morning.

The question is to find out whether these needs could be satisfied by driving aids covering the PROMETHEUS functions, as defined in the reference documents (PROMETHEUS, 1989a).

PROMETHEUS functions

This programme comprises 23 functions, 14 of which (see Table 15.2) are linked directly to safety. These functions are not aids in themselves. They represent a basis for a specification sheet covering devices which ensure one or several of these functions. This applies particularly to Common European Demonstrators (CEDs) which the various manufacturers taking part in the programme (PROMETHEUS, 1989b; Clarke, 1990) are now beginning to develop.

The actual performances of these functions will only be known when the demonstrators are in operation. At the present time, performance can only be judged in relation to the brief description given of their operating mode. Nevertheless, the lack of precision of this information prevented us from assessing functions 9, 10, 12 and 13. The following hypotheses were also put forward:

- a device will be developed for each function. This device will attain the set objectives in a reliable manner,
- drivers will use them to their full extent, thus pre-supposing that the ergonomics of these systems will be adequate, and more particularly, that the use of these systems will not have any adverse effect,
- that drivers will not take advantage of this accrued safety to take additional risks, e.g. driving faster, particularly in situations where visibility is poor.

Having set out these far-reaching hypotheses, an attempt was made to determine which accidents could have been avoided by putting each of these

Table 15.2 PROMETHEUS safety functions. A description of these functions and their operating modes can be found in the document: 'FUNCTIONS', or how to achieve PROMETHEUS objectives, Stuttgart, 1989a.

F1: Obstacle detection	F8: Dynamic vehicle control
F2: Monitoring environment/road	F9: Supportive driver information
F3: Monitoring driver	F10: Intelligent manoeuvring and control
F4: Monitoring vehicle	F11: Intelligent cruise control
F5: Vision enhancement	F12: Intelligent intersection control
F6: Safety margin determination	F13: Medium-range pre-information
F7: Critical course determination	F14: Emergency warning

functions into practice. To do so, each accident in the file was examined case by case, to determine whether the identified needs could have been satisfied by the different functions. It should be remembered that for all users for whom at least one need would have been satisfied by an aid, the accident would probably have been avoided. Moreover, the results shown here assume that all 4-wheelers could be equipped, but that pedestrians and 2-wheelers will not have access to these aids. Figures 15.2 and 15.3 show the percentage of needs which could have been covered, with a distinction being made between built-up areas and the open country.

Outside built-up areas, it can be seen that needs are relatively well provided for, with the exception of prediction (no aid could predict the intentions of another user) and status diagnostics. This is because it was assumed that these aids would only be informative, i.e. they would not dissuade all drivers who have been drinking alcohol, or who are overtired, from taking the wheel. This is seen as one of the most difficult problems to solve. Users would not accept a system that was too restrictive, and a system that was merely informative could have little effect on driver habits. It could even have adverse effects; drivers who have been drinking alcohol or who are overtired could feel over confident if driving aids were available. The high percentage of 'no need' drivers, in general those who are run into, from the rear, or by a vehicle out of control, and for whom no aid is possible, should also be noted.

In built-up areas, the percentage of needs which can be satisfied is lower. This is due to the greater number of 2-wheelers and pedestrians which, it was considered, could not be equipped with driving aids. In view of this, their needs could not be covered. In the same way, they cannot be included in cooperative systems. This means no aid could detect a pedestrian obscured by

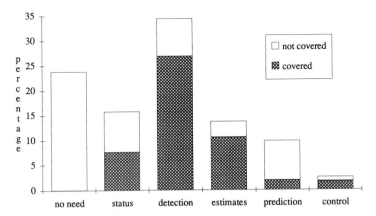

Figure 15.2   Outside built-up areas, the percentage of needs which could be covered by aids. These needs have been grouped together in broad categories. The blackened sections represents the percentage of each category which could have been covered by aids providing PROMETHEUS functions 1, 2, 3, 4, 5, 6, 7, 10, 11 and 14.

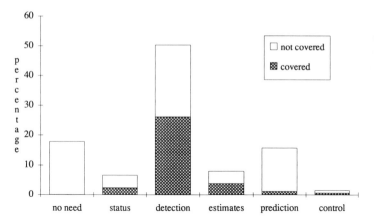

Figure 15.3   In built-up areas, the percentage of needs which could be covered by aids. These needs have been grouped together in broad categories. The blackened section represents the percentage of each category which could have been covered by aids providing PROMETHEUS functions 1, 2, 3, 4, 5, 6, 7, 10, 11 and 14.

a stationary vehicle. Notwithstanding, as there are more accidents in built-up areas, the number of users whose needs were satisfied is on the whole greater, despite a lower rate of effectiveness. This evaluation does not take into account the seriousness of the accidents. To do so would give greater significance to accidents which occur in the open country.

The effectiveness of the PROMETHEUS functions listed above were also compared. In Table 15.3 the percentage of all the identified needs which could have been satisfied (excluding no need users), is given opposite each function.

Table 15.3   Percentage of total needs which could be covered for each of the PROMETHEUS functions. To calculate the sum of these percentages would be meaningless, as certain needs could be covered equally well by several functions.

Functions	% of needs covered outside built-up areas	% of needs covered in built-up areas
F1	7.8	18.8
F2	2.2	1.8
F3	2.3	1.5
F4	0.4	0.2
F5	4.5	6.1
F6	4.8	2.3
F7	9.1	19.0
F10	1.5	.
F11	2.4	.
F14	1.0	0.2

It can be seen that function F7, i.e. determining critical courses (collisions with other users or driving off the road) heads the list. It makes it possible to cover 28 per cent of all needs. A more detailed analysis of these needs show, that they are above all simple detection needs and are the same as those covered by F1. More complex needs dealing with course calculation represent only 29 per cent of the effectiveness. As for functions F1 and F5, they only cover (by definition) detection needs. Function F3 monitors driver status. In this instance the hypothesis (highly arbitrary) was formulated that it would dissuade 50 per cent of drivers who had been drinking alcohol (60% of N1.1 needs) or those who were overtired (14%) from taking the wheel. On the other hand, this would not be effective for distraction, whether from inside the vehicle (passengers) or outside. However, concerning intelligent cruise controls (F11), this only confirms their relatively low importance with regard to safety. At the present time, we do not know which devices will be actually set on vehicles and when. We can only predict that the more rapidly available will be those:

- which are autonomous, that is to say that do not require sophisticated vehicle-vehicle communication but rather rely on radars, lidars, sensors, or any optical systems,
- which will not require expensive equipment of the road network,
- which will not raise liability problems, such as deciding who is responsible for the accident in case of failure in an automatism or in a communication.

But conversely, systems providing drivers only with information will be more prone to unsafe changes in behaviour, such as speeding or closing up. In effect, some drivers will be tempted to wait for a warning before reducing their speed, even in poor visibility conditions.

Avoidable accidents

The question here is to express the satisfaction of individual needs in terms of accidents which have been avoided. It should be remembered that any user for whom at least one need could be satisfied, would, as a consequence, avoid an accident. These are given in terms of avoidable accidents and no longer in terms of personal involvement. In an accident involving several vehicles, the use of an aid by only one of those involved, could often be sufficient to avoid an accident. The effectiveness expressed in avoidable accidents is therefore higher than in terms of individual needs satisfied, except of course when only one vehicle was involved.

Should PROMETHEUS be applied only outside built-up areas, approximately 22 per cent of all accidents could be avoided. These improvements apply particularly to loss of control (6.6%), accidents at junctions (5.3%), rear-end collisions (2.9%), or head-on collisions (2.2%). In cases of loss of control, the aid provided involves monitoring driver status, giving warnings with regard to alignment, skid resistance and recommended speeds. At

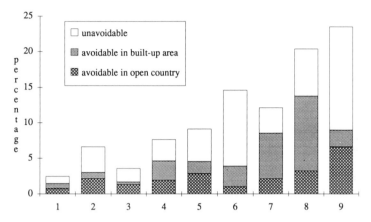

Figure 15.4 Accidents which could have been avoided using PROMETHEUS functions, depending on the type of accident. Avoidable accidents outside built-up areas have been distinguished from those in built-up areas. This figure clearly shows the importance of junctions in urban areas.

1 changing lane	*6 collisions involving a pedestrian*
2 head-on collisions	*7 junctions (going straight-on)*
3 overtaking	*8 junctions (turning)*
4 merging	*9 loss of control*
5 rear-end collisions	

junctions, the proportion of avoidable accidents is very high (approximately 2/3), and thus could have been even higher if function F12 (intelligent junction control) had been evaluated. This, however, was not possible, due to a lack of adequate performance data. If PROMETHEUS also operated in built-up areas, the overall effectiveness would be appreciably increased, by preventing in particular accidents at junctions (16.9%), and accidents involving pedestrians which, although avoidable only to a very small degree, represent a considerable improvement (2.9%). *All in all, nearly one out of two accidents could have been avoided, which corresponds to the targeted objectives.* To be more realistic, however, estimates should take into account the number of vehicles on the road which should be fitted with these aids.

Implications as to equipping the vehicle fleet

Although care should be taken, it is possible to try to calculate the impact PROMETHEUS could have on safety, depending on the different ways in which the vehicle fleet could be equipped. Assuming that the automobile fleet structure (breakdown in terms of age and power) remains relatively stable, and that all new vehicles are equipped with a full range of functions, the following results would be obtained (see Table 15.4). Depending on the hypothesis

Table 15.4   Percentage of accidents which, with PROMETHEUS, could have been avoided, according to different hypotheses as to the equipping of the French vehicle fleet. The evaluation of equipment effect was made at the end of a 5 and 10 year period, assuming that all new vehicles in the categories considered were equipped. At present, 82 per cent of passenger cars involved in accidents are less than 10 years old, and 15 per cent have a power/weight ratio greater than 90 Hp/tons (metric tons). The Table below was obtained by calculating the percentage of avoidable accidents for each category, and by supposing that the fleet structure would remain constant over the next 10 years.

effectiveness	top of range ($\geqslant$90 hp/ton)	top + middle ($\geqslant$70 hp/ton)	all 4-wheelers (including HGV)
after 5 years	4%	11%	24%
after 10 years	6%	18%	41%
after ... 30 years	7%	21%	51%

considered, it can be seen that PROMETHEUS could reduce the number of accidents from 4 per cent to 51 per cent. It is known, however, that this objective will never be reached due to possible changes in behaviour and a probable increase in mobility. The risk of involvement per vehicle-kilometre for equipped users should nevertheless fall, except in cases where user motivations, which will depend both on economic factors and the media coverage given to these aids, are directed more towards technology and high performance for its own sake, rather than to a search for improved safety.

Conclusion

It is extremely difficult to predict the effectiveness of a range of driving aids which have not yet been developed, and forecast to what extent the vehicle fleet will be equipped. The functions, as defined, will not all result in effective aids. Even the CEDs will only be demonstrators. They are only designed to display know-how, to evaluate performance and limitations, together with any improvements to be expected in real traffic conditions. They in no way anticipate the systems which will be made available to the vehicle fleet. The a priori method of evaluating effectiveness cannot be used to make accurate long term predictions. It assumes that all things will remain equal, which is of course never the case. On the other hand, this method makes it possible to correctly categorize implications. Needs analysis and implication categorization would therefore seem to be the two strong points of this research work.

When the needs and the systems studied are compared, it would seem a pity that the choice of demonstrators seems to have been motivated more by technological considerations than by taking actual needs, supported by accident analysis or on-the-spot studies, into account. It is to be feared that real driver practices have been neglected in favour of a theoretical even legalistic vision of the driving task (Malaterre and Saad, 1984, 1986). In these conditions, the evaluation phase of demonstrators in actual driving situations may be rather disappointing, with regard to both acceptability and their real impact on safety. Furthermore, the reliability of some functions could be questionable. A great number of problems seem to be very difficult to overcome. These include:

- the allocation of wavelengths required for vehicle-vehicle liaison. We are at present experiencing a shortage in this field, which the rapid expansion of telecommunications will do nothing to improve (David and Hane, 1990).
- the reliability of sensors and transmissions.
- the cost and maintenance of ground-based installations required for certain functions.

The cost of installing these aids on vehicles should also be taken into consideration. Most systems could not be fitted to vehicles already on the road. This is due to both financial considerations and the technical problems involved in fitting them into the vehicle, in particular, onto the dashboard. Two questions arise with regard to new vehicles:

- will all new vehicles be equipped; which will considerably increase the price of bottom-of-the-range models?
- if only top-of-the-range models are to be equipped, will this be standard practice or only an option?

Should they be an option, effectiveness may be limited by changes in behaviour; users who buy these aids could be motivated by reasons other than a quest for greater safety (improved performance?). This phenomenon is probably the cause of the disappointing results shown by ABS (Biehl *et al.*, 1988). This is why, from a safety stand-point, it would be better to set more reasonable objectives, by focusing on aids which correspond to the most pressing needs (driver status, detection). Aids should be as simple and as cheap as possible and could therefore be fitted to standard vehicles, at least from middle-of-the-range vehicles upwards.

Acknowledgments

The authors would like to thank Pierre Van Elslande who took part in a study phase which was not dealt with in this paper, and Clare McDonald who provided the translation. This research was sponsored by RENAULT.

References

Biehl, B., Aschenbrenner, K. and Wurm, G., 1988, Einfluss der Riskokompensation auf die Wirkung von Verkehrssicherheitsmassnahmen am Beispiel ABS, in *Schriftenreihe Unfall und Sicherheitsforschung Strassenverkehr*, Heft, 63, 65–70, Bergisch Gladbach.

Clarke, N.J., 1990, Common European Demonstrations, in PROMETHEUS *Research Newsletter* no 9, ed PROMETHEUS Office, Stuttgart.

David, Y. and Hane, B., 1990, Perspectives de développement des aides à la conduite et à la gestion de la circulation, *Revue RTS*, no 26, 7–14.

Elslande, P. van, and Malaterre, G., 1987, Les aides à la conduite: analyse des besoins en assistance des conducteurs, *Rapport de recherche INRETS*, no 23.

Fell, J. C., 1976, A motor vehicle accident causal system: the human element. *Human Factors*, **18**, 1, 85–94.

Ferrandez, F., Fleury, D. and Malaterre, G., 1986, L'étude détaillée d'accidents (EDA). Une nouvelle orientation de la recherche en sécurité routière, *Revue RTS*, no 9–10, 17–20.

Fontaine, H., Malaterre, G. and Elslande, P. van, 1989, Evaluation de l'efficacité potentielle des aides à la conduite, *Rapport de recherche INRETS*, no 85.

Fontaine, H., Gourlet, Y., Jurvillier, J-C. and Malespert, J-P., 1990, Constitution d'une base de données sur les accidents corporels, année 1989, Rapport de convention INRETS.

Kurucz, C. N., Morrow, B. W., Fogarty, W. J., Janicek, A. and Klapper, J., 1977, Multidisciplinary accident investigation, in, *Single vehicle accident study*, Volume 2: technical report, Coral Gables, Florida: University of Miami.

Malaterre, G., 1990, Error analysis and in-depth accident studies, *Ergonomics*, **33**, 10–11, 1403–21.

Malaterre, G., Fontaine, H. and Elslande, P. van, 1991, Analyse des besoins des conducteurs a partir de procès-verbaux d'accidents: évaluation a priori des fonctions PROMETHEUS, *Rapport INRETS* no 139.

Malaterre, G. and Saad, F., 1984, Contribution à l'analyse du contrôle de la vitesse par le conducteur: évaluation de deux limiteurs, *Cahier d'études ONSER* no 62.

Malaterre, G. and Saad, F., 1986, Les aides à la conduite; définitions et évaluation. Exemple du radar anti-collision, *Le Travail Humain*, tome 49, no 4, 333–46.

PROMETHEUS, 1989a, 'FUNCTIONS', or how to achieve PROMETHEUS 'objectives', Stuttgart.

PROMETHEUS, 1989b, PROMETHEUS is rolling: Pre-Dinner Speeches—Thematic Projects—Demonstrator Projects, Presentation Board Members, Munich.

16

Information presentation in in-car navigation systems

J. M. C. Schraagen

Introduction

Finding one's way in an unfamiliar city can be a difficult task, particularly when one has to attend to other traffic at the same time. Getting to know the city beforehand, by studying maps or asking people familiar with the city to point out the way, may make the task somewhat easier. Still, one frequently has to stop to check one's position, since information studied beforehand is often partially forgotten. The task would be somewhat easier if the driver could have permanent access to several kinds of information necessary for navigating in an unfamiliar city. One solution would be a simple piece of paper, where the driver has written down names of roads to be followed, or has drawn a sketch map of the city. This solution may still be very effective, but it is limited in the information it offers. Another solution, which is now technically feasible, would be a navigation system, an electronic aid that offers the driver various kinds of information upon request. Currently, two major alternatives are under development:

1. Map displays, possibly with current location and desired destination indicated. Commercial examples of map-display systems include the American ETAK Navigator system, the European TRAVELPILOT system, and various Japanese systems.
2. Displays using simple arrows or voiced instructions. These systems are still under development and are not commercially available. One of the best-known examples of these systems is the system under development in the Autoguide project (Rees, 1989).

These two types of system may get their information from various sources, for example, satellites, road transmitters, or by autonomous means (the so-called dead reckoning and map matching technique, in which the direction of the vehicle and the distance moved are measured by the vehicle itself). We are not concerned here with the question of where in-car navigation systems get their information from, but rather with the question what, from the driver's point of view, is the most safe and reliable way of presenting navigation information. Hence, it is critical to ensure that consulting the navigation system does not seriously interfere with the driving task and thus make the navigation task even more demanding than it already is. Therefore, it is important to know the types and amount of information to present to drivers, taking into account the driver's capabilities and limitations.

This section will proceed as follows. We will review a number of studies that have been carried out comparing conventional maps with various electronic means of presenting navigation information. Based on these studies, and our previous work on the strategies drivers use when navigating, we will develop some guidelines on what navigation information to present to drivers. Based on these guidelines, we have developed three experimental navigation systems that have been tested in a field experiment. Finally, implications for in-car navigation systems will be discussed.

Previous studies on in-car navigation information

We have been able to find seven studies that compared some form of navigation system with the use of a conventional map. Since all these studies used the conventional paper map as the control condition, we will be able to compare the seven studies in terms of the effectiveness of the navigation system compared with the paper map. The effectiveness will be expressed in terms of the average number of navigation errors per person per route. We will first give a brief description of each study before summarizing and comparing the results of all studies.

1. Streeter *et al.* (1985). In this study, customized route maps were compared with (i) vocal directions on tape alone, and with (ii) tape and map combined. The average number of errors in the map condition was 1.87, in the tape alone condition 1.11, and in the combined map and tape condition 1.61. Unfortunately, Streeter *et al.* only carried out significance tests on a subset of the navigation errors, leaving out errors they did not consider 'true' errors.

2. van Winsum (1987). This study compared a map on display with verbal 'left-right' instructions provided by the experimenter. In the map condition, the average number of navigation errors was 2.2, whereas in the verbal instructions condition no navigation errors were made. The relatively high number of errors in the map condition was due to the miscounting of side roads that subjects had to count in order to update their position on the map (van Winsum, personal communication).

3. Verwey and Janssen (1988). In this study, a memorized map was compared with (i) auditory system-generated left-right instructions, and (ii) visual system-generated arrows pointing 90 degrees to the left or right. The average number of errors in the memorized map condition was 1.92, in the auditory

condition 0.42, and in the visual condition 0.77. The pairwise differences between the three systems were all significant. The results further showed an interesting interaction with traffic density and route complexity: the number of navigation errors in the map condition was much higher than in the navigation system conditions when traffic density was high and routes were complex. The higher number of errors in the visual as compared with the auditory condition may have been due to the types of arrow used in the visual condition (Verwey, personal communication). Since all arrows extended under 90 degree angles, subjects expected the actual turns also to be 90 degrees. However, in a laboratory study, Verwey (1989) showed that even with a verbal presentation format subjects tended to expect perpendicular or straight turnings. It remains to be seen whether arrows that are isomorph with the actual turnings in the environment can change these expectancies.

4. Wierwille *et al.* (1988). In this study, a conventional paper map was compared with the ETAK navigation system, which provides the driver with a CRT-drawn map of the roadway network. The number of navigation errors was not reported, but the 'navigation effectiveness' was rated by the experimenters on a scale from 1 to 5 for each of three routes. The navigation effectiveness rating means were 3.06 for both the paper map and the navigation system. This indicates that there were no differences in the directness or quality of routes selected when using either the paper map or the navigation system.

5. Parkes (1989). In this study, a paper map on display was compared with a simple text display on an LCD screen. The text display showed left-right instructions and street names (e.g. 'left at end: Tennis Court Drive', or 'First right: Ambassador Road'). The average number of navigation errors in the map condition was 0.85, and in the text display condition 0.45. This difference was significant.

6. Pauzie and Marin-Lamellet (1989). In this study, a memorized map was compared with arrows displayed on an LCD screen. The average number of errors in the memorized map condition was 2.5, and in the 'arrow' condition 0.41. This difference was significant.

7. Schraagen (1990). In this study, a conventional memorized paper map was compared with a customized memorized paper map. The customized paper map contained stickers with names of road signs on them. These names could be memorized by subjects and used to make navigation decisions more easily. The average number of navigation errors in the conventional map condition was 2.15, and in the customized map condition 1.80. This difference was marginally significant.

Summary of navigation studies

Table 16.1 presents the summary results of all studies discussed above. Since all studies differ in numerous aspects, the average number of navigation errors cannot be used to compare studies. What we have done instead is, for each

Table 16.1   Percentage reduction in navigation errors.

Study	Navigation system	% reduction
Streeter *et al.*	tape	41%
	map + tape	31%
Van Winsum	auditory left-right	100%
Verwey & Janssen	auditory left-right	78%
	visual (arrows)	60%
Wierwille *et al.*	ETAK	0%
Parkes	left-right + street names	47%
Pauzie	visual (arrows)	84%
Schraagen	road signs on map	16%

study, calculate the percentage reduction in errors caused by a particular type of navigation instruction compared with the conventional map as control condition. Hence, this percentage reduction is calculated as:

$$\frac{(\text{control} - \text{experimental})}{\text{control}}$$

where: control = average number of errors with conventional map, and experimental = average number of errors with navigation system.

Guidelines

The following guidelines may be derived from Table 16.1:

1. Map-like displays should not be used in navigation systems, if we want to improve upon conventional paper maps. This is shown by the study by Wierwille *et al.* (1988), in which the ETAK system did not lead to improved performance compared with a conventional paper map.
2. The amount of information presented should be kept as small as possible. *Simple left-right instructions*, whether auditorily or visually presented, led to the largest reduction in errors. The taped instructions used by Streeter *et al.* (1985) contained too much information, hence did not lead to a large reduction in errors.
3. Street names should not be used. In Parkes' (1989) study, street names were used together with left-right instructions, but these instructions led to a reduction in errors of only 47 per cent, which is low compared to the 60–100 per cent reduction in errors with only left-right instructions. Moreover, from the study by Schraagen (1990) it was clear that subjects who heavily used street names when navigating made more navigation errors than subjects who attended to road signs, landmarks, and topological characteristics. Street names are probably not appropriate for navigational purposes because they are either difficult to locate, altogether

absent, or sometimes too far away when navigation decisions have to be made and therefore hardly visible.

Although most navigation systems that are commercially available use map displays, the empirical results clearly show that map displays do not lead to a reduction in errors when compared to conventional paper maps. The reason for this is not apparent from Table 16.1. However, one of the reasons may be that maps simply contain too much information not directly relevant for navigational purposes. The driver therefore has to devote a great deal of attention to extracting the appropriate information from a map. This is shown in a study by Gordon (1981) who found that simple instructions such as 'Alexandria' could be located much faster on freeway routing signs than when subjects had to make their own navigation decision based on a survey map. Furthermore, several studies have shown that maps severely detract attention away from the traffic situation to the in-car display (Wierwille *et al.*, 1988; Noy, 1989; Parkes, 1989; Ashby *et al.*, 1991). Whenever the situation outside the car demands a great deal of attention (e.g., when traffic density is high or when roads are narrow and winding), the time spent consulting the navigation system will decrease (Wierwille *et al.*, 1989). Since relevant navigation information is difficult to extract from map-based navigation systems, more navigation errors will be made in these demanding conditions (cf. Verwey and Janssen, 1988). Map-based displays may therefore be possible safety hazards. The study by Schraagen (1990) showed that when information directly relevant for navigational purposes, in this case road signs, was prominently displayed on a map, drivers made 16 per cent fewer navigation errors than when no such information was provided. Detailed analysis of the errors showed that road signs were particularly useful when choice points were complex. This indicates that drivers have problems relating two-dimensional map-based representations to three-dimensional real-world situations, particularly when those situations are complex. In conclusion, maps are perfectly useful for determining where one is, but are less useful for determining in a short amount of time where one has to go to. Simple left-right instructions are much easier to understand than maps, particularly on complex routes and with busy traffic, and provide information directly relevant for navigational purposes.

Use of the existing infrastructure for navigation

Currently, in-car navigation systems hardly use existing forms of routing by means of road signs. Still, use of the existing infrastructure in the form of road signs may be a very elegant form of navigation support (Godthelp, 1990). Instructions such as 'Follow A 28' or 'Take exit Utrecht' seem more natural on highways than 'left' or 'right'. Reference to road signs on highways by in-car navigation systems also opens up the possibility of flexible routing systems. For instance, when traffic density is very high on a particular highway, the driver may be advised to take another highway. This advice may be transmitted both to dynamic road signs and to the in-car navigation system. Even on main roads within cities, there is often an elaborate system of road signs that are there for people to find their way. The use of road signs in actual navigation systems has not been investigated before. We do know that people make use of prominent environmental features or landmarks when learning navigational decisions or when describing routes to others (Lynch, 1960; Carr and Schissler, 1969; Allen *et al.* 1978; Heft, 1979; Schraagen, 1989, 1990). Therefore, road signs, being one example of prominent environmental features, at least match the way people think of routes, that is, as a succession of discrete choice points. The main difference with left-right instructions is that road signs indicate both where to turn and what direction to turn to, whereas left-right instructions only indicate what direction to turn to. With left-right instructions, the driver has to decide where to turn. This may make the driver feel uncertain, particularly when navigation instructions are given close to a choice point. Use of road signs, on the other hand, may help the driver anticipate a change of direction long before the actual choice point.

A potential disadvantage of road signs is that they require the driver to look for the road sign in the environment first, before being able to make a navigation decision. Left-right instructions do not require such an extensive search in the environment. Road signs may be hard to detect when other distracting signs, for instance, advertisements, are placed nearby (Holahan *et al.*, 1978; Cole and Hughes, 1984; Shoptaugh and Whitaker, 1984; Boersema *et al.*, 1989).

Empirical studies

Field experiment 1: strategy differences in map information use

In order to find out how drivers navigate under 'normal' circumstances, and what difficulties they encounter, a field experiment was conducted in which 24 drivers, half of them male and half of them female, unfamiliar with a city, had to follow four prescribed routes indicated on maps. Since we did not know beforehand what different kinds of information drivers attend to, we opted for the situation where drivers could choose the different kinds of information for themselves. We used maps instead of verbal instructions because maps provide a large amount of different kinds of information and allow for a study of different strategies. We asked subjects to think aloud while studying the maps and while driving, in order to find out what types of information they attended to. Further details can be found in Schraagen (1990).

Categorization scheme

The subject's verbalizations were categorized into one of four types of spatial knowledge. In our study, we adopted as a framework Kuipers' theory of navigation and mapping in large-scale space (Kuipers, 1978; Kuipers, 1982; Kuipers and Levitt, 1988). Kuipers distinguishes between four types of spatial knowledge:

1. Sensorimotor knowledge: the knowledge that supports recognition of landmarks from a strictly egocentric point of view.
2. Procedural knowledge: knowledge of how to find and follow routes, stored in procedures. One may view these procedures as 'travel plans' (Gärling *et al.*, 1984), hierarchically organized around goals and subgoals.
3. Topological knowledge: a description of the environment in terms of fixed entities, such as places, paths, landmarks, and regions, linked by topological relations, such as connectivity, containment, and order. At this level of description, the traveller is able to go beyond strictly egocentric sensorimotor experience. Instead, he or she is able to recognize places as being the same, despite different viewpoints; identify places as being on a single path, in a particular order; define boundary regions to the left or right of a path. Using topological knowledge, a driver is able, for instance, to identify a street as being a main street that goes left of the centre.
4. Metric knowledge: a description of the environment in terms of fixed entities, such as places, paths, landmarks, and regions, linked by metric relations, such as relative distance, relative angle, and absolute angle and distance with respect to a frame of reference. Using metric knowledge, a driver is able, for instance, to infer that place A is south of place B, that a turn should be made with a sharp angle, and that a particular route is two kilometres.

Results

The average number of navigation errors across the four routes was 1.6 for males, and 2.4 for females. This was a marginally significant difference, $F(1,8) = 4.68$, $p < 0.10$. The effect of sex remained significant, even when age and yearly kilometrage were controlled for, $F(1,21) = 3.28$, $p < 0.10$ and $F(1,21) = 4.87$, $p < 0.05$, respectively. Driving experience did not have a significant effect on the number of navigation errors, $F(1,8) < 1$. In an unfamiliar environment, driving experience does not seem to benefit navigation performance. The verbal protocols were literally transcribed. Partly based on Kuipers' theory, five categories were distinguished:

1. street names
2. road signs
3. landmarks (e.g., school, church, railroad)
4. topological knowledge
5. metric knowledge

All the relevant terms that subjects mentioned were put into one of these five categories. In this way, a reference list resulted with 81 street names, 17 road

signs, 25 landmarks, 62 topological knowledge items (e.g., road characteristics, road types, counting streets, recognition of places from various angles),
and 21 metric knowledge items (e.g., compass directions, distance, angle). A
computer programme was written that compared the subject's verbalizations
(stored in ASCII format) with this reference list. For this purpose, the four
routes were considered as replications and taken together. In this way, we
could determine the relative emphasis a subject put on the various categories.
The distribution of the subjects' verbalizations during driving across the
different types of knowledge was as follows:

1. street names 42%
2. road signs 14%
3. landmarks 15%
4. topological knowledge 25%
5. metric knowledge 4%

There was a strong relationship between the type of information used during
driving and the number of navigation errors. If we subtract, for each subject,
the number of street names from the sum of the four other categories, we end
up with a score indicating the relative emphasis put on the other categories,
such as topological and metric knowledge. This score has a high negative
($r = -0.61$, $p < 0.001$) correlation with the number of navigation errors. Thus,
the more subjects attended to items other than street names, the fewer navigation errors they made. A more elaborate representation of the route leads
to more robust navigation performance, since whenever one type of information is forgotten or cannot be found in the environment, another type of
information may be retrieved from memory. If we enter this score as a
covariate in a separate ANOVA, the main effect of sex on number of navigation errors disappears, $F(1,21) = 1.02$, $p > 0.10$. This means that the effect
of sex was mainly due to differences in the information attended to. Women
used street names more exclusively, while men focused more on landmarks and
road signs, and used topological and metric knowledge to a larger extent.

In order to look at individual differences, 'good navigators' were compared
with 'poor navigators'. The top eight navigators made 4.5 navigation errors
on average, the bottom eight 12.5. This cannot be attributed to differences in
map study times (2.38 min versus 2.06 min, $p > 0.10$ by Mann-Whitney U-test),
but could be attributed to the different way of looking at maps by males and
females: the top eight were 4 men and 4 women, the bottom eight 1 man and
7 women. The difference between good and poor navigators could largely be
attributed to the greater attention of the poor navigators to street names (51%
versus 36%, $p < 0.05$ by Mann-Whitney U-test), mainly at the cost of attention
to topological characteristics (21% versus 27%, although this difference failed
to reach significance). Again this fits well with a corresponding difference in
map study behaviour. The poor navigators looked more at street names (57%
versus 46%, $p < 0.05$ by Mann-Whitney U-test), again at the cost of attention
to topological characteristics (18% versus 27%, $p < 0.05$ by Mann-Whitney

U-test). In order to look more closely at the causes of navigation errors, three types of navigation errors were distinguished:

1. Errors due to insufficient map inspection (this could very often be determined from the subjects' own verbalizations during driving, e.g. 'I must have overlooked that on the map').
2. Errors due to memory failures (forgetting of parts of the route to be driven, e.g. 'I can't remember whether it was left or right').
3. Errors due to insufficient visibility of the environment (in this case, subjects actively searched for the correct landmark or sign or intersection, but could not find these).

Each navigation error was assigned to one of the three types. The distribution of errors across these three categories for the two groups is shown in Table 16.2. An interesting aspect is that the difference in type of navigation error between good and poor navigators lies only in memory failures, and not in map reading errors or errors caused by the environment. Thus, the top navigators memorized well, the poor ones admitted that they had forgotten the relevant information. Correspondingly poor navigators consulted maps during driving twice as often as good navigators (9 versus 3.9, $p < 0.01$ by Mann-Whitney U-test). On average, the poor navigators consulted the map after 3.1 correct navigation decisions, versus 6.5 for the top ones ($p < 0.001$ by Mann-Whitney U-test).

Summary of main results

The main findings from this study were:

1. Females do worse than males since they focus more exclusively on street names.
2. Poor navigators (mostly females) fail in particular because they cannot memorize more than three items.
3. Driving experience does not matter for navigation.

Field experiment 2: experimental comparison between different types of in-car navigation information

A second field experiment evaluated the use of road signs and left-right

Table 16.2 *Distribution of errors across three categories for good and poor navigators.*

	Map reading	Memory	Environment
Good	2.4	0.2	1.7
Poor	2.6	8.3	1.9

instructions in a simulated in-car navigation system. Since we were particularly interested in how much navigation systems enable the driver to anticipate a choice point, we asked the drivers to indicate their subjectively experienced workload after each navigation instruction. We assumed that the more drivers can anticipate, the lower their mental workload will be. Mental workload was assessed by a subjective estimate on a single dimension, which is generally regarded as a satisfactory measure (cf., Verwey, 1990).

Subjects

Forty-two subjects participated in the experiment, half of them male and half of them female. Subjects were recruited by an advertisement in a local newspaper. They had no knowledge of the area where the experiment took place.

Routes

All routes were situated in an urban area of the medium-sized Dutch city of Amersfoort (c. 100 000 inhabitants). There were three experimental routes. Route I was 22 km long, was driven mainly on highways, and contained eight decision points (relevant intersections); Route II was 13.3 km long, was driven on main roads within Amersfoort, and contained nine decision points; Route III was 4.1 km long, was driven on secondary roads in a residential area, and contained 10 decision points. The end point of one route was the beginning point of the next route. If negotiated without navigation errors, route I took about 25 minutes to drive, route II took about 30 minutes, and route III took about 10 minutes to drive.

Conditions

The route guidance while driving was given by navigation instructions printed on postcards and presented by the experimenter. There were three types of navigation instructions on the cards. The first type of instruction was a left-right instruction in the form of an arrow. The arrows not only extended over perpendicular, but also non-perpendicular (i.e., 45 degree), angles. Non-perpendicular angles were used when they corresponded to the actual turnings in the environment. The second type of instruction consisted of messages such as 'Follow Utrecht' and 'Take exit Amersfoort'. These are referred to as 'road sign instructions', because they corresponded to names of cities on road signs placed at the decision points. The third type of instruction consisted of two or three messages presented simultaneously on one card. These instructions are referred to as 'multiple instructions'. An example of such an instruction is: 'Turn right, Galvanistraat; Turn left; Turn left, Leusderweg'. In this example, the subject is given feedback by means of street names that were clearly visible. On highways, instructions such as 'Follow Amsterdam; Follow Amersfoort' were given.

Procedure

After each route, subjects had to fill in a questionnaire with rating scales. The questionnaire asked about the number of times subjects had looked at the navigation instruction, the clarity and reliability of the instructions, whether the instructions were presented in time, whether subjects felt they had devoted sufficient attention to other traffic, and the general difficulty of the route. After all three experimental routes were completed, subjects had to compare the three different types of navigation systems by indicating, in pair-wise comparisons, which one they preferred. Subjects could at this point also make any comments and suggestions for improvement. Further details about this study can be found in Schraagen (1991).

Results

Type of navigation instruction had a significant effect on several dependent variables. Compared to multiple instructions, both road signs and arrows led to fewer navigation errors, a lower workload estimate, fewer number of looks at the navigation instructions, a higher judged clarity of instructions, and more attention to other traffic. Overall, routes driven with multiple instructions were considered more difficult than routes driven with arrows and road signs. Table 16.3 shows the average score on the dependent variables that showed a significant difference among the three types of navigation. Post-hoc tests did not reveal any significant differences between road signs and arrows on any of the dependent measures shown in Table 16.3. With one exception, the significant differences are between road signs and arrows on the one hand, and multiple instructions on the other hand. The exception is that subjects did not make fewer navigation errors with road signs than with multiple instructions, $F(1,86) = 2.53$, $p = 0.11$. Subjects did make fewer navigation errors with arrows than with multiple instructions, $F(1,86) = 4.72$, $p = 0.03$. Further analyses of the error types revealed that multiple instructions were particularly

Table 16.3 Average score on the dependent variables that showed a significant difference among the three types of navigation instructions (road signs, arrows, multiple instructions). The fourth column shows the F-values and the corresponding significance level ($^* = p < 0.05$; $^{**} = p < 0.01$; $^{***} = p < 0.001$).

	Road signs	Arrows	Multiple	F(2,34)
Navigation errors	0.57	0.53	0.92	3.26*
Workload estimate	1.29	1.34	1.84	11.64***
Number of looks	1.86	1.65	2.80	22.51***
Clarity of instructions	5.79	5.43	4.78	5.81**
Attention to traffic	5.82	5.63	4.94	5.96**
Difficulty of route	5.14	5.14	4.00	4.52*

likely to result in navigation errors on complex decision points involving multiple lane choices. Road type had a significant effect on subjects' judgement of whether instructions were given too late, $F_{(2,34)} = 5.87$, $p = 0.006$. Subjects expressed their judgement on a scale ranging from 0 (far too late) to 6 (not too late at all). Subjects thought instructions given on main roads were more often presented too late (average: 4.81) than instructions given on highways (average: 5.60) or secondary roads (average: 5.83), $F_{(1,86)} = 8.92$, $p = 0.004$ and $F_{(1,86)} = 12.24$, $p = 0.001$, respectively. The questionnaire given at the end of the experiment showed that 48 per cent of the subjects preferred the arrows, 12 per cent preferred the road signs, 17 per cent showed no preference for either of the two, and 24 per cent preferred arrows in cities and road signs on highways. Quite a few subjects remarked that multiple instructions took too much attention away from the traffic situation.

Summary of main results

The two major results of this study were, first, the finding that multiple navigation instructions resulted in more rather than fewer navigation errors compared with single instructions in the form of arrows. Second, use of the existing infrastructure in the form of road signs as navigation instructions did not result in fewer navigation errors than when simple arrows were used for navigational purposes.

Implications for in-car navigation systems

The results from this study confirm our guidelines presented in the introduction on the basis of a literature review of navigation experiments. The most important guideline is that navigation instructions should be kept as simple as possible: simple left-right instructions should be used instead of map-like displays (a similar conclusion can be drawn from the studies by Parkes and Martell, 1990; and Ashby *et al.*, 1991). Moreover, only one navigation instruction should be presented at a time. This number may be increased to two when the driving time between two choice points is less than 10 seconds (Alm, 1990). Navigation instructions in the form of arrows or in the form of verbal commands with reference to road signs, such as 'Follow Amsterdam', seem both equally effective, but further research is needed on this issue. If we look at the overall preference of drivers for the different types of navigation instructions used in this experiment, we may conclude that almost half of the drivers preferred arrows to use of road signs. A quarter of the subjects suggested the use of arrows in cities, and the use of road signs on highways. No one preferred the multiple instructions to the single instructions. On the basis of these results, a sensible guideline may be to use road signs on highways and arrows within cities.

Finally, it should be noted that our experiments have focused on relatively young drivers with a mean age of approximately 30 years. The oldest driver in the experiments was 48 years of age. We have not excluded elderly drivers because they may be of less interest, but primarily for reasons of experimental control and to keep the number of conditions within bounds. Thus, although the question whether elderly drivers have different information profiles than younger drivers is highly interesting, it has to be answered in future experiments.

Acknowledgments

The work reported in this publication is part of the project 'Generic Intelligent Driver Support Systems' (GIDS) carried out under contract DRIVE V1041 of the European Community, in which the Traffic Research Centre, University of Groningen acts as prime contractor and Delft University of Technology, INRETS-LEN, Saab/Scania, TNO Institute for Perception, YARD Ltd., MRC-Applied Psychology Unit, Tregie Groupe Renault, Swedish Road and Traffic Research Institute VTI, Universität der Bundeswehr München, and University College Dublin are represented as partners. The opinions, findings and conclusions expressed in this report are those of the author alone and do not necessarily reflect those of the EC or any organization involved in the project.

References

Allen, G. L., Siegel, A.W. and Rosinski, R. R., 1978, The role of perceptual context in structuring spatial knowledge, *Journal of Experimental Psychology: Human Learning and Memory*, 4, 617–30.

Alm, H. and Berlin, M., 1990, *What is the optimal amount of information from a verbally based navigation system*? Technical Report, Swedish Road and Traffic Research Institute (VTI), Linköping, Sweden.

Ashby, M. C., Fairclough, S. H. and Parkes, A. M., 1991, A comparison of route navigation and route guidance systems in an urban environment, *Proceedings of ISATA Conference*, Florence, Italy.

Boersema, T., Zwaga, H.J.G. and Adams, A.S., 1989, Conspicuity in realistic scenes: an eye-movement measure, *Applied Ergonomics*, **20**, 267–73.

Carr, S. and Schissler, D., 1969, The city as a trip: perceptual selection and memory in the view from the road, *Environment and Behavior*, **1**, 7–36.

Cole, B. L. and Hughes, P. K., 1984, A field trial of attention and search conspicuity, *Human Factors*, **26**, 299–313.

Gärling, T., Böök, A. and Lindberg, E., 1984, Cognitive mapping of large-scale environments: The interrelationship of action plans, acquisition and orientation, *Environment and Behavior*, **16**, 3–34.

Godthelp, H., 1990, Naar een beheerst wegverkeer, *Verkeerskunde*, **41**, 3, 112–6.

Gordon, D. A., 1981, The assessment of guide sign information load, *Human Factors*, **23**, 453–66.

Heft, H., 1979, The role of environmental features in route-learning: two exploratory studies of way-finding, *Environmental Psychology and Nonverbal Behavior*, **3**, 172–85.

Holahan, C. J., Culler, R. E. and Wilcox, B. L., 1978, Effects of visual distraction on reaction time in a simulated traffic environment, *Human Factors*, **20**, 409–13.

Kuipers, B. J., 1978, Modeling spatial knowledge, *Cognitive Science*, **2**, 129–53.

Kuipers, B. J., 1982, The 'map in the head' metaphor, *Environment and Behavior*, **14**, 202–20.

Kuipers, B. J. and Levitt, T. S., 1988, Navigation and mapping in large-scale space, *AI Magazine*, **9**, 2, 25–43.

Lynch, K., 1960, *The image of the city*, Cambridge, MA.: MIT Press.

Noy, Y. I., 1989, Intelligent route guidance: Will the new horse be as good as the old? in *Conference record of papers presented at the First Vehicle Navigation and Information Systems Conference*, pp. 49–55, Toronto, September 11–13, 1989.

Parkes, A. M., 1989, *Changes in driver behaviour due to two modes of route guidance information presentation: a multi-level approach*, Report No. 21, DRIVE Project V1017, HUSAT Research Institute, UK.

Parkes, A. M. and Martell, A., 1990, A usability analysis of a PC based route planning system, in E. J. Lovesey (Ed.), *Contemporary Ergonomics*, pp. 50–5.

Pauzie, A. and Marin-Lamellet, C., 1989, Analysis of aging drivers behaviors navigating with in-vehicle visual display systems, in *Conference record of papers presented at the First Vehicle Navigation and Information Systems Conference*, pp. 61–7, Toronto, September 11–13, 1989.

Rees, N., 1989, Autoguide: policy and practice in Great Britain, in *Conference record of papers presented at the First Vehicle Navigation and Information Systems Conference*, pp. 244–9, Toronto, September 11–13, 1989.

Schraagen, J. M. C., 1989, *Navigation in unfamiliar cities: a review of the literature and a theoretical framework*, Report IZF 1989-36, Soesterberg, The Netherlands.

Schraagen, J. M. C., 1990, *Strategy differences in map information use for route following in unfamiliar cities: implications for in-car navigation systems*, Report IZF 1990 B-6, Soesterberg, The Netherlands.

Schraagen, J. M. C., 1991, *An experimental comparison between different types of in-car navigation information*, Report IZF 1991 B-1, Soesterberg, The Netherlands.

Shoptaugh, C. F. and Whitaker, L. A., 1984, Verbal response times to directional traffic signs embedded in photographic street scenes, *Human Factors*, **26**, 235–44.

Streeter, L. A., Vitello, D. and Wonsiewicz, S. A., 1985, How to tell people where to go: comparing navigational aids, *International Journal of Man-Machine Studies*, **22**, 549–62.

Verwey, W. B. and Janssen, W. H., 1988, *Route following and driving performance with in-car route guidance systems*, Report IZF 1988 C-14, Soesterberg, The Netherlands.

Verwey, W. B., 1989, Simple in-car route guidance information from another perspective: modality versus coding, in *Conference record of papers presented at the First Vehicle Navigation and Information Systems Conference*, pp. 56–60, Toronto, September 11–13, 1989.

Verwey, W. B., 1990, *Adaptable driver-car interfacing and mental workload: a review of the literature*, Report IZF 1990 B-3, Soesterberg, The Netherlands.

Wierwille, W., Antin, J. F., Dingus, T. A. and Hulse, M. C., 1988, Visual attentional demand of an in-car navigation display system, in Gale, A. G., Freeman, M. H., Haslegrave, C. M., Smith, P. and Taylor, S. P. (Eds) *Vision in vehicles II*, pp. 307–16, Amsterdam: North-Holland.

Wierwille, W., Hulse, M.C., Fischer, T.J. and Dingus, T., 1989, Visual adaptation of the driver to high-demand driving situations while navigating with an in-car navigation system, *Abstract for Vision in vehicles*, third international conference, Aachen, September 12–15, 1989.

Winsum, W. van, 1987, *De mentale be lasting van hut navigeren in het verkeer*, Rapport Verkeerskundig Studiecentrum VK 87-30.

17

Route navigation. Deciding driving information needs

H. Alm

Introduction

By providing drivers with new types of in-car information, modern information technology can make traffic more efficient and rational. One risk associated with this possibility is that drivers may be distracted by the increased amount of new information. New information may, in the worst case, come from different uncoordinated systems. This internal distraction may increase the risk of an accident. To minimize that risk, different information technology devices should be adapted to drivers' needs for information, and to drivers' information processing abilities, not only to what is possible from a technological point of view. To simply let the market decide what are, and are not, safe products can be seriously questioned. If the market possesses this ability, then it is hard to understand the existence of dangerous products today. Furthermore, this belief assumes a high degree of rationality among drivers. Given the finding that most drivers consider themselves to be more competent that the average driver (Svenson, 1981), this high degree of rationality can be questioned. This overestimation of driving ability may, in the worst case, increase the number of accidents for some drivers. Looking at other safety measures we can also note that the safety belt was not generally accepted by the driving population in Sweden, despite its documented safety effects. Also, as pointed out by Rumar (1988), most drivers do not experience the risks as they can be seen in accident statistics. On the individual level the experience of risk is normally very low. The average driver can drive many years without being injured in an accident. On an aggregated National or European level, on the other hand, the number of killed and injured in traffic gives quite another impression. This difference in available information may, according to Rumar, explain the less than rational behaviour of individual car drivers.

To understand drivers' *need for information* in different driving tasks, it is necessary to analyse these tasks in more detail. To adapt systems to drivers' *information processing abilities*, it is often necessary to perform research. The purpose of this paper is to deal with the question: 'How to adapt a navigation system to drivers' need for information, and to drivers' information processing abilities'. It will start with an analysis of the concepts of navigation and route guidance, and from that deduce drivers' needs. Finally, it will discuss areas where research is needed to adapt the system to drivers' information processing abilities.

Navigation and route guidance, clarification of the concepts

Navigation and route guidance can be defined in many ways (see for instance Parkes, 1990), and what follows is one way to do it. Navigation is, in this context, defined as a type of spatial problem solving. It starts with a need to travel to, or be in a different spatial position than the present. The reason(s) behind the need to travel may be of many different sorts, from business and job related to more pleasure seeking. Navigation can be broken down into a number of subproblems. One subproblem is to locate the present position (where am I?), another to plan the trip (where is my destination?), and a third is route guidance (how do I drive to reach it?). Using a time dimension to analyse navigation, we can identify three stages.

Before the trip

The driver needs information about the destination(s). A number of pretrip or strategical questions can be listed, and the following list is probably not exhaustive.

Identification of present position
● Where am I right now?
Identification of destination(s)
● Where is my destination(s) located?
Identification of transportation modes
● In what way(s) can I reach the destination(s)?
Evaluation of transportation modes
● What is the time of departure for the different transportation modes?
● What is the total travel time for the different transportation modes?
● What is the cost associated with the different transportation modes?
● Where is the place of departure for the different transportation modes?
● Given the choice of one way to travel, how many times do I have to change train/bus/plane etc.?
● What is the predictability of the different transportation modes?

- Which activities are possible to perform when travelling by the different transportation modes?
- What are the risks associated with the different transportation modes?
- What is the nature of the social environment associated with the different transportation modes?
- What is the nature of the physical environment associated with the different transportation modes?
- How do the different transportation modes interfere with other activities?
- How do the different transportation modes enhance the performance of other activities?

To make a decision about how to travel to a certain destination, people may need information about some, or all of these questions. The nature of the trip (business or pleasure) may have an influence upon people's need for information. To simplify the situation, we will assume that the person in question has decided to use a car to reach the desired destination.

During the trip

The driver needs information that will make it possible to follow a chosen route. This type of information will be called route guidance information, simply because it guides the driver along a chosen route. The driver will also need information about the possibility to proceed along the chosen route.

After the trip

The driver needs information about different options to park the car, and the associated costs and benefits. To summarize, a driver's need for information from a navigation system depends upon what stage (before, during, after the trip) the driver is in. The ideal navigation system should be able to give information relevant for all stages. In the following section we shall discuss how to adapt the information for a route guidance system to drivers' information processing abilities, and research questions that demand an answer.

Route guidance system and drivers' information processing abilities

A basic assumption is that a route guidance system must transmit information to the driver. It is possible to pose a number of basic questions about this information. What follows is an example of some important questions.

1. Which sense modality, or sense modalities, should be used as receivers for the route guidance information? Looking at the driver as an information processor, s/he has five different channels, with different properties, to the outside world. The most likely candidates for route guidance information are

probably the visual and the auditory channels. This is not to deny the possibility of using other sense modalities in future vehicles. Future research may tell us more about that.

Research efforts have been oriented towards comparison of the effectiveness of visual and auditory information during route guidance. Streeter *et al.* (1985) concluded that carefully designed verbal directives seem to have an advantage over carefully designed visual directives. Somewhat surprisingly, they also found that the combination of visual and verbal information did not lead to the optimal performance. In a more recent study Verwey and Janssen (1988) also found that auditory information seems to be superior to visual information. Thus auditory information may have some advantage over visual information. From a traffic safety point of view it can also be argued that the use of auditory or verbal information has some advantages. It is commonly assumed that the feedback a driver gets from driving is visual to a very large degree (90 per cent according to Rockwell, 1972). Therefore, it seems wise to avoid equipment that makes demands on the drivers' visual processes, especially if the traffic situation is complex, or the driver is unskilled. On the other hand, it can be argued that very simple visual information has some strong positive features. The advantage with visual information is that it can be 'self-paced', that is, the driver can pick up the information when the traffic situation so allows. Auditory information can be missed, and takes some time to repeat. So, instead of a repeating function for auditory information, a very simple visual symbol could possibly be used. Since Streeter *et al.* found negative effects of a combination of verbal and visual information, we obviously need more research in order to see if visual and verbal information can be combined (in terms of content and temporal aspects) in some optimal way.

2. Another basic question concerns the type of information that should be used by a route guidance system. A route guidance system can show a graphic description of something in the environment, talk about, or point at something. In all these cases a decision must be taken concerning what the system should represent, talk about, or point at. In order to adapt the system to the drivers' representation of the driving environment, we need basic knowledge about that. Studies within DRIVE (GIDS) have given some insight into drivers' representation of routes (Alm, 1990), and also found some 'landmarks' that drivers seem to use frequently. Traffic lights, orientation and traffic signs, petrol stations, and shops were frequently used as indicators of a route. Within PROMETHEUS one study has looked at the effect of using landmarks for a route guidance system (Alm *et al.*, 1991). It was found that landmarks had a positive effect upon drivers' route following ability. These are positive indications, but also in this case more research is needed.

3. A third basic question concerns the amount of information a route guidance system should offer to the driver. The maximum amount of information is to inform the driver about every choice leading to the destination,

before starting the trip. The minimum information is to guide the driver from choice point to choice point, without any overall information. It seems reasonable to assume that the amount of information from a route guidance system should be dependent upon the complexity of the driving environment, and one study within GIDS (Alm and Berlin, 1991) has indicated this. In the study twenty-four subjects, randomly assigned to three experimental conditions, were given the task to drive to an unknown destination. The amount of information was varied for the different groups. The results from the study indicated that information about one and two choice points seems to be optimal. A route guidance system should also be able to vary its amount of information depending upon the driving time between choice points. Exactly when the system should present information about one or two choice points respectively, must be further investigated.

4. A fourth basic question has to do with the timing of route guidance information. The optimal system presents information so that the driver is able to drive safely. This means that the driver must be given time to see if the traffic situation allows a certain manoeuvre, and time to implement the manoeuvre. The timing of the information must, on the other hand, not be made too early. Information should ideally be presented when the driver starts to experience a need for route guidance information. Too early presentations may put an extra demand on the drivers' memory capacity, and increase the drivers' mental workload. Too late presentation may increase the temporal demands upon the driver, as demonstrated in the study by Ashby *et al.* (1991). This is obviously a difficult question, and so far we do not have any empirical studies that can support educated guesses.

Conclusions

Through task analysis and empirical research it seems possible to give guidelines for the construction of navigation and route guidance systems, adapted to the drivers' needs, and information processing abilities. In this context it is important to stress the fact that we are only in the beginning of this process. Much more analysis and empirical research is needed to really reach safe conclusions. The work performed in DRIVE and PROMETHEUS will hopefully lead to the construction of both effective and safe information technology devices.

References

Alm, H., 1990, Drivers' cognitive models of routes, in van Winsum, W., Alm, H. and Schraagen, J. M., *Laboratory and field studies on route representation and drivers' cognitive models of routes*, Deliverable GIDS/NAV2, Traffic Research Centre, University of Groningen.

Alm, H. and Berlin, M., 1990, *What is the optimal amount of information from a verbally based navigation system*, Contribution to deliverable GIDS/NAV3, VTI, Linkoping, Sweden.

Alm, H., Nilsson, L., Järmark, S., Savelid, J. and Hennings, U., 1991, *The effects of landmark presentation on driver performance and uncertainty in a navigation task—a field study*, Swedish PROMETHEUS IT-4, VTI, Linkoping, Sweden.

Ashby, M. C., Fairclough, S. H. and Parkes, A. M., 1991, *A Comparision of Two Route Information Systems in an Urban Environment*, Report No. 49, DRIVE Project V1017, HUSAT Research Institute, Loughborough University of Technology, U.K.

Parkes, A. M., 1990, *Route guidance devices: Too many, too soon*? Paper presented at SICS Workshop on Navigation and Planning, Nässlingen, Sweden, Stockholm University Press.

Rockwell, T. H., 1972, Skills, judgement and information acquisition in driving, in Forbes, T. W. (Ed.) *Human factors in highway traffic safety research*, pp. 133–64, New York: Wiley-Interscience.

Rumar, K., 1988, Collective risk but individual safety, *Ergonomics*, **31**, 4, 507–18.

Svenson, O., 1981, Are we all less risky and more skilful than our fellow drivers? *Acta Psychologica*, **47**, 119–33.

Streeter, L. A., Vitello, D. and Wonsiewicz, S. A., 1985, How to tell people where to go: Comparing navigational aids, *International Journal of Man-Machine studies*, **22**, 549–62.

Verwey, W. B., and Janssen, W. H., 1988, *Route following and driving performance with in-car guidance systems*, Report IZF C-14, Soesterberg (The Netherlands): Institute for Perception TNO.

18

Selection of routes in route navigation systems

W. van Winsum

Introduction

In electronic route guidance systems the route selection algorithm is of great importance. The system is required to select the best route from an origin to a destination. In order to ensure a high level of user acceptance, the route selection logic should be the same as the decision logic of the user of the system. It is assumed the system works best if the driver is led over routes he would have selected with a full knowledge of all available routes. To develop the input for an algorithm for route selection, it is paramount to know the driver needs. The driver needs, in this context, relate to the reasons why car drivers prefer some routes over other routes. A number of potential benefits of electronic navigation systems are mentioned in the literature. They reduce total kilometrage and travel time (Armstrong, 1977; Jeffery, 1981a, 1981b), enable drivers to circumvent congested routes (Hounsell et al., 1988), result in fewer drivers getting lost (King and Lunenfeld, 1974), leading to safer behaviour (Färber et al., 1986). Furthermore, traffic safety will be increased by diminishing the required amount of controlled attention by navigation (van Winsum, 1987).

There are numerous studies on route choice to be found in the literature. None of these studied the decision rules of car drivers in route choice (van Winsum, 1989). In addition to time savings, distance is frequently reported as a relevant route choice criterion (Trueblood, 1952; Mortimer, 1955; Moskowitz, 1956; Freeman et al., 1971; Radcliffe, 1972; Koning and Bovy, 1980). Safety of the route appears to be important in some studies, but not in others. The same can be said of criteria relating to congestion and self pacing (Anderson and Peterson, 1964; Ueberschaer, 1971; Benshoof, 1970; Huchingson et al., 1977; Janssen, 1985). Scenery seems to be important in non-business related trips, but not in business related trips. Costs appear to be unimportant as a route choice criterion. Travel time, distance, congestion, safety and the possibility to keep moving all seem to be important. A problem, however, is that these criteria are not independent. Previous studies on route choice by car drivers have not been able to unravel these factors. The influence of network type (urban, inter-urban), of distance between origin and destination, of personal and demographic characteristics and of trip motive on route choice criteria have not received much attention in the literature. The investigations that included these variables generally showed no effects on route choice criteria.

For the development of a route selection algorithm it is imperative to know how several attributes of routes are traded off by car drivers. Also, the determining factors of travel time, from the car driver's perspective, should be known. To establish the weighting of route choice criteria by car drivers, a decision analysis experiment was designed.

Method

General

Route choice is assumed to be determined by a number of attributes of routes. The implied psychological process is compensatory in nature. This means that poor levels on some attributes can be compensated by good levels on others. In decision analysis, the appropriate methodology for estimation of the importance of multiple attributes is Multi Attribute Utility Theory (MAUT). The assessment of alternatives, in this case routes, is simplified by expressing the overall value of an alternative as a decomposed function of the separate outcome attributes (Fischer, 1979). The evaluation task is broken down into single attributes of which evaluations are constructed. Trade offs among attributes are quantified as importance weights. Finally, formal models are applied to reaggregate the single attribute evaluations (von Winterfeldt and Edwards, 1986). A decision analysis computer programme was developed for this purpose. This programme offers a number of different options for weight estimation. Only the results from the best weight estimation method will be reported here. A comparison of results from different methods and methodological considerations can be found in van Winsum (1990).

Subjects

The data from 49 subjects were analysed. These subjects volunteered for participation in the experiment after an invitation in two local newspapers. Between subjects groups were gender, age, level of education, trip motive (home-work or home-school, social trip, recreative trip) and driving experience (inexperienced v. experienced). Inexperienced drivers were defined as:

years of driver's licence $\leqslant 5$ AND driving $\leqslant 100000$ kilometres in total OR years of driver's licence >5 AND driving $\leqslant 10000$ kilometres a year OR years of driver's licence <1.

Experienced drivers were defined as:

years of driver's licence $\leqslant 5$ AND driving >100000 kilometres in total OR years of driver's licence >5 AND driving >10000 kilometres a year.

Gender, age, education and driving experience were used as between subjects variables in order to study whether the parameters for the route selection in the navigation system should be different for different driver categories. Trip motive was included as a between subjects variable to study whether route selection should be based on different parameter values for different trip motives.

Procedure

Four routes were selected by each subject. They received a large map of the city of Groningen in the Netherlands with all streets and street names (scale 1:10000) and a smaller topographical map (scale 1:20000). The subjects were asked to think up one destination, within the range of the map, they visited often by car. The origin was always the home address. The destination had to be chosen far enough from the home address to make a variety of different routes possible. The subjects were first asked to draw the most frequently used route on the small map with a red felt pen. Subsequently they drew a green, a blue and a yellow route on the small map. These were routes they used less frequently. It was, however, required that they were familiar with these routes. In the remainder of the experiment the routes were identified by the colour. The subjects were asked to think about the routes in terms of the type of trip the route was used for, on a certain time of the day, for ordinary use of the route (and not for the purpose of, e.g. making a detour to pick up a person).

As a criterion variable for the correctness of the estimated weights, the global preference was used. The subject selected the best route on a global basis. This route was assigned a value of 100. Then the subject selected the least preferred route. This route was assigned a value of 0. The other two routes were scaled in between the extremes on a horizontal bar, such that the differences between the routes on the bar reflected the relative preferences as accurately as possible. From a restructuring of attributes which appeared promising from a literature review (van Winsum, 1989), the following nine attributes were derived:

- time loss by giving right of way
- time loss by traffic lights and queues

- number of lanes and maximum speed (road type)
- adaptation of speed caused by other traffic
- number of turns
- certainty of arrival in time
- concentration required
- scenery
- distance

The attributes were the same for all subjects for the sake of comparability. Subjects were, however, free to delete attributes with a severely constricted range within the choice set. The values of each route on every attribute were then estimated. The scaling of the routes on every single attribute was done in the same way, as in the global preference estimation method. The subjects then determined the attribute importance weights by means of an indifference procedure. They were presented with two horizontal bars located above each other. The upper bar represented the units of global preference, the lower bar represented the units of value for a specific attribute. The subjects were asked how much a specific route would increase in global preference (relative to the other routes), if the current attribute would improve from the level on that route to the best value of that attribute. The best value was represented by the route with the best value on that attribute. The two bars were scaled such that the target route on the upper bar was directly above the same target on the lower bar. The weights were determined by the increase in global preference divided by the increase in preference of the respective attribute. Since all weights are expressed in the same units (global preference) they are comparable. In this way, the strength of preference of each attribute was matched by the strength of preference of the global preference. Since one unit increase in global preference is worth the same as an increase of 1/'weight' units in an attribute, the indifference values are $1/W_i$, where W_i is the weight of the attribute i. This is because the values and the global preferences have the same scale (0–100). If we know the 'objective' range of the attributes, the indifference values are calculated as $RANGE(i)/W_i$.

Since there are four routes there are, in principle, three possible routes which are not best on a single attribute. Therefore there are, for each attribute, three measurements to reduce measurement error and to get enough data points for a multiple regression analysis for each individual. The number of cases for regression analysis on the indifference measurement is (number of attributes) × (number of alternatives − 1), i.e. 27. The dependent variable of the regression analysis is (global preference of route A + increase in global preference caused by an increase in preference in one attribute). The independent variables are the values on the attributes on that route, where the attribute which caused the increase in global preference has a value of 100 (increase to best level). So, the regression analysis is based on the indifference measurements and filters out irrationalities by the subject and measurement errors.

Determination of aggregated values

The weights were normalized to sum to one.
Aggregated values were calculated as:

$$V(x) = Sum(Wi \times Vi(xi)), \text{ summed for all attributes } i = 1 \text{ to } n.$$

where x   = route x
  Wi   = weight of attribute i
  Vi(xi) = value of route x on attribute i

Results

Environmental dependency

A positive correlation between attributes in the real world forms no threat to the validity of the MAUT model. It conveys something of the meaning of the attributes to the subjects. A factor analysis (principal component analysis with varimax rotation) was applied on the estimated values of the objective attributes. The criterion for insertion of a factor was a value greater than 1. Three factors were selected with a total explained variance in the set of variables of 64.2 (see Table 18.1). The factor coefficients, after varimax rotation, are listed in Table 18.2.

The first factor concerns the ability to keep driving with as little interference as possible. 'Road type' and 'in time' have negative coefficients because these (objective) attributes are scaled the other way around (more is better, while with the other attributes in this factor more is worse). In the value measurement procedure the attributes were all scaled in the same direction to insure conditional monotonicity. It also appears that the reliability of the route (in time) is related to this factor. The second factor concerns travel time. Travel time relates to the two types of waiting time and distance, but not to road type. The third factor consists of scenery. This attribute is not related to the other attributes.

So, 'road type', 'adaptation of speed', 'number of turns', 'in time' and 'concentration' are related attributes which covary in the real world, and probably have as a common factor the ability to keep driving with little interference, or ease of driving.

Table 18.1   Explained variance by factors with value > 1.

Factor	Value	Variance (%)	Cumulative percentage variance
Factor 1	3.40	34.0	34.0
Factor 2	2.01	20.1	54.1
Factor 3	1.01	10.1	64.2

Table 18.2   Factor coefficients after varimax rotation.

	Factor 1	Factor 2	Factor 3
A-travel time	0.14	*0.89*	−0.05
1-right of way	0.27	*0.68*	0.09
2-traffic lights	0.18	*0.70*	0.25
3-road type	*−0.76*	0.22	−0.28
4-adaptation speed	*0.84*	0.10	0.28
5-number of turns	*0.60*	0.05	−0.27
6-in time	*−0.66*	−0.28	0.14
7-concentration	*0.79*	0.21	0.23
8-scenery	0.10	0.07	*0.88*
9-distance	−0.26	*0.64*	−0.13

'Distance', 'travel time', 'waiting time to give right of way' and 'waiting time because of traffic lights' are also related attributes which covary in the real world because they have travel time as a common factor.

Effect of weights on convergence with the global preference

The correlation between global preferences and aggregated values was 0.91, indicating that the decision model predicted the preferences of routes very well. If on the other hand it was assumed that all attributes were equally important (equal weights) the correlation between aggregated values and global preference was only 0.66. This indicates that a differential weighting did make a difference and that the extracted weights reflected the decision process of the subjects.

Attribute weights

Table 18.3 presents an overview of the average weights of the attributes. This indicates that 'distance', 'in time' (reliability of the route), 'road type' and 'traffic lights' are the most important route choice criteria.

Simulations

In decision analysis, sensitivity analyses are used to study the implications of varying the weights in a systematic way. The robustness of the model can be tested by measuring the effect of weight variations on the aggregated values.

In this study a comparable procedure has been followed. For every subject, the effects on the aggregated outcomes of deletion of the least important attribute from the set of attributes has been studied. The effects on the aggregated outcomes were measured by the correlation between the aggregated outcome and the global preference. This can be regarded as the search for the

Table 18.3   Average weights of the attributes.

Attribute	Weight
1-distance	0.21
2-in time	0.13
3-road type	0.12
4-traffic lights	0.13
5-adaptation	0.10
6-right of way	0.10
7-concentration	0.09
8-number of turns	0.08
9-scenery	0.07

smallest subset of attributes with the highest predictive power of the global preference. It also allows us to see if there is a relation between the number of attributes and the correlation between aggregates and global preference. Subjects are assumed not to take all attributes into consideration when evaluating routes. The preference for routes might be determined, or at least explained equally well, by a smaller subset of attributes. The weights in the reduced set have to be rescaled such that the relative magnitudes of the weights are the same as they were, and the sum of weights equals one.

In the value estimation the range of every attribute is 100 (from 0 to 100). Indifference between attributes can be calculated as range/weight, i.e. 100/weight. Suppose the weight of 'distance' is 0.30 and the weight of 'waiting time' is 0.20. This means that 100/0.30 units of distance can be compensated by 100/0.20 units of waiting time. So 333.33 units of distance can be compensated by 500 units of waiting time. This means that if there are two routes which are the same on all attributes except on distance and waiting time, these routes are equally preferred if route A is 333.33 units worse in distance than route B, but 500 units better in waiting time than route B.

To translate the value units to objective units of the attributes, the subjects were asked to give estimations of the real objective value (distance in metres and waiting times in seconds and so on). The attribute with value 0 corresponds with the worst route on that attribute (e.g. 5000 metres) and the attribute with value 100 corresponds with the best route on that attribute (e.g. 3000 metres). The range of 100 in value then corresponds to the range in metres of 2000. These objective ranges can be divided by the weight as well, to derive indifference values. This was done according to:

$$1000 \times (\text{range}(i)/W(i))/(\text{range}(\text{dist})/W(\text{dist})),$$

where i   = attribute i,
       dist = distance,
       range = worst level − best level.

In this way all units of the attributes (seconds or scale points or percentages) were expressed in the number of kilometres distance that is equal in preference. The averages of these values were determined over all subjects with the exclusion of outliers (deviations of more than 2 standard deviation from the mean).

A good route selection algorithm has to use clearly measurable objective attributes, use as few as possible attributes, assign car drivers to routes which would be chosen by the subjects themselves if they would know these routes well. The attributes which are network characteristics and thus measurable by a navigation system (integrated with stations along the road) are 'time loss by giving right of way', 'time loss by traffic lights and queues', 'road type', 'number of turns' and 'distance'. The indifference values, worth one kilometres distance, are listed in Table 18.4. The number behind road type indicates the number of extra metres of higher order road that would compensate for an increase of 1000 metres in total length of the route.

These parameters were used as input for a simulation programme for route selection. The data of all subjects (global preferences of routes and objective characteristics of routes) were used as input as well. So the parameters were the same for all subjects. The programme determined the powerset of these five attributes. The powerset consists of all possible subsets of the five attributes, e.g. only 'right of way' and 'number of turns' or only 'traffic lights', 'number of turns' and 'distance'. For every subset the programme calculated:

• The correlation between the global preferences of routes and aggregates of routes based on the parameters for the attributes in the subset.
• The percentages of the routes that were the best (most frequently used) according to the subjects that were assigned to a certain preference rank order by the algorithm.

The second route in preference, according to the subject, is an acceptable route most of the time. So, an algorithm which would lead the subject as much as possible over the best route according to the subjects and to a lesser extent over the second best route, but not over the third or the fourth route, seems to be acceptable. If only distance was used as a selection criterion, only 40.8 per cent of the routes selected by the algorithm were the subjects' first choice. This confirms Koning and Bovy (1980) who found (with a completely different

Table 18.4   Average indifference values that are worth one kilometre distance.

Right of way (in seconds)	112.2
Traffic lights (in seconds)	102.0
Road type (in metres)	− 3568.0
Number of turns (number)	5.7
Distance (in metres)	1000.0

method) that distance as an optimalization criterion predicted around 38 per cent of routes used.

If we select the attributes with not only a high correlation, but also a high percentage predicted routes, which are also the best according to the subjects, then the attributes 'right of way', 'traffic lights', 'road type' and 'distance' appear to be a good choice. The aggregated values for those four attributes correlated 0.78 with the global preferences. These four attributes resulted in computed best routes of which 81.6 per cent were also best routes according to the subjects, and 14.1 per cent of the computed best routes were second best according to the subjects. Only 4.1 per cent of the computed best routes were second worst according to the subjects, and worst routes were never computed as best by the simulation programme.

Conclusions and discussion

A decision analysis study on route choice by car drivers was designed to estimate parameters for a route selection algorithm for navigation systems. It was argued that a good route selection algorithm has to result in leading car drivers over routes they would have chosen themselves, if they had complete knowledge of the characteristics of the routes.

Between subjects factors were gender, age, education, trip motive and driving experience. These factors did not influence the subjective importance of route choice criteria in a meaningful fashion. It should, however, be noted that the trip motive 'recreational trips' in this study is not representative for all networks. Recreational trips in this study were time restricted, meaning that the subjects had to reach the destination (the recreation spot) in time. Moreover, the recreation was located on the destination and not on the route. That means that scenery of the route may still be important in non-urban trips where the trip motive is recreation. Total travel time of the route strongly influences the preference of the route. This is not surprising in the light of numerous other studies on route choice. Total travel time appears to be determined by distance, waiting time to give right of way and waiting time for traffic lights. Of these attributes distance appeared to be the single most important determinant of route choice, but waiting time at traffic lights was experienced as important as well. Waiting time to give right of way was in general experienced as less important. Yet, not only travel time is important as a decision criterion in route choice. Road type (whether a larger part of the route consists of higher order roads) is an important route choice criterion as well. This attribute has much more influence on the reliability of the route (whether the driver is certain to arrive in time) than on travel time. Reliability of the route is an important determinant of route choice as well. Scenery, the number of turns and the concentration required appear unimportant attributes in general.

When the weights were used as input for a route selections algorithm, it appeared that the attributes 'waiting time for giving right of way', 'waiting time for traffic lights', 'road type' and 'distance' resulted in selection of routes which would have been the first choice for the car drivers in 82 per cent of the cases. Only 14 per cent of the selected routes were second best according to the subjects. Worst routes were never selected and only 4 per cent of the selected routes were second worst according to the subjects. If only distance had been used as a selection criterion, as in a number of conventional route selection algorithms, 41 per cent of the selected routes would have been first choices by the car drivers. Since this figure conforms with the literature, this strongly suggests that the fact that people often choose routes which are sub-optimal on distance is not caused by insufficient knowledge, but by the fact that criteria other than distance are important as well. Since there were hardly any differences between groups of gender, age, trip type, educational level and driving experience, the route selection can be the same for all groups.

References

Armstrong, B. D., 1977, *The need for route guidance*, TRRL supplementary report 330.

Anderson, J. L. and Peterson, G. D., 1964, *The value and time for passenger cars: further theory and small scale behavioral studies*, SRI, Menlo Park.

Benshoof, J. A., 1970, Characteristics of drivers' route selection behavior, *Traffic Engineering and Control*, **11**, 604–9.

Färber, B., Färber, B. and Popp, M. M., 1986, Are orientated drivers better drivers? *Fifth International congress ATEC 86*, Paris.

Fischer, G. W., 1979, Utility models for multiple objective decisions: do they accurately represent human preferences? *Decision sciences*, **10**, 451–79.

Freeman, Fox, Wilbur Smith and associates, 1971, Choice of route by car driver, Report prepared for Department of the Environment [UK], Mathematical Advisory Unit.

Hounsell, N. B., McDonald, M. and Wong, C. F. S., 1988, Traffic incidents and route guidance in a SCOOT network, PTRC summer annual meeting, 1988.

Huchinson, R. D., McNees, R. W. and Dudek, C. L., 1977, Survey of motorist route-selection criteria, *Transportation Research Record*, **643**, 45–8.

Janssen, W. H., 1985, Routevoorbereiding, een enquete onder bestuurders, IZF c2.

Jeffery, D. J., 1981a, *The potential benefits of route guidance*, TRRL Report 997.

Jeffery, D. J., 1981b, *Ways and means for improving driver route guidance*, TRRL Laboratory Report 1016.

King, G. F. and Lunenfeld, H., 1974, Urban guidance: perceived needs and problems, *Transportation Research Record*, **503**, 25–37.

Koning, G. J. and Bovy, P. H. L., 1980, Routeanalyse, een vergelijking van model-en enqueteroutes van autoritten in een stedelijk wegennet, TH Delft, memorandum nr. 24.

Mortimer, W. J., 1955, Trends in traffic diversion on Edens expressway, *HRB Bulletin*, **119**.

Moskowitz, K., 1956, California method of assigning diverted traffic to proposed freeways, *HRB Bulletin*, **130**.

Ratcliffe, E. P., 1972, A comparison of drivers' route choice criteria and those used in current assignment processes, *Traffic Engineering and Control*, **13**, 526–9.

Trueblood, D. C., 1952, Effect of travel time and distance on freeway usage, *Public Roads*, **26**, 241–50.

Ueberschaer, M. H., 1971, Choice of routes on urban networks for the journey to work, *Highway Research Record*, **369**, 228–38.

Winsum, W. van, 1987, De mentale belasting van het navigeren in het verkeer, VK 87-30.

Winsum, W. van, 1989, *Route choice criteria of car drivers: a review of the literature*, Traffic Research Centre, University of Groningen.

Winsum, W. van, 1990, *A validation of MAUT decision analysis techniques*, Technical report GIDS/NAV/R5, Traffic Research Centre, University of Groningen.

Winterfeldt, D. von and Edwards, W., 1986, *Decision analysis and behavioral research*. New York: Cambridge University Press.

19

Conflicts between high efficiency and high acceptance due to the way information is provided to the driver

K. Lorenz

Introduction: the urban driving task as a challenge for new technologies

In all major cities, two of the basic tasks in getting from A to B, in an urban road network, are becoming more and more difficult for the single driver:

(a) finding an appropriate route from A to B knowing that the shortest one is not necessarily the best in terms of travel time,
(b) leaving the car near the destination (somewhere cheap, legal and safe).

New technologies such as route information and route guidance systems, as well as parking guidance and management systems, are offering new ways of providing conventional information for the driver. Furthermore, some systems include new sources of information which had never been available for the driver before. They can even replace the driver as captain on board, leaving him as the man at the wheel. Three major European systems currently under development are EUROSCOUT (the advanced follow-up to the LISB-System tested in Berlin; Siemens AG), TRAVELPILOT (Bosch Blaupunkt) and CARMINAT (PHILIPPS). EUROSCOUT and CARMINAT are route guidance systems (beneath some extra applications, which are of no importance here). They recommend a full route to the destination. Advanced versions also recommend mode of transport (at P + R interchange points), which can be seen as a first step into a real integrated traffic management system. Routes are chosen in a central computer, using a constantly updated travel time database, and the result is the shortest route in terms of travel time predicted for the time of the journey. The method of route calculation and travel time prediction depends on the actual system, but it is of no importance here (as it is invisible for the driver anyway). The distinctive feature of guidance systems, in principle, is that they provide the driver with manoeuvring information (displayed as standardized icons) on every point of decision. Contrary to those systems, route information systems, such as TRAVELPILOT, leave the route choice completely to the driver who can use a small screen with an electronic map display which is constantly aligned with vehicle direction and moving according to the actual vehicle position. This map can be zoomed, from a simplified overview of the whole city, to a very detailed road map of the near surrounding of the vehicle's position. This up-grade of the conventional paper map is meant to ease the driver's task in finding his way through a complex network, but leaving him the full decision making process. Because the displayed map is visible and zoomable all the time, the driver is completely free to decide if, and when, he wants to use the information. No traffic data are used and no route recommendation is given (except that major roads appear highlighted on the screen).

The question remains however, which, and to what extent the difficulties encountered in tasks (a) and (b) are really caused by a lack of information, and what part of the problems is due to routines, personal preferences and prejudices and non-acceptance of already existing sources of information. If there is an information deficit, advanced on-board information systems as described above can be the key to solve current problems. If other aspects in driver's problem awareness and decision making are more dominant, then there have to be other measures to deal with road traffic problems. Furthermore, it has to be investigated whether this new kind of driver information is likely to be accepted by the driver. Only this can really change driver behaviour from the heuristic, routine based and intuitive driving of today to the informed, but increasingly 'remote controlled' driver of the future. As part of the DRIVE Project BERTIE (V1017) three field studies have been carried out to assess the potential for changes in drivers' strategic behaviour due to the introduction of new route and parking guidance systems.

Selected results from BERTIE field experiments

First a brief summary of the BERTIE experiments TUB was involved in. A parking questionnaire and a car-following survey were carried out in Berlin to investigate 'before RTI' parking search behaviour (Lorenz, 1990a). On the basis of these ex-car flied trials another experiment was realized in Nottingham with in-car data collection. Searching behaviour was investigated with and without a parking guidance system. For this small-scale experiment only a simple simulation of a guidance system (acoustic commands) was used in order to compare guided from unguided driving/searching (Lorenz, 1990b). The experiences from both approaches were integrated within the next field trials in Berlin. This time two real RTI applications were made available in-car for experiments. The experiments were designed to investigate changes in driver behaviour due to the way information is provided and displayed:

- autonomous route information system (the system chosen for that category was the Bosch TRAVELPILOT IDS),
- infrastructure based route guidance system (the system chosen for that category was the Siemens LISB System).

Details of this experiment can be found in Ashby *et al.* (1991) and Lorenz (1991). In the following, findings of this third field experiment will be explained and compared against the results of the two other studies.

Route choice behaviour: a comparison between upgraded conventional driving (with an electronic map) and driving with a route guidance system

Twenty-four subjects were recruited from the age groups 20–30 and 55 +, both male and female (the age-groups were chosen for the simultaneous collection of physiological data). They had two tasks; in both driving to a given destination and locating a parking place. In one task the subjects had to follow the instructions of LISB; in the other task they had to choose a route on their own, using the map of Bosch TRAVELPILOT and their own experience—they were all residents of Berlin. The two tasks were carried out in different areas of the city and counterbalanced in order of tasks, order of systems, order of destinations etc. The hypothesis was that a guidance system cannot be significantly more efficient than local drivers assisted by an electronic map.

However, it was found that even local drivers with the help of an electronic map could not 'beat' the travel times of drivers following the route guidance system. Therefore, it can be said that improving travel times through route choice is a problem of information. These advantages, however, will only change driver behaviour (trusting a guidance system more than individual experience) if they can be felt by the driver. Under normal non-experimental conditions no driver will keep records on travel times and alternative routes. Therefore, the correctly measured travel times of the experiment are not the right source for analysing individual evaluation and the potential for the acceptance of guidance systems. Acceptance in that respect means the willing-ness to use the guidance system regularly and follow the recommended routes (= a real change in driver behaviour), and not the acceptance to purchase a system and use it from time to time (= just another status symbol and no change in driver behaviour). The subjective estimation of travel time com-pared to a subjective experience will be the basis for individual evaluation instead. For that reason, drivers have been asked to evaluate travel time and parking search time immediately after a task had been completed. The results are shown in Figure 19.1 (for both tasks together). The difference of the

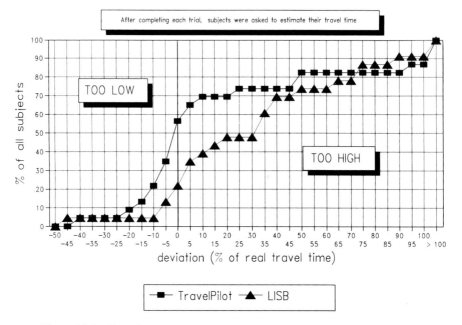

Figure 19.1   Travel time estimation deviation: estimation/real travel time.

estimated travel time from the real one has been calculated for all experimental tasks, and the deviation from the real value is displayed as relative error (in per cent of the real travel time) and shown in a cumulative percentage distribution. Though the majority of subjects arrived at the destination earlier with LISB than with TRAVELPILOT, subjects tended to overestimate travel times with LISB: 80 per cent of all subjects felt a travel time longer than the measured one. There is a wide range of deviation from the real travel time value on the 'too long'-side of the scale. The same subjects estimated their travel times with TRAVELPILOT more precisely. Too short and too long-estimations were given with a proportion of about 50 per cent each and the deviation from the real value is much smaller than with LISB.

This result decreases the advantage of a dynamic route guidance system found in the real travel time analysis. Subjects are not fully aware of increased efficiency of their task. This, however, is a requirement for a successful implementation of guidance systems unless these systems are not compulsory. Two aspects may be the reason for this over-estimation of travel times:

1. A dynamic route guidance system is choosing routes on a purely mathematical way using a dynamic database and a route search algorithm. Many times the result is a route which is uncommon and seems to be not logical at first sight (without knowing the traffic conditions on all sections of a network at that time). That was the case for task 1 of the experiment. Most

times LISB chose a route which was remarkably longer than the 'normal' straight route(s) into the city centre because LISB knew the travel time advantages of the urban motorway ring. Furthermore this recommended route was going to a completely different direction for a long time, which gave subjects the impression of following the wrong way or at least a major detour. This uncertainty, and the incompatibility of the recommended route with the individual cognitive map and common sense of direction, can produce the feeling that one is 'driving for ages in the wrong direction' and lead to an over-estimation of travel time. The uncertainty not to know whether this is an error or really the optimum route is increased by the fact that LISB does not show the full recommended route in advance, which would give the driver a more comfortable feeling that this recommended route is actually leading to the destination.

2. Another reason for over-estimating travel times can be found in the relatively passive part the driver has to play. With LISB the driver is free of the navigation-part of the driving task. All he has to do is steer the vehicle safely and stand by for instructions. Subjective assessments show that the tasks with LISB have been generally evaluated as easier and less demanding. This is an advantage of the route guidance concept, which relieves the driver from mental workload, but it may have the effect that the driver now has more 'spare time' which leads to a longer impression of travel time.

For the first aspect application designers have to think about a different way of presenting guidance information to the driver. The potential for increasing efficiency with a route guidance system is the recommendation of alternative or non-standard routes; in other words, under heavy traffic conditions a selection of routes that drivers do not normally follow, because they do not know them or they are convinced of the efficiency of their own preferred route. The guidance system has to be highly reliable and the way information is being displayed to the driver should contribute to the impression of reliability. For instance, a preview of the (currently) recommended route should be given before the start on a TRAVELPILOT-like display.

On the other hand, a route information system, which leaves the navigation to the driver did not produce a higher efficiency, because drivers preferred straight routes to the destination. These routes are standard routes for many drivers and therefore not free from congestion. TRAVELPILOT does not provide the driver with any route recommendation, except the fact that major roads appear brighter on the screen and minor roads are not displayed in the larger scales of the scale menu. Subjects missed most information on traffic restrictions, for example not every link on the display is a link in reality due to one-way-roads, which are not marked. For destinations in a completely unfamiliar environment, a route information system is a useful tool, especially for the final approach, and can be easily done with the detailed and complete road network display. For local drivers it is of limited value because it is not providing the driver with new information. Most subjects chose routes by combining their experience and the map display; a few subjects misused the electronic map for rat-running strategies and trying new routes.

Parking: is parking guidance to off-street facilities what drivers really want?

For parking the key question is basically the same: is the general acceptance that searching for long and inefficient on-street parking is caused (a) by a lack of information about alternatively available parking facilities; or (b) due to the rejection of commercial parking?

(a) Parking guidance systems, combined with in-car route guidance are the appropriate way to improve the parking situation.
(b) In-car parking guidance systems (which have to be bought by the drivers) will fail because they will offer the same parking spaces rejected by the driver today.

Interviews and observations carried out in Berlin showed that the reality is more a combination of both extremes: because free on-street parking is so much more attractive (despite uncertainty and stress), commercial parking facilities are not (or only a vague) part of the drivers' 'cognitive map'. Therefore, this is a problem of acceptance.

If a parking search in the favourite parking area is unsuccessful after the maximum acceptable searching time, the driver has to think about a substitute strategy. Because of the widespread exclusion of commercial car-parks in drivers' minds, most drivers decide on illegal parking. But even if they accept commercial parking as an 'emergency' strategy, very often they do not know about the next car-park and its occupancy. This is a problem of information, and this is where parking guidance systems will show their benefits even for drivers who are not using them right from the beginning of their trips.

How did drivers deal with the problem in the in-car experiment? Both guidance systems had no significant influence on parking search strategies. Subjects were not prepared to use the systems (showing the destination, or the distance and direction of the destination) for parking purposes. The road network in Berlin is relatively simple: a fairly right-angular network, few one-way systems and turning restrictions. Contrary to the experiences from Nottingham, subjects did not lose their sense of direction during the parking search. The questionnaire revealed, however, the need for integrated parking guidance either as 'P's shown on the TRAVELPILOT display, or guidance to a car park close to the destination instead of guidance to that destination itself. Subjects argue that if they accept a guidance system then they want to be guided completely.

In line with the results from the Berlin parking interviews, the parking search did not take very long in Berlin. Figure 19.2 shows the distribution of

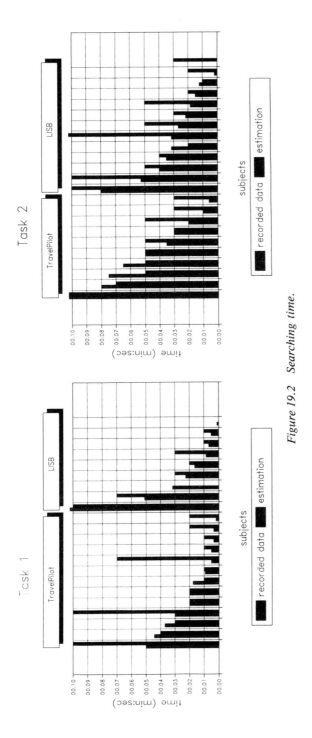

Figure 19.2 Searching time.

parking search duration, both the measured duration and the subjective estimation. The majority of subjects needed less than 5 minutes to find a parking place, many of them less than 2 minutes. The Berlin parking interviews had revealed a similar situation. Task 1 ended in a busy area of the city centre with department stores and a wide range of commercial off-street parking facilities. In this area on-street parking is known to be difficult, and therefore many drivers accepted to park in a multi-storey car park right from the beginning and not as a result of an unsuccessful on-street search.

Task 2 led to an area which attracted more drivers to search for on-street parking than in task 1. The search durations were still acceptable for most subjects, and the search result was very good, especially for those who found an on-street space. The distribution of found spaces in the destination area showed that on-street spaces are not less convenient than off-street spaces. In terms of subjective acceptability they are even better, because they are free of charge and walking distances are not remarkably longer, especially if one takes into account the walking distances (both horizontal and vertical) inside a multi-storey car park. In that respect, the advantage of being guided to a car park is limited to the elimination of searching time and the comfortable feeling of avoiding uncertainty and stress. This experiment is in line with the interviews which showed that the success rate of conventional on-street parking was not so low that driver behaviour would change after new guidance systems have appeared on the market.

Evaluation of parking problems as a potential for changes in future driver behaviour

Changes in driver behaviour can only happen if two basic requirements are fulfilled:

- internal: the driver has to be able to handle the new system properly (MMI),
- external: the driver has to be aware of the unacceptable state of the road transport system and accept the new system as a key to solve the problem.

While the experiments showed that the first requirement can be fulfilled with some minor modifications of recently developed systems, the external requirement is still not reached in all urban areas.

The general public is very aware of parking problems in the cities, but is not, in general, prepared to change its parking behaviour and accept other modes of transportation or new concepts. Only extreme and singular traffic conditions (for example big sporting events or worst traffic and parking restrictions in the destination area, e.g. wheel clamp areas in London) are creating an environment for behavioural changes. In city centres like Berlin the situation is still far from those extremes. On-street parking is becoming more difficult, but is still the rule, off-street parking remains the exception.

Therefore the general strategy drivers are using today can be described as follows:

1. The driver is aware of parking problems, the reaction is an extended time budget for parking search.
2. The driver knows some areas for free and unlimited on-street parking.
3. The driver uses multi-storey car parks as an emergency strategy only, unless this car park is connected with his destination directly (department stores for example with parking fee refund).

That was the result of both the Berlin parking interviews and the driving experiment. Figure 19.3 shows the distribution of searching times. It can be seen that their absolute values are not unacceptably long though they cover a relatively large proportion of a short inner city trip. The individual estimation of searching time tends to over-estimate the real searching time, but still the majority of subjects were not disappointed. A simplified calculation illustrates the effect of an increased efficiency, supposing that an integrated parking guidance system will reduce searching time completely and guide directly to a car park. Eighty per cent of all subjects searched less than 5 minutes for a parking space (both on- and off-street). Due to the over-estimation of searching times, a subjective feeling of 5 minutes searching can be seen as a realistic value for individual evaluation. Route guidance in task 1 for instance improved travel times by another 5 minutes on average. This adds up to a large

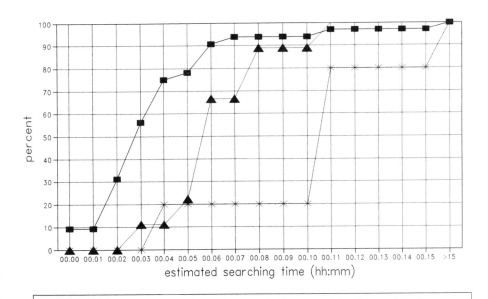

Figure 19.3 Parking search evaluation in relation to estimated searching time.

travel time saving of 10 minutes. Fifty per cent of all subjects, however, over-estimated travel-times with route guidance by more than 30 per cent. For the average travel time of task 1 (about 19 minutes), this means an error of nearly 6 minutes. Therefore, for the individual evaluation, there was only a 50 per cent improvement. That was the case for task 1 with high travel time improvements, for task 2 the subjective benefits are even shorter due to smaller improvements of travel time. The benefits referred to here are only individual benefits of car drivers. The general benefit of a (even small) reduction of unnecessary traffic in city centres is not part of the analysis here. The concept of RTI-systems is (still) to sell applications on a voluntary basis and not as compulsory systems, therefore individual benefits will be the main incentive for the purchase of such systems.

The searching time for on-street space was longer than the time to look for a multi-storey car park (Figure 19.4) for those who chose a multi-storey car park as their first choice. Forty-four per cent reached their car park entrance in less than a minute while only 22 per cent got their on-street space in the same amount of time. On the other hand, 22 per cent is nearly a chance of 1 in 4, which gives a good reason for the attraction of on-street parking. Another reason is the remaining walking distance. Off-street parking, though (on average) closer to the destination, is not necessarily closer in terms of walking time because these distances are only representing horizontal distances. Figure 19.4 shows also that the on-street-line ends at 7.5 minutes, while the off-street line continues until 15 minutes. This can be interpreted as the threshold for on-street parking search. After an unsuccessful search of about eight minutes the strategy is changed and a multi-storey car park will be considered as an emergency alternative. These search procedures are extremely inefficient and will be the strongest argument for new parking guidance strategies. If the rate of such unsuccessful on-street search routes could rise from 20 per cent now up to 60 or 80 per cent, then this would mean an environment appropriate to create behavioural changes. Two decisive factors will change drivers attitudes towards on-street parking:

1. to know that after a long search, the probability of having to park finally in an off-street car park is very high, and the unsuccessful search was a complete waste of time,
2. if resulting from (1) the price for parking is accepted, then a quick and comfortable guidance to the nearest free car park could be seen as an incentive. ('If I have to pay for parking anyway, then I want maximum comfort and quality.')

Nearly all subjects in Berlin said that they would appreciate parking information, or a straight guidance to the nearest car park, if they were using a route guidance system. From the interviews, previously carried out, drivers preferring on-street spaces did not know about other car parks at all, or had only a vague impression about the nearest one. The result of the Nottingham experiment was that drivers started highly inefficient trial-and-error-search for

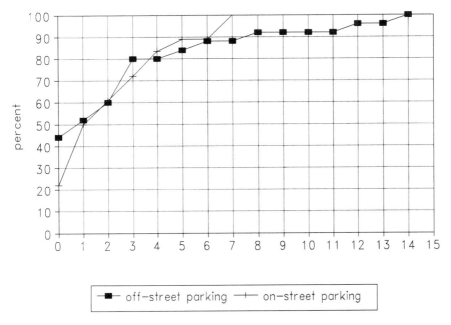

Figure 19.4 Parking efficiency: searching times.

a substitute car park if their first choice (and standard) car park was full/closed, and only one of the 10 subjects found the shortest route to the next car park in the 20 tasks.

The key to assessing the potential for behavioural changes is the distinction between individual benefits and global assessment of efficiency. It would be a major success if we could reduce the amount of (parking) search traffic in city centres. Although in Berlin we found only a relatively small proportion of drivers who did not find an on-street space after 8 minutes, it would still mean a success due to the size of city traffic. The global evaluation of efficiency of integrated parking guidance systems would be positive. The accumulated searching times, of all vehicles per day, would end up as an extremely high figure of wasted time, because in about 80 per cent of the cases in our experiment, drivers did not find an on-street space immediately (< 1 minute searching). The individual evaluation however is different, a success rate of one in four for ideal on-street parking encourages drivers to try this option. Even some time spent searching is well accepted and still more appreciated than off-street spaces as the questionnaire revealed. About two-thirds of those who parked off-street evaluated this mode of parking as non-optimum. Eighty per cent of the subjects parking on-street found there was no better parking despite the search. The advantage of parking in front of, or near the destination, legally, free of charge, with unlimited time, and without stairs to climb and levels to remember, is so attractive that the risk of an unsuccessful search is taken.

Conclusions

Following the analysis of the experimental data, the following key sentences can be taken as the results regarding efficiency and the potential for changing driver behaviour:

1. Route guidance systems improve travel times even for local drivers with an electronic road map, due to the integration of traffic-information inaccessible to the single driver.
2. The improvements of travel times are not fully recognized by the drivers.
3. The route guidance principle (commands and standardized icons) is generally accepted as comfortable, easy to understand and hardly distracting.
4. Existing systems are not efficient for the final part of a journey to the city centre. Parking search is not eliminated or even assisted by the systems. Thus the increased efficiency of the approach to the destination area is smaller if the efficiency of the whole journey is considered.
5. Changes in parking behaviour and the acceptance of parking guidance systems require several external factors:

   (a) On-street parking has to be not only difficult, but impossible or more expensive than off-street parking.
   (b) The individual chance of an on-street space has to be so extremely low, that drivers do not start on-street searching strategies first. The experimental results show that a success rate of 22 per cent is still not low enough.
   (c) The road network structure has to be unsuitable for intuitive searching procedures, especially close loops around the destination.
   (d) Off-street parking procedures (manoeuvring, paying, finding parking position, etc.) have to be as easy and comfortable as possible. A central location of the car park alone is not sufficient if the other factors are producing high resistances for the driver. These resistances are together influencing driver behaviour more than the discomfort and pressure of an 8-minute on-street search. Task 1 produced high rates of off-street parking due mainly to aspect (b), but also due to a centrally located car park, easy to drive and directly connected to a department store with smaller fees for customers.

6. For cities with capacity problems in off-street facilities parking guidance has to be combined with route guidance in order to start optimum routes right from the starting point.
7. System design has to assure that uncertainty is avoided at any stage and advantages of the system are visible to the driver. Uncertainty about the route choice can influence the individual estimation of travel time and decrease the acceptance of the system.

References

Ashby, M. C., Fairclough, S. H. and Parkes, A. M., 1991, A Comparison of Two Route Information Systems in an Urban Environment, *Consortium Report to the Commission of the European Communities Directorate General XIII DRIVE programme*: BERTIE Report 49, The HUSAT Research Institute, Loughborough, Leics. U.K.

Lorenz, K., 1990a, Localised Parking Behaviour—Interviews and Observations in Berlin, *Consortium Report to the Commission of the European Communities Directorate General XIII DRIVE programme*, report reference: BERTIE Report 27, Technische Universität Berlin, Institut für Verkehrsplanung und Verkehrstechnik, Fachgebiet Strassenplanung und Strassenverkehrstechnik.

Lorenz, K., 1990b, Real Parking Behaviour Study in Test Vehicles, *Consortium Report to the Commission of the European Communities Directorate General XIII DRIVE prograrmme*, report reference: BERTIE Report 50, Technische Universität Berlin, Institut für Verkehrsplanung und Verkehrstechnik, Fachgebiet Strassenplanung und Strassenverkehrstechnik.

Lorenz, K., 1991, A Comparison of Route Guidance and Route Information Systems in an Urban Environment, *Consortium Report to the Commission of the European Communities Directorate General XIII DRIVE programme*. BERTIE Report 49A, Technische Universität Berlin, Institut für Verkehrsplanung und Verkehrstechnik, Fachgebiet Strassenplanung und Strassenverkehrstechnik.

20

Voice communications in vehicles

A. M. Parkes

The introduction of communication devices into vehicles has the immediate effect of changing the possibilities of the driving experience. Whereas previously the driver was restricted to information given by the vehicle, or from the immediate traffic environment, access to a communication channel opens up potential for a large range of point to point voice and data links. The vehicle can no longer be regarded as an object within the traffic system, it becomes a complex system in itself. The driver need no longer base strategic and tactical decisions on information from the immediate environment; many information sources can be accessed and interrogated. The initial market demand has been for an equivalent to the normal telephone service to be available in the vehicle, with mainly business users being those most willing to pay the relatively high costs. However, once the carrier technology is in place, and a certain critical mass of market users has developed, it becomes possible for additional service providers to enter the market offering a wide range of products. Because many of these 'value added' services will be similar to those offered by the fixed telephone network, it is reasonable to assume that the driver will use many of them in a voice-input auditory-output fashion. The vehicle of the future may become quite a noisy environment, with in-car entertainment systems being joined by telephone systems, and a number of auditory displays and controls. Research into the safety and acceptability issues surrounding mobile telephone conversations, or indeed any voice input and output system for use whilst driving, though ranging over twenty years (e.g. Brown *et al.*, 1969) is less than comprehensive. Research into the effects of in-vehicle telephone conversations is important because it has implications not only for that application, but for the nature of future interface design in general.

Allowing the driver to communicate with the outside world has the potential to make journeys more acceptable: if better use can be made of the time by making arrangements by carphone; they can be more efficient, if real time-task relevant information can be received; and safer if stress can be reduced for the driver. Alternatively, certain auditory displays might be aggravating in the vehicle, certain information might be more appropriately presented visually, and there might be a safety decrement if drivers find that carphone conversations distract them from their driving control.

Voice interactions in the vehicle, whether with the on-board systems, or with others via the carphone, introduce an additional task to the driver. It might be argued that driving and holding a carphone conversation is no more difficult than driving and listening to the radio; but this ignores the important interactional component which requires an activity level by the driver beyond that necessary in the passive situation of listening to a radio broadcast.

We need to know if drivers are able to hold carphone conversations in the same way that they would use a fixed system, without reducing their safety margin in the driving task. If they cannot, there are implications not only for carphone usage, but for other proposed systems that intend to use the voice channel in some way.

Traditional models of psychological behaviour have often been based on the assumption of the brain being a limited channel for processing information (e.g. Wickens, 1980). A task may take up a proportion of the capacity, leaving a certain 'spare capacity'. The introduction of a second task will make demands on this limited channel and take up spare capacity of the system. If necessary it will impinge on the capacity originally occupied solely by the primary task. Driving and communication system usage can be considered within the context of this model, the driving element being the primary task, and conversation the secondary.

Driving is a primary task which inherently has high risk associated with it. As such, it may be reasonable to suppose that a decrement in driving performance (if it exists) would only continue to a certain floor level in normal circumstances. Beyond this, the subjective risk associated with paying more attention to the conversation and less to the driving, would reach an unacceptable level. Thus at this floor level, any further decrement in performance, associated with increasing task difficulty, would be transferred to the secondary task, and an element of stress would be imposed upon the driver as a consequence. This floor level will of course vary between individuals, as will the level of spare capacity involved in reaching it.

We have the position that Europe is making a heavy investment in cellphone systems to support carphone usage. It will be possible to provide a wide range of services on this cellphone system, and enable drivers to spend significant portions of their driving time engaged in conversations with people remote from the vehicle. It has become important to know the answers to two questions:

- does talking on a carphone affect driving performance?
- does driving make it difficult to use a carphone in the same way as a standard fixed telephone?

As early as 1961, Brown and Poulton used addition and attention secondary tasks and found a decrease in driving performance in heavy traffic, though the nature of the performance measure is not indicated. Quenault (1968) used mental addition, counting auditory blips and giving antonyms in his experiment. This study investigated three types of road condition; closed track, open road and 30 mph restricted zone with three types of verbal problem solving tasks. No effect on driving speed in the three types of road condition was found. Brown *et al.* (1969) used a verbal reasoning task based on grammatical transformation to assess the effect of telephoning whilst driving. They found increased errors in judgement of gaps, decreased skill in driving through narrow gaps and decreased speed. The effect of driving on the verbal reasoning tasks was manifested in increased errors and increased response times.

For some reason there seems to have been a lull in research in the 1970s, with an upsurge in the 1980s. Wetherell (1981) found that under six different task conditions (verbal reasoning, short term memory, mental addition, memory search, attention and random number generation) drivers showed no significant changes in their driving performance. The criteria in this experiment being trip time, deceleration rate, gear changes and brake operation. He concluded that task priority was maintained as intended. Drory (1985) measured driver behaviour when using mobile telephones in a driving simulator. The subjects' verbal task involved reporting their position, and reading from a simple visual display. No serious decrement was found in driving behaviour, except when the subjects actually dialled the telephone and were distracted from the roadway view. Mikkonen and Blackman (1988) studied the influence of telephone conversation on driving behaviour whilst driving around familiar routes in a city environment. They found that drivers increased in alertness and anticipation which decreased the need to use their brakes.

In an interesting study, Brookhuis *et al.* (1991) looked at subjects driving an instrumented vehicle and performing carphone tasks for one hour each day for three weeks, in three traffic conditions. They found that the carphone conversation itself had little effect on the quality of the driving task at the operational (control) level. In accord with Mikkonen and Blackman, they report suggestions of an alerting effect in quiet traffic conditions. When considering the task at the strategic and tactical levels, they found several parameters affected. In motorway conditions, subjects checked the rear-view mirror significantly less often whilst telephoning. The subjects did not decrease average speed, but reaction time to braking lights increased by 6.5 per cent, and reaction times to lead car speed changes increased by 22.6 per cent, with a potentially significant effect on stopping distances. They found that performance on the carphone task (a paced serial addition task) improved over time, in accord with what might be expected from the power law of practice.

As part of the BERTIE project within DRIVE, Alm and Nilsson (1990) used a moving-base driving simulator to assess drivers' behavioural responses to the use of handsfree carphones. It was found that a carphone conversation (in the guise of a working memory span test) had a negative effect on drivers' reaction times, when the driving task was easy, less so when the primary task was difficult. Speed also decreased with the onset of a call. The carphone conversation had a negative effect on drivers' lane position, worse so when the tracking component of the driving task was hard. Mental workload was rated higher when driving whilst using the carphone. In a follow-up study (Alm and Nilsson, 1991), an elegant experimental design was used to investigate the effects of carphone conversations on drivers' choice reaction time, headway, lateral position, and workload in a simulated car following situation. Young and elderly experienced drivers took part in the study. The carphone conversation was found to have a negative effect on choice reaction time, and that the effect was more pronounced on elderly drivers (an increase in mean reaction time of around 1.5 seconds). No compensation for the prolonged reaction time could be detected. The subjects did not increase their headway during the carphone conversation, nor was it sufficient to accommodate the prolonged reaction times produced.

Of the studies cited above, four (Quenault, 1968; Wetherell, 1981; Mikkonen and Blackman, 1988; and Brookhuis *et al.*, 1991), indicate no primary driving task performance change when using a carphone, and four (Brown and Poulton, 1961; Brown *et al.*, 1969; Alm and Nilsson, 1990, 1991) indicate change. The most typical change is a reduction in speed, but also a decrease in ability to control the vehicle, and an increase in driver reaction time. Of the real road studies, changes in driving behaviour were associated with a driving task of high complexity, i.e. complex manoeuvres or urban traffic conditions. If a consistent pattern can be gleaned from the above studies, it would seem that priority can be given to the primary task of driving whilst talking on a carphone, without observable decrements in performance, so long as some threshold point is not reached. The report of Mikkonen and Blackman indicates that drivers may increase their levels of activation to cope with the increased workload of doing two things at once. However, the studies of Alm and Nilsson are interesting because they indicate that even in low workload conditions in a driving simulator, very fine detail changes in performance could be detected.

In association with the DRIVE programme HUSAT has conducted four experiments that are relevant to this area. The first study (Parkes, 1991a) examined whether the style and structure of negotiation dialogues held over a carphone differed from those held face to face, via a fixed telephone, or from those held between a driver and a passenger in a car.

Conversations can be considered in two ways. There are the ideas conveyed in the conversation, and also the words and structures employed to convey them. Previous research has looked at the efficiency of the traditional telephone as a conversation medium, and the experimental literature has focused on three main areas:

- transmission of information and problem solving,
- persuasion, including negotiation, conflict and opinion change,
- person perception.

From a theoretical viewpoint a major influence has been the 'intimacy' model of Argyle and Dean (1965). Though alternatives have developed, notably the 'formality' model (Morley and Stephenson, 1969), the 'social presence' model (Short *et al.*, 1976) and the 'cuelessness' model (Rutter *et al.*, 1981). These models make similar assumptions about the way that a conversation may be influenced by the medium in which it is conducted. The thrust of the argument being that a lack of natural cues when the participants are separated visually, leads to a more impersonal style of conversation.

It is clear that in normal direct conversation, participants use a style of talking to each other which is different to that used on a telephone link. There is, however, very little difference in terms of the acceptability of the outcome of the negotiation. People seem able to reach satisfactory outcomes even without the benefit of the feedback provided by the range of facial and bodily cues available face to face. Indeed some people prefer to negotiate by telephone, where they perceive the impersonal nature of the medium to be of benefit in formal discussions.

In the first study (Parkes, 1991a) subjects were given role play scenarios, and asked to negotiate with an experimenter in similar role play. Transcripts were made of the negotiations made in each of the experimental conditions (face to face, via fixed telephone, driver to passenger, or via carphone in normal traffic) and subjected to a detailed analysis, based on Chapanis (1972) of conversation style and efficiency. The outcomes of the negotiations were also rated by the subjects themselves, and independently scored by an observer in terms of appropriateness of outcome. The results showed that the separation of the two speakers in the telephone and carphone conditions resulted in shorter conversations than when the two speakers were together (the 'telephone effect') as would be predicted by the psychological distance theory. Subjects ranked the two conditions where the driving task was involved as more difficult than the speaking only condition (the 'driving effect').

If the 'telephone effect' has the major influence on conversation style, it seems that the 'driving effect' has the greater effect on negotiation content and outcome. The interaction of the two, in the carphone condition, appears to present the worst case scenario. Conversations were of medium efficiency (time taken, number of pauses etc.), but led to the least appropriate outcomes with negotiation tasks. The conversations tended to involve quite long pauses, and require more confirmation (e.g. did you say?) questions, but fewer new information (e.g. can you do this?) questions. Subjects reported feeling pressured when attempting to complete the two tasks simultaneously, and often made the decision to foreclose the conversation early, before a reasonable conclusion was reached.

In a related experiment (Parkes, 1991b), the same experimental conditions were used (face to face, fixed telephone, driver to passenger and carphone), but a tightly controlled decision making test was given to the subjects. It was necessary to develop a hybrid test that not only tapped the essential components of decision making, but which could be both administered and responded to verbally. The resulting test instrument consisted of seven sections; numerical and verbal memory, simple arithmetic, numerical reasoning, inference, deduction and interpretation. These tests although artificial, addressed the components of real decision making in a readily quantifiable format, allowing ready comparison across conditions. Of particular importance were the three components of critical thinking which measure the ability to draw a sound inference from a statement of facts, to recognize assumption implied by a statement, and to reason logically by interpretation. These being the very abilities business people are likely to need as the basis for decision making whilst using a carphone.

Results showed that subjects scored significantly lower when using the carphone than in the other experimental conditions. Looking at the seven elements of the test in detail, it is clear that the difference in total scores is attributable to three components, but not others. It seems that the interaction of the driving task, with the telephone medium results in difficulty in remembering verbal or numerical data, and in making correct interpretations from background information. The type of memory test employed was to ask subjects to repeat increasing strings of digits both forwards and backwards, or important points from short (around 50 word) paragraphs. Interpretation tasks involved the subject listening to a short statement, and then being asked if any of three subsidiary statements necessarily followed from the first. In this type of test scores in the carphone condition were typically over 20 per cent worse than in the other conditions. The combination of demanding primary task (driving) with the remote conversation medium (carphone) appear to produce difficulty in remembering complex information. The lower scores cannot be solely attributed to the dual demands of the driving task, as scores in the 'driver to passenger' condition were not significantly different to those produced in the laboratory in single task conditions. From a videotape analysis this seemed to be due to the fact that the experimenter sitting in the front passenger seat, naturally, even unconsciously, made some allowance for the temporal demands of the driving task when administering the test.

This may indicate that some reservation should be placed on notions of the development of the future 'office on the move'. Whilst it will be technically feasible to provide many of the features of the office (e.g. telephone, fax machine, personal organizer) in the vehicle, it must be remembered that it is not possible for the human component, the business person, to operate in the same way whilst they simultaneously drive a car.

The third experiment conducted by HUSAT (Fairclough *et al.*, 1991) again used negotiation role play, in a real road environment, but this time the focus was on driving behaviour. As such two speaking conditions were considered; using a carphone, and talking to a front seat passenger, and one control, no speaking condition was considered. Driving behaviour was measured both in terms of objective data, e.g. time to complete the route, heart rate, eye movement behaviour; and also subjective data, using both NASA-TLX and other questionnaires. The results showed the expected finding that speaking and driving is associated with a higher level of mental demand than driving alone. In both the speaking conditions subjects took longer to complete the route. Although the reduction in speed was slight (around 10 per cent in a mixed urban and rural route) it does seem indicative of some strategic level decision to reduce the primary task loading to allow more concentration on the secondary task. There were no differences in drivers' eye movements between the speaking and no speaking conditions. Looking at these results in isolation, one might think that there were no important differences revealed between using a carphone and the other two conditions, i.e. visual behaviour seemed unaffected and any increase in demand might be compensated by a decision to drive more slowly. However the data could be looked at in more detail. The increased difficulty associated with the speaking tasks was also shown by the subjective data. An analysis of the individual Task Load Index dimensions show specific increases in mental demand, task effort and frustration. It was possible to discriminate between speaking via a carphone and speaking to a passenger at the physiological level, as evidenced by heart rate data. Heart rate was significantly higher in the carphone condition than either the comparative passenger condition or the control.

The final study of interest, relates to carphone use and motorway driving (Parkes *et al.*, 1991). The focus was the decision making of drivers in a real road environment. It was decided to investigate low complexity driving in the relatively constrained environment of a three lane motorway with moderate traffic flow. Each subject drove two, twenty minute journeys on the motorway. One journey took place in silence, during the other the subject received four telephone calls. The verbal tasks during the conversations took the form of rudimentary mental arithmetic and memory tasks. Subjects' driving behaviour was captured via accelerator depression, steering wheel reversal and speed. Subjective mental workload was measured via modified versions of the NASA-TLX and the MCH (Modified Cooper Harper scale). Observational data of driving style were also recorded; for example, the number of lane changes, overtaking behaviour and the proportion of time spent in each of the three possible lanes.

This study revealed no evidence for a change in driving behaviour during carphone conversations. Strategic level choices of speed; tactical level choices of lane occupancy; and operational level activity such as accelerator depression, were consistent across the experimental conditions. The average speed in both control and carphone conditions was around 80 mph, slightly above the legal limit, and consistent with normal driving patterns. Thus it would appear that talking via a carphone in moderate traffic conditions on a motorway might impose an additional load on the driver, but it cannot be detected via the simple metrics of driver behaviour. Analysis of the subjective responses revealed an increase in perceived workload; for example MCH scores for the control condition suggest acceptable driver effort for adequate safe driving, but moderately high effort to attain safe driving in the carphone condition. This semantic categorical difference is not large, but is a clear indication of the extra load imposed by the relatively straightforward tasks involved. The combination of MCH and TLX results indicate that mental workload is increased by the carphone task to levels where it is judged to be high but not approaching levels where maximal levels of driver effort are required to maintain safe driving. However, it is worth noting that each call in this study lasted for only two minutes. Sustaining the higher mental workload for more protracted periods, might lead to a greater demand on resources. It should also be noted that this study did not attempt to replicate other studies' findings (Alm and Nilsson, 1990, 1991; Brookhuis *et al.*, 1991) of increased reaction and choice reaction times in similar low complexity situations.

What do these studies tell us about the future of voice communications in vehicles? It certainly seems that there are problems associated with conversations held over carphones; both in terms of the effect on the driving task, and on the conversation itself. However, there is not sufficient evidence to proclaim hands free carphone operation unsafe, nor to assume that 'routine' conversations involving little complex information are beyond the capabilities of the normal driving public. It seems that the operational skill level component of the driving task is reasonably robust and only likely to show deterioration at times of high primary task difficulty. It might be hoped that the driver might take appropriate action in such situations and either not accept incoming calls, or close down current ones.

Drivers certainly seem aware that holding carphone conversations is involving them in increased workload and a certain amount of stress. In most real life cases this increased workload appears to be accompanied by an appropriate increase in the activation level of the driver. The popularity of carphones in Europe certainly attest to their perceived utility, and it is difficult to see the likelihood of any future recommendations or legislation curtailing their use, though this has been suggested in some quarters (e.g. Zwahlen *et al.*, 1988). However, the evidence presented above does point to a serious problem. Both the studies of Alm and Nilsson, and of Brookhuis *et al.* from recent research, and the much older work of Brown *et al.* (1969) point to significant increases in response times of drivers when they are engaged in carphone conversations. The drivers themselves are largely unaware of this phenomenon, and it may be difficult to convince them of the need for the appropriate level of cautionary driving, e.g. increased headway distance to vehicles in front, or reduced speed in urban areas. The analogy with the public perception of the level of impairment produced at low levels of blood alcohol is clear.

The use of carphones will continue to increase, and refinements to hands-free facilities will mean that an increasing number of controls and commands in the system will be activated by voice. Already many carphones have sophisticated voice recognition facilities, and a large number of commands for call set-up, number dialling and call termination are activated by voice alone. Such voice input, allowing truly handsfree and more importantly, eyes-free operation, is to be encouraged if it can be demonstrated that a superior manual alternative does not exist. Current speech recognizers do not operate at the level of speaker independent natural language. Voice driven systems require consistency in speech characteristics of the driver, and the ability to remember sometimes complex or convoluted command sequences, to ensure that there is not inappropriate activation of the system in response to other noises or conversations in the vehicle. It seems that the optimal design will in fact be a sensible combination of very simple manual and verbal operations.

The functions available on carphones will continue to improve. New systems typically offer divert, scratchpad and answer phone facilities. Manufacturers and recommendations for the use of the systems are also making an increasing emphasis on safety of operation. A combination of intelligent system intervention and customer usage could go a long way to remove the pacing aspect that appears to be the major problem experienced in current carphone conversations. Because the other person is distant from the driving task they are not aware of the immediate task demands experienced by the driver (Parkes, 1991a) and do not 'punctuate' their speaking in the same way that a front seat passenger would. The driver often not only has to deal with the complexity of the conversation, but can feel pressured to keep up a flow of conversation, even at times when full concentration on the current driving task would be preferred. It is easy to imagine future carphone systems would be able to intervene with appropriate messages to the caller, if able to detect that the driver was engaged in a complex driving task at that time, or if the on-board dialogue manager decided that the driver needed to attend to other more urgent system information, for example from a route guidance or collision avoidance system.

Voice communications will increase in vehicles, as will the number of voice inputs and auditory displays. The integration of such systems is one of the greater challenges facing the designer of future vehicles.

References

Alm, H. and Nilsson, L., 1990, *Changes in Driver Behaviour as a Function of Handsfree Mobile Telephones: A Simulator Study*, DRIVE Report No. 47 for DRIVE Project No. V1017, VTI Linkoping, Sweden.

Alm, H. and Nilsson, L., 1991, *The Effects of a Mobile Telephone Conversation on Driver Behaviour in a Car Following Situation*, DRIVE Report No. 73 Project V1017, VTI Linkoping, Sweden.

Argyle, M. and Dean, J., 1965, Eye-contact, distance and affiliation, *Sociometry*, **28**, 289–304.

Brookhuis, K. A., de Vries, G. and de Waard, D., 1991, The effects of mobile telephoning on driving performance, *Accident Analysis and Prevention*, **23**, 4, 309–16.

Brown, I. D. and Poulton, E. C., 1961, Measuring the spare 'mental capacity' of car drivers by a subsidiary task, *Ergonomics*, **4**, 35–40.

Brown, I. D., Tickner, A. H. and Simmonds, D. C. V., 1969, Interference between concurrent tasks of driving and telephone, *Journal of Applied Psychology*, **53**, 419–24.

Chapanis, A., Ochsman, R. B., Parrish, R. N. and Weeks, G. D., 1972, Studies in Interactive Communication I. The Effects of Four Communication Modes on the Behaviour of Teams during Co-operative Problem Solving, *Human Factors*, **4**, 487–510.

Drory, A., 1985, Effects of Rest Versus Secondary Task on Simulated Truck Driving Performance, *Human Factors*, **27**, 2, 201–7.

Fairclough, S. H., Ashby, M. C., Ross, T. and Parkes, A. M., 1991, Effects of handsfree telephone use on driving behaviour, *Proceedings of the ISATA Conference*, Florence, Italy, ISBN 0 947719458.

Mikkonen, V. and Blackman, M., 1988, Use of Car Telephone While Driving. *Technical report no. A39*. Department of Psychology, University of Helsinki.

Morley, I. E. and Stephenson, G. M., 1969, Interpersonal and interparty exchange: a laboratory simulation of an industrial negotiation at the plant level, *British Journal of Psychology*, **60**, 543–5.

Parkes, A. M., 1991a, The effects of driving and handsfree telephone use on conversation structure and style, *Proceedings of 24th. Annual Conference of the Human Factors Association of Canada*, Vancouver.

Parkes, A. M., 1991b, Driver decision making ability whilst using carphones, in, Lovesey, E. J. (Ed.), *Contemporary Ergonomics 1991, Proceedings of the Ergonomics Society Annual Conference*, Southampton, 16–19 April 1991, London: Taylor & Francis.

Rutter, D. R., Stephenson, G. M. and Dewey, M. E., 1981, Visual communication and the content and style of conversation, *British Journal of Social Psychology*, **20**, 41–52.

Short, J., Williams, E. and Christie, H., 1976, *The Social Psychology of Telecommunications*, Chichester: John Wiley.

Wetherell, A., 1981, The efficiency of some auditory-vocal subsidiary tasks as measures of the mental load on male and female drivers, *Ergonomics*, **24**, 3, 197–214.

Wickens, C. D., 1980, The structure of attention resources, in Nickerson, R. (Ed.) *Attention and Performance VIII*, Hillsdale, N.J.: Erlbaum.

Zwahlen, H. T., Adams, Jr. C. C. and Schwartz, P. J., 1988, Safety aspects of cellular telephones in automobiles, *Proceedings of the ISATA Conference*, Florence, Italy.

21

Violation detection and driver information

J. A. Rothengatter

Introduction

The efficiency and safety of the road transport system depends to a large extent on the willingness of road users to keep to the traffic rules. Unfortunately, in most EC countries traffic rules are violated on a massive scale, with, as a consequence, an unnecessary high level of accident occurrence, congestion and fuel consumption. The effects of police surveillance on road-user behaviour have been studied rather extensively. It appears that it is the subjective probability of detection which determines the level of traffic law violations. Moreover, increased objective probability of detection is a necessary, but as such not a sufficient, requirement for influencing the subjective probability of detection (Rothengatter, 1990). Often, the objective probability of detection that can be achieved in everyday practice is not sufficiently high to reduce the number of violations to an acceptable level. Lack of manpower, in particular, is mentioned as one of the limiting factors (De Bruin *et al.*, 1989). Considering the increase in traffic volumes and the changes in the priorities of police duties, it is to be expected that this problematic situation will only worsen (Compston, 1992).

The task of surveillance and law enforcement can be, in principle, virtually completely automatized. Several different stages can be distinguished. The first step, the *detection function*, concerns the detection of the fact that a violation is being committed. When a violation has been detected the road user can be informed about this so that he can adapt his or her behaviour. This is the *warning function* of the system. If the road user does not adapt his or her behaviour, the violation can be registered, and, in the case of serious violations or crimes, a surveillance unit can be alerted so that the driver may be stopped. The registered violations can be the basis for further legal action.

of the DRIVE project AUTOPOLIS this approach has
ith particular attention to:

that are suitable for automatic police surveillance,
specifications,
requirements for such systems.

Prioritizing traffic violations

The question of priority for automatic enforcement systems in terms of selecting the most relevant law violations has been approached in two different manners. Firstly, both road users and police officers have been asked to indicate which violations they considered to be serious. In addition, the subjective probability of detection has been investigated. The violations investigated concern: (1) speeding violations; (2) traffic light and priority violations; (3) non-usage of safety belts; (4) vehicle-related offences; (5) driver-related offences (such as alcohol usage, fatigue); (6) illegal manoeuvres; (7) failing to stop after an accident; and (8) violations related to improper documentation, no valid licence, etc. (see De Bruin *et al.*, 1989). The results of a postal survey carried out in four EC countries revealed that, in general, there is a high degree of agreement between road users and police officers with regard to the seriousness of the different violations. In particular, driving under the influence of alcohol, driving with defective brakes and failing to stop after an accident are considered to be serious violations. In Norway and the Netherlands, drivers consider the violations more seriously than do police officers. There are, furthermore, clear differences between the different countries. Dutch drivers consider driving without a licence more seriously than others, while in Spain overtaking where this is prohibited is considered to be particularly serious. Norwegian drivers, who in general consider violations to be more serious, are particularly concerned about speeding violations. In general there is a remarkably strong relationship between seriousness and subjective probability of detection; with the exception of parking violations on the one hand, which are considered to have a high probability of detection but rank low in seriousness, and insufficient following distance and vehicle defects on the other, which are considered to be serious but of which the probability of detection is estimated to be low.

The role of violations in accident occurrence has also been investigated by means of an analysis of the different accident data bases available in the EC, and by means of a detailed analysis of the French 'REAGIR' data (see Lemaitre and Rhodes, 1989; Van Opheusden, 1990). Although there are large differences in the methods of data collection and storage in the different databases, and although the data bases often do not contain sufficiently detailed information, a number of violations can be identified as being particularly accident related:

1. speed violations, in particular insufficiently adapting speed to the prevailing road, traffic or weather conditions;
2. priority violations, in particular inside a built-up area in combination with motorized traffic and vulnerable road users, and outside a built-up area with high vehicle speeds;
3. alcohol usage, fatigue or other factors momentarily influencing driver performance;
4. following distance;
5. not staying in lane, overtaking, using restricted lanes.

System specifications

A review of presently used semi-automatic systems (Harper and Nauwelaerts, 1990) revealed that such systems are exclusively used as on-site systems dedicated to the detection of speeding violations and jumping red lights. Different detection technologies are used, including radar, induction loop configurations and infra-red detection. Systems, which are still in an experimental stage, are being developed for the detection of parking violations, stop signs violations and restrictions for heavy goods vehicles. As registration carrier, both photo and video registration are used. It is noteworthy that none of the systems incorporate any form of warning or feedback to the driver.

Location-specific systems can, in principle, be used for virtually all relevant violations. The warning function can be realized by means of variable message signs. The automatic identification of the vehicle or the driver still poses problems with the registration carriers presently in use. Even though video image analysis systems are in development, this does not seem to be the most appropriate short-term solution. Introduction of electronic vehicle identification will provide a considerable improvement as this enables the fully automatic processing of the detected violations. This is, of course, even more so with systems that permit electronic driver identification.

In-vehicle systems have the advantage of functioning independently from an—expensive—electronic infrastructure and that they can analyse driver behaviour in more detail. Moreover, these systems can take the specific characteristics of the driver into account if these are communicated to the system by means of a smart card. The same smart card can then be used to register the violations. The main advantage of vehicle systems is that the driver can continuously be provided with individual feedback and, in this way, can also be given the opportunity to correct his or her behaviour. The application of in-vehicle systems is limited to those violations that can be inferred from vehicle parameters alone. Violations, such as crossing solid lines (i.e. illegal overtaking) and insufficient following distance, can be detected when the vehicle is equipped with the required sensors. Systems that can detect the drivers' state (i.e. fatigue and alcohol usage) are under development (Brookhuis, 1990).

For a full application of the possibilities, interaction between the environment and the vehicle is a requirement. In the case of automatic surveillance of speeding behaviour, for example, the in-vehicle system can, in a relatively straightforward manner, be informed about the local speed limit, by 'electronic tagging' of the present speed signs. Moreover, prescribed speeds can be adapted to the situational and momentary circumstances, e.g. the speed limit can be lowered during fog, while when the traffic intensities are low the speed limit can be raised. This has as an advantage that the driver is forced to adapt his speed to the prevailing circumstances and will be warned that he is committing an offence if he does not comply. Introduction of such a system can contribute considerably to the consistency of traffic behaviour dependent upon momentary and local circumstances. In the comprehensive system configuration, the behaviour required is determined on the basis of vehicle, driver and environmental conditions, and required behaviour and realized behaviour are compared. If discrepancies occur, the driver is informed and is given the opportunity to adapt his behaviour. Should he fail to do so, this is registered in such a manner that it can result in prosecution.

Legal considerations

Considerable attention has been devoted to the legal requirements the system has to fulfil in order to function effectively (see Van Opheusden, 1990). In some EC countries, over 65 per cent of the court cases concern traffic law violations. The possibility to deal with violations administratively has already been created in several countries and is expected to gain more extensive application in the future. A major issue is what will be admissible as legal evidence. In other EC countries, a police officer still is legally required to establish that a violation or crime has been committed, even where automatic detection equipment has been used to register the violation committed. Substantial changes in the legal system are required therefore before computer data can be accepted as irrefutable proof.

Introduction of automatic policing systems

The possibilities of automatic policing are to a larger extent determined by societal factors rather than by technological factors. Given the expected introduction of RTI systems in the transport system, and given the fact that automatic policing will almost exclusively use applications that are either in development or already exist, no serious technological problems are to be expected in the introduction of automatic policing. However, the social considerations are of great importance. The attitudes and opinions of road users, which are manifestly highly ambivalent, have to be examined and considered. Surveys carried out to date indicate that on-site systems are considered more acceptable than in-vehicle or interactive systems. In particular, registration of violations in-vehicle is associated with negative consequences such as infringement of personal freedom. In addition, there are large discrepancies in the acceptability of such systems between different EC countries. Societal organizations will emphasize the procedural aspects. Privacy aspects and considerations of 'proportionality' of enforcement efforts are major factors to be considered. In addition, the user-friendliness of the equipment for the police officers is a factor to be considered. A questionnaire indicated that police units may, in principle, view the envisaged systems positively, but also indicated that the complexity and reliability of the equipment determines to a large extent whether it will be employed in everyday police work.

Another question is whether the road administration will accept that the RTI infrastructure can be used for police surveillance purposes. It will be necessary to check carefully that the different data structures cannot be confused even when the same infrastructure is used. As outlined above, law changes are necessary to optimally use the envisaged automated systems.

Introduction of the described systems will not be possible without first completing field experiments. Obviously, these will concern systems that (1) have the largest traffic safety effects, (2) use existing applications and (3) do not pose insurmountable legal problems. This probably implies that limited applications directed at specific target groups (such as repeated offenders) or at violations that are already subject of semi-automated surveillance (speeding) are the most feasible.

Although the introduction of automatic enforcement will certainly not be without problems, there are good reasons to pursue this. Based on empirical data concerning speed-accident relations (Nilsson, 1990) and pilot automatic enforcement sites in Norway (Muskaug, 1992), even the most conservative estimate (considering speed reductions only) indicates that an accident reduction of at least 25 per cent seems feasible. At present, there are no other RTI applications that could reasonably attempt to equal this.

References

Brookhuis, K. A, 1990, *Summary report of the DREAM project*, Submitted to the EC DRIVE Office, Brussels.

Bruin, R. A. de, Østvik, E. and Vaa, T., 1989, Opinions of drivers and police officers about the seriousness of traffic violations, in Rothengatter, J. A. (Ed.), *The identification of traffic law violations*, Report 1033/D1, Haren: Traffic Research Centre, University of Groningen.

Compston, J. J., 1992, Urban traffic enforcement, candidate for endangered species list, *Preprint 920735, 71st Annual Meeting Transportation Research Board*, Washington, DC.

Harper, J. J. and Nauwelaerts, T., 1990, *Current technology: reliability and implications for automatic policing*, Report 1033/D5, Haren: Traffic Research Centre, University of Groningen.

Lemaitre, G. and Rhodes, S., 1989, A qualitative approach to accident-contributory violations and the construction of a video database, in Rothengatter, J. A. (Ed.), *The identification of traffic law violations*, Report 1033/D1, Haren: Traffic Research Centre, University of Groningen.

Muskaug, R., 1992, Results of pilot automatic enforcement sites in Norway, Paper presented at the First DETER workshop, Oslo.

Nilsson, G., 1990, *Reduction of the 110 km/h speed limit to 90 km/h during summer 1989: Effects on personal injury accidents, injured and speeds*, Report 358A, Linköping: Swedish Road and Traffic Research Institute.

Opheusden, P. van, 1989, *Analysis of accident databases*, Report 1033/R5, Haren: Traffic Research Centre, University of Groningen.

Opheusden, P. van 1990, *Legal requirements for automatic policing information systems*, Report 1033/D4, Haren: Traffic Research Centre, University of Groningen.

Rothengatter, J. A., 1991, Normative behaviour is unattractive if it is abnormal: relationships between norms, attitudes and traffic law, in *Proceedings OECD international road safety symposium Enforcement and Rewarding*, Leidschendam: SWOV.

22

How can we prevent overload of the driver?

W. B. Verwey

The introduction of low-cost, high speed computers as a tool to affect traffic behaviour has just started. One of the potential dangers coming along with such systems is that they convey too much information to the driver at the wrong moment, that is, when the driver needs to attend to the traffic environment, or to more important information presented concurrently by other in-car systems. The reality of this danger is indicated by conclusions that even without sophisticated in-car displays driver inattention plays a role in about 30 to 50 per cent of all accidents (Treat *et al.*, 1979; Sussman *et al.*, 1985). If we do not want this figure to increase care should be taken that drivers can not be overloaded by information from in-car devices.

This paper will give a brief account of work that has been carried out in the Generic Intelligent Driver Support (GIDS, V1041) project on interfaces that adapt to driver workload. It will address issues relevant for the design of adaptive interfaces, and give results of a study that attempted to find determinants of driver workload as required for the construction of adaptive interfaces. It will be argued that the solution to the overload problem is actually quite simple: a system should be developed which schedules information presentation while taking the driver's capacities and limitations into account. This implies the design of an integrated interface between the driver and the car, i.e. a dialogue controller, which presents only one attention demanding message at the time and which gives priority to more important messages over less important messages. In addition, it should adapt message presentation to the current demands exerted upon the driver by the driving task (Figure 22.1). That is, the system should behave as 'an executive's assistant who zealously guards the superior's time and resources' (Rouse *et al.*, 1987, p. 96). When, for example, a driver gets very busy with manoeuvring his/her car in heavy city traffic the interface might adapt to the increased demands of driving by blocking phone calls and automatically turning the wipers on when it starts raining.

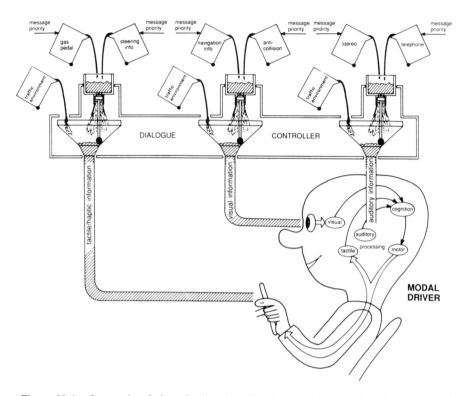

Figure 22.1   One task of the adaptive interface is to take care that the perceptual channels of the driver will not be overloaded.

Ideas about such adaptive interfaces are not new as they have been described before in general terms, and with regard to aviation (e.g. Chignell and Hancock, 1985; Hancock and Chignell, 1988; Rouse, 1988). However, with the exception of reports related to the GIDS project (Smiley, 1989; Verwey, 1991b) no reports have been published on the application of adaptive interfaces in the car. Given the current efforts to develop intelligent in-car systems, clearly now is the time to put efforts in the development of adaptive interfaces.

What is workload?

The term workload is used quite often in the area of interface design. However, it is often used with very different meanings. Therefore, a brief review of the concept of workload will be given in order to come to a definition of workload adaptivity.

The reason for measuring workload is the wish to predict performance of a man-machine system before it gets into production, to decide about task alterations, and to assess individual differences in task performance (Wickens, 1984a). But according to Sanders (1979) the concept of workload is defined in common sense terms and, therefore, not consistently dealt with in models of human performance. Examples of workload definitions are: 'Workload is the extent to which an operator is occupied by a task' (Jahns, 1973); '... the term workload will be used to refer to the integrated effects on the human operator of task-related, situation-related, and operator-related factors that occur during the performance of a task' (Hart, 1986); and '... the rate at which information is processed by the human operator, and basically the rate at which decisions are made and the difficulty of making the decisions' (Moray, 1979). Indeed, these definitions are abstract and not very illuminating. This is probably caused by the fact that the usage of workload measures have been of more concern to human factors practitioners than to theoretical psychologists, and from the beginning the emphasis was on how and when to measure workload rather than on the theoretical and empirical foundation of the assessment methodology.

There are at least two basic problems with the concept of workload (Sanders, 1979). First, when speaking about workload different researchers have different things in mind. Some refer merely to perceptual motor workload. Others incorporate the task environment with its physical, social and emotional components. Then the concept becomes almost equivalent to stress or strain. However, more recently an attempt has been made to define the term workload more formally by stating that workload involves the load on perceptual, central and output processing resources (Wickens, 1984b, 1989). This indicates that workload is multidimensional. The second problem is the absence of a standardized measure of workload. Basically, workload measures can be divided into four categories: primary task measures, secondary task measures, physiological measures, and subjective measures (e.g., Verwey, 1990). Yet, these measures often correlate less than moderately and it is difficult to equate the different measures of workload.

In addition to these workload assessment methods, one can also predict workload from an analysis of the task or a—usually highly simplified—model of the human operator. The classical method is time-line analysis which assumes that workload is hundred per cent when the operator is busy for hundred per cent of the available time. Reliable operator models are very complicated to make because they have to include effects of practice, they sometimes have to consider the human as a multiple channel processor and sometimes as a single channel processor, and the effects of prior workload, self-paced flexibility and strategies have to be considered (see Wickens, 1984a; Schneider and Detweiler, 1986). Subsequently, it has been stated that the more complex a task, the more difficult it is to reliably model human behaviour (Rouse, 1981). However, recent developments in cognitive modelling (e.g. Anderson, 1983; Laird *et al.*, 1987; Schneider and Detweiler, 1988) seem very promising for predicting driver workload, because these models are based on modern views of human information processing and incorporate effects of practice and strategies. The relevance of such models for workload is that when we know what the operator does and how he or she processes the information, we are also able to estimate workload.

In this paper, we will define workload in terms of the load on perceptual (visual, auditory, and tactile), central (cognitive) and output (hand, foot, and vocal) resources (Wickens, 1984b). This definition indicates that we should obtain workload by utilizing workload measuring techniques that can differentiate between loads on the various resources. Most appropriate, then, is to use a battery of secondary tasks, each of which indicates a load on one particular resource.

Workload and interface adaptation

Given this conception of workload, we now can state that the task of the adaptive interface is to prevent overload of any of the perceptual, central, and output resources of the driver. Basically, two issues evolve. First, what should the system do to adapt? Second, how does the system know that it has to adapt?

How to adapt?

There are three techniques to alleviate driver workload in situations where the driver becomes overloaded. First, messages can be postponed or completely cancelled. Criticality of the messages should indicate which option is required for a particular message. Less critical messages can be postponed or cancelled, whereas highly critical messages should be presented immediately. Second, modality, format and content of messages can be changed. For example, a specific message may contain additional details when workload is low, but may be excluded when workload is high. Another modality also may be used. The problem with this approach is that in situations of high workload a driver is suddenly confronted with a message that may be simple but unfamiliar. A more complex but highly familiar counter message may, therefore, yield less workload. Third, tasks may be allocated to the driver in low workload situations and be taken over by the system in high workload situations. This has been called dynamic task allocation (Verwey, 1990). Dynamic allocation has the advantage that the drivers are supported only when they require it. Also, the risk of underload and drowsiness lessens because, under regular conditions, they will have to do the task by themselves. It is probable that dynamic task allocation may lead to confusion about who is in control of the task. Therefore, dynamic task allocation should, in principle, be restricted to tasks which are either not critical for traffic safety, or tasks which keep the driver in control by automatically having the system present support information, without taking further action, when the driver is overloaded. The latter situation requires special attention of the designer because driver work-load must be reduced and not increased by such, possibly unexpected, support.

When to adapt?

Before the driver-car interface can adapt to driver workload, the system should know when to do so, that is, it should have an indication of the driver's workload. Workload measurement utilizing subjective measures, secondary tasks, or physiological methods are clearly not applicable outside the laboratory. The only way out is to develop a model which describes driver behaviour and indicates workload (Rouse *et al.*, 1987). Sophisticated techniques have been used to model the driver (e.g. Aasman, 1988), but simpler models can be developed, as well. For example, if we could find the parameters that determine the level of driver workload, we might be able to model workload by constructing a static look-up table in which the levels of the determining parameters are used as input. Such parameters may be driving experience, age, road situation and familiarity with the road. Given the development of intelligent infra-structures and in-car sensors the system may, in future, have this information at its disposal. The difference between the sophisticated and the simple approach is that the latter yields less accurate estimates, but is much easier to develop.

In the light of the earlier definition of workload, the system should be able to estimate load on the various processing resources. A system capable of differentiating between load on various resources would, for example, distinguish between driving on a curvy road (only loading visual and manual resources) and negotiating a dangerous intersection (loading central resources as well). In the first situation, the system might still allow the driver to make a difficult telephone conversation on a (handsfree) phone, whereas in the latter situation it should not. In addition, workload caused by interaction with the interface should be known to the system so that the system knows that presentation of a specific message, in a specific situation, will not overload the driver given his current workload.

A preliminary study for determinants of driver workload

If we want to develop an adaptive interface that relies on a simple driver model we should investigate exactly which factors determine driver workload. There-fore, a preliminary study was carried out in which the levels of visual and cognitive load were determined as a function of road situation, driving experience and traffic density (see Verwey, 1991a for a more extensive description). In designing this study it was assumed that load on the output resources will yield less problems in designing car interfaces because driver actions are self-paced. In the study, 24 subjects drove a particular route while performing one of three secondary tasks. The route involved six road situations at which performance was analysed. Three involved a four lane motorway with a speed limit of 100 km/h: performing a sequential merging and exiting manoeuvre (800 m), driving the right lane of a straight motorway (500 m), and driving a two lane roundabout (500 m, from 100 m before to 20 m after the cross-road) with traffic entering the roundabout having right of way. The other three situations involved rural roads with a speed limit of 80 km/h. These situations were turning right on a cross-road (120 m, from 100 m before to 20 m after the junction) and turning left just before a curve to the right (120 m, from 100 m before to 20 m after the turn). The latter turn was more complicated than the first because it involved crossing a bicycle lane and the outlook toward oncoming traffic was limited. Finally, a straight rural road was driven where bicycles were allowed as well (500 m). In an attempt to evaluate performance under varying conditions of traffic density, each subject performed at three different times of the day, one of which included the rush hour.

All subjects drove an instrumented car in four secondary task conditions: one no-task control condition and three task conditions. Responses to the three secondary tasks were always vocal. In the condition with the visual detection task subjects had to vocally indicate that they had detected a number on a dashboard-mounted display by saying 'yes'. In the visual addition condition the subjects' task was to add twelve to the number presented on the display and to utter the answer. Auditory addition also involved adding twelve, but now the numbers were presented auditorily. The subjects were explicitly instructed not to pay attention to the secondary task when that would affect driving performance. Hence, secondary task performance is assumed to indicate variations in resource load. Base-line secondary task performance was acquired before and after driving. For each subject, these measures were used to express secondary task performance in the driving conditions as percentage deterioration. In addition, the number of glances at the display was analysed.

The effects of the secondary tasks on driving were measured by counting eye glances left and right outside the car and at the interior mirror. Also, pedal usage and driving speed were measured. Secondary task performance and driving measures were analysed only when driving speed exceeded 1 km/h.

Twelve male and twelve female drivers between 21 and 28 years of age served as subjects. Six female and six male subjects were experienced, in that they had their driving licence for five or more years and had driven more than 10 000 km per year. The other six males and six females were inexperienced drivers who had possessed their driving licence for less than one year and had driven less than 10 000 km.

Driving speed and pedal usage were not different when comparing conditions with and without a secondary task. This suggests that driving performance did not deteriorate when a secondary task was being performed. Only the number of mirror fixations was smaller when a secondary task was carried out—especially when this involved the visual addition task.

In Figure 22.2, performance on the three secondary tasks in the six road situations is presented as a function of driving experience. The figure shows secondary task performance as a percentage of performance in the base-line condition. Interactions between secondary task type and road situation, and

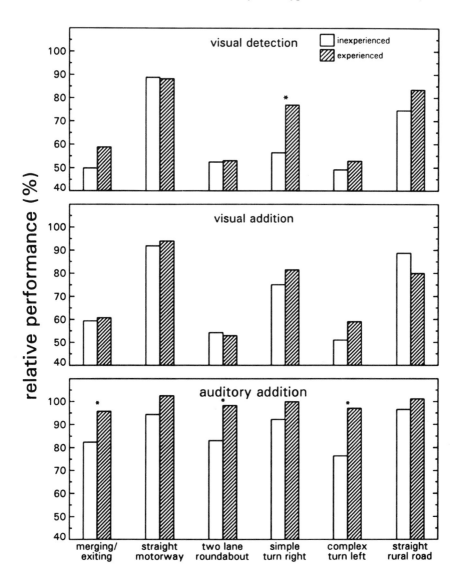

Figure 22.2 *Relative secondary task performance while driving as a function of secondary task type, road situation and driving experience. Asterisks indicate significant differences between inexperienced and experienced drivers.*

secondary task type, road situation and driving experience indicated that performance on the secondary tasks deteriorated differently in the different road situations, and that performance on auditory addition deteriorated less in experienced drivers than in inexperienced drivers. In fact, performance of experienced drivers on auditory addition did not come below 95 per cent of base-line performance, whereas it dropped to below 85 per cent during merging and exiting, the roundabout, and the complex turn left for inexperienced drivers. Display glance frequency was lowest on the roundabout and greatest while driving straight ahead on the motorway. Inexperienced drivers glanced less at the display than experienced drivers. The difference between inexperienced and experienced drivers was greatest while driving straight on the motorway and smallest at the roundabout. Finally, with increasing traffic density the display was glanced at less in visual addition, but not less in visual detection.

Together, the data suggest that the visual resources of inexperienced drivers are more loaded than those in experienced drivers. Yet this difference was rather small and display glance data and performance on the visual secondary tasks were not consistent. By contrast, pooled over both groups of drivers, the visual demands varied largely between road situations. Adaptive interfaces should, therefore, use the road situation as a factor to estimate visual load. Likewise, for inexperienced drivers load on cognitive resources varied between road situations, whereas the effect of traffic situation on cognitive load in experienced drivers was small and not significant. With regard to the adaptive interface this finding suggests that road situation should be taken into account when messages, requiring cognitive resources, are presented to inexperienced drivers. Finally, effects of traffic density were limited to display glance frequency. Basically, this would suggest that traffic density is not an important determinant of driver workload. One should, however, keep in mind that traffic density was only varied by manipulating time of driving. Also, the road situations were highly familiar to the subjects. Future research should indicate whether traffic density may still be of importance when measured otherwise and in interaction with familiarity and transparency of the situation at hand.

Conclusions

Given the potential danger of overloading the driver with information from various in-car devices, there is a need for an interface that integrates information from these devices, that schedules the messages according to their priority, and that adapts information presentation to the current demands imposed upon the driver. Adaptation may involve postponing or cancellation of messages, changing the nature or content of the messages, and taking over non-critical tasks. In order to do this the interface must contain a workload estimator. Preferably, the workload estimator is based upon a model of the driver which indicates what the driver is currently doing or will be doing in the near future. A more pragmatic approach is setting-up a multidimensional static look-up table. This is simpler than building a driver model, but will probably yield less accurate estimates. In order to construct that table, studies are required to find determinants of driver workload. The study described in this chapter suggests that, at least, traffic situation and driver experience should be used for workload estimation. Future research should investigate the role of other factors on driver workload like age, traffic density, and familiarity with the driving environment, as well as the effect of variable road signs and in-car messages.

Adaptive interfaces in the car are not just a fantasy. A preliminary version has actually been incorporated as a core element in the GIDS system. In order to test the GIDS system, both in terms of system operation as well as in terms of effects on driving performance, it has been built into an instrumented car at the TNO Institute for Perception in Soesterberg (NL) and into a driving simulator at the TRC in Groningen (NL). The scope of the adaptive interface is extended in the successor of the GIDS project called ARIADNE[1].

Whether or not adaptive interfaces will be introduced into the car of the future will depend on the accuracy of workload estimation. No driver will accept a system which, as a result of inaccurate workload estimation, blocks in-coming phone calls which can actually be handled by the driver and which gives irrelevant information when the driver is busy negotiating a roundabout. The perspectives of the adaptive interface will depend on the availability of reliable information supplied by intelligent sensors (e.g. radar), an intelligent infra-structure (e.g. up-to-date road lay out), and vehicle-to-vehicle communication (e.g. location of other traffic participants). Without such information, the scope of the adaptive interface will be limited. But when, on the other hand, workload estimation can be made reliable by proper modelling of the human driver and ample availability of information required for workload estimation, adaptive interfaces will and should be part of the regular outfit of any future car.

Note

[1] Application of a Real-time Intelligent Aid for Driving and Navigation Enhancement, V2004.

References

Aasman, J., 1988, Implementations of car-driver behaviour and psychological risk models, in Rothengatter, T. and de Bruin, R. (Eds) *Road User Behavior: Theory and Research*, pp. 106–18, Assen: van Gorcum.

Anderson, J. R., 1983, *The architecture of cognition*, Cambridge, MA: Harvard University Press.

Hancock, P. A. and Chignell, M. H., 1988, Mental workload dynamics in adaptive interface design, *IEEE Transactions on Systems, Man, and Cybernetics*, **18**, 647–58.

Hart, S. G., 1986, The relationship between workload and training: An introduction, *Proceedings of the 30th Annual Meeting of the Human Factors Society*, pp. 1116–20, Santa Monica, CA: Human Factors Society.

Jahns, D. W., 1973, *A concept of operator workload in manual vehicle operations* (Tech. Rep. No. 14), Meckenheim, FRG: Forschungsinstitut für Anthropotechnik.

Laird, J. E., Newell, A. and Rosenbloom, P. S., 1987, SOAR: An architecture for general intelligence, *Artificial Intelligence*, **33**, 1–64.

Moray, N., 1979, Models and measures of mental workload, in Moray, N. (Ed.) *Mental Workload. Its Theory and Measurement*, pp. 13–22, New York: Plenum Press.

Rouse, W. B., 1981, Human-Computer Interaction in the control of dynamic systems, *Computing Surveys*, **13**, 71–99.

Rouse, W. B., 1988, Adaptive aiding for human/computer control, *Human Factors*, **30**, 431–43.

Rouse, W. B., Geddes, N. D. and Curry, R. E., 1987, An architecture for intelligent interfaces: Outline of an approach to supporting operators of complex systems, in Moran, T. P. (Ed.) *Human-Computer Interaction*, pp. 87–122, Hillsdale, NJ: Erlbaum.

Sanders, A. F., 1979, Some remarks on mental workload, in Moray, N. (Ed.) *Mental Workload. Its Theory and Measurement*, New York: Plenum.

Schneider, W. and Detweiler, M., 1986, Changes in performance in workload with training, *Proceedings of the 30th Annual Meeting of the Human Factors Society*, pp. 1128–32, Santa Monica, CA: Human Factors Society.

Schneider, W. and Detweiler, M., 1988, The role of practice in dual-task performance: Toward workload modelling in a connectionist/control architecture, *Human Factors*, **30**, 539–66.

Smiley, A., 1989, Mental workload and information management, in Reekie, D. H. M., Case, E. R. and Tsai, J. (Eds) *Proceedings of the First Vehicle Navigation and Information Systems Conference (VNIS'89)*, pp. 435–8, Toronto: IEEE.

Sussman, E. D., Bishop, H., Madnick, B. and Walter, R., 1985, Driver inattention and highway safety, *Transportation Research Record*, **1047**, 40–8.

Treat, J. R., Tumbas, N. S., McDonald S. T. *et al.*, 1979, *Tri-level study of the causes of traffic accidents: Final report* (Vol. 1 and 2) U.S. DOT HS-805-086, (NTIS PB 80-121064).

Verwey, W. B., 1990, *Adaptable driver-car interfacing and mental workload: A review of the literature* (GIDS, V1041, deliverable DIA1), Haren, The Netherlands: University of Groningen, Traffic Research Centre.

Verwey, W. B., 1991a, *Towards guidelines for in-car information management: Driver workload in specific driving situations* (Report IZF 1991 C-13), Soesterberg, The Netherlands: TNO Institute for Perception.

Verwey, W. B., 1991b, Adaptive interfaces based on driver resource demands, in Quéinnec, F. and Daniellou, F. (Eds) *Designing for Everyone, Proceedings of the 11th Congress of the International Ergonomics Association*, Vol. 2, pp. 1541–4, London: Taylor & Francis.

Wickens, C. D., 1984a, *Engineering Psychology and Human Performance*, Columbus, Ohio: Merill.

Wickens, C. D., 1984b, Processing resources in attention, in Parasuraman, R. and Davies, D. R. (Eds) *Varieties of Attention*, pp. 63–102, London: Academic Press.

Wickens, C. D., 1989, Attention and skilled performance, in Holding, D. H., (Ed.) *Human Skills*, pp. 71–105, Chichester: John Wiley.

PART IV
DRIVER ORIENTED DESIGN

23

Driver oriented design

S. Franzén

The total traffic system can be seen as a multiperson-multimachine system running under a great variety of environmental and other situational characteristics. The system is organized by implicit and explicit rules, where the key system elements, humans (drivers) and machines (vehicles) interact. Several interactions can be identified and are performed through the driver-vehicle interface and the driver-environment interface. The design of the driver-vehicle interaction is not a question of specification of displays and controls. The design must be related to the goal of the total man-machine system, i.e. to the context in which the system will operate and to the content of the information flow between user and technical system. This ideal top-down analysis path is presented in the PROMETHEUS MMI Checklist (Franzén *et al.*, 1991). The basic elements of the approach cover the problem and the user, the tasks and subtasks, the interaction and the interface. Aggregated questions like: 'why a technical system?'; 'a system for whom?'; 'which user tasks are supported?'; 'what should be the content of the inter-action?'; 'when will the system be activated?'; 'how should the system functions be realized technically?'; and 'where should the device be located?' exemplify the top-down approach.

However, the ideal design process of man-machine interactions and related interfaces is not to be found in real life, i.e. in the practical driver-vehicle interface development and design work. The reality is, in fact, a combination of the ideal top-down systems approach and analysis, and bottom-up technology-oriented synthesis work. If given ample time the design process is iterative, with a series of re-designs and system evaluations in-between. The evaluation criteria being both acceptable system performance and technical feasibility as well as user acceptance of the product or the system provided, i.e. the potential or real success in the market place.

The industrial elements of the design process which have to be incorporated in the necessary human factors work, are the practical realization of ideas which have to be turned into functional concepts and the following implementation of these concepts into user accepted solutions. Acceptance being the trade-off or a cost-benefit analysis between system features and system costs. The features are subdivided into usability, utility and likeability, whereas the effort to learn and use, the loss of skills and new elements of risk introduced and the financial costs are the cost elements. The difficulty to attach variables and measures of these elements is evident, but the principle of user acceptance is an approach that clearly highlights all the diverging elements that could, would and should influence the design process. It can simply be stated that if the features are graded higher than the costs, in the weighted criteria/cost function, the solution is accepted, purchased, and used.

The industrial and economic reality of enterprises and society, and the judges of success, i.e. the market place actors such as the end users, the customers, the consumers, must be somehow incorporated in the design work. How is that taken care of? In industrial work systems engineering is accepted as an engineering approach to solve problems, either to develop products or to arrange the production of products in an efficient way. Systems engineering can be described as 'a phase-oriented methodology' related to the cycle of problem solving and the corresponding stages of system life. However, as with any methodology, one cannot solve problems by solely applying systems engineering ideas. Other factors such as expertise, creativity, situation knowledge etc. have also to come into play. Systems engineering can only help to improve the efficiency of these other factors and help optimize system performance (Franzén and Reichart, 1991). Part of the systems engineering approach is the inclusion of human factors, competence and inputs from the very start of the problem solving industrial process.

The development of new information technology has also stimulated the introduction of new design tools, e.g. rapid prototyping tools. Such tools will be interesting for design departments, as they will provide the means to speed up the design phase, as well as reduce the needs to build several 'full-size' prototypes for the testing and selection of basic concepts for the application studied. Rapid prototyping will, in a laboratory or simulator environment, make it possible to test concepts at a reasonable cost and at a much higher speed than by building mock-ups etc. However, designers should be aware of the limitations of such a design tool. It can not grasp all the features of an application concept and it is difficult to use in a realistic environment. Rapid prototyping tools can be used to quicken the very time-consuming process to choose between many first-hand alternatives, and to help design teams to concentrate on testing only the most promising alternatives.

Other important design aspects to consider are that new products or solutions are very seldom developed to solve or to meet completely new problems and needs. Very often a better performance of already existing solutions is sought. It is also clear that old solutions and products will exist side-by-side with the new ones. The penetration of new technology in society is very slow and starts with people that can afford to be 'modern' and most up-to-date. That new products have to co-exist with old ones indicates that the design must allow parallel operation, and that step-by-step development could be the market-oriented approach, i.e. would increase the group of potential customers substantially. Other elements to consider are that the long term goals of systems for traffic and transport very often are societal, while the short term (market-oriented) are individual, i.e. to meet an instant demand and that there is an inherent conflict to be resolved as well.

The standardization of man-machine interfaces has always been a controversial issue. From an industry point of view the less standardization the better would be the first standpoint. The societal interest in standardization would be greater especially when it concerns areas like traffic and transport. The introduction of new systems in traffic, based on advanced information and communication technologies would even further strengthen this wish. There could be standardization of the design process itself, i.e. the design team could be asked to follow given design regulations in the design, involving human factors competence and well designed testing procedures. In short it could be said that the company should be licensed to design driver-vehicle interfaces. Another level of standardization is that of the performance. Based on knowledge about the limitations and capacities of human beings in certain tasks and situations, the expected performance level could be stated as a threshold level to overcome before the design could be accepted. The third level is, of course, the product standard, but in this case there is a risk that an established product standard could seriously hamper future possible enhancement of product performance. The reason being that a product standard is hard to free from the technology involved in the design and, because it would be impossible to foresee future possible technological development, the product standard would block the use of these elements in the design.

The contributions in Part 4 can be presented as part of the driver oriented design process as follows:

Top-down approach
- information handling/dialogue
  management   (Brown and Höök)
- copilot concepts structure/monitoring
  concepts   (Nirschl and Geiser)
- navigation/guidance information concept   (Ashby and Parkes)

Bottom-up approach
- voice I/O   (Leiser)
- active gas pedal   (Schuman and Godthelp)
- adaptive interfaces (technologies/solidly
  computer science)   (Piersma)

Standardization
- standardization   (Ross)

References

Franzén, S. and Reichart, G., 1991, *Human Factors and Systems Safety—Two Aspects of Systems Engineering*, presented at the 11th Congress of the International Ergonomics Association (IEA), July 15–29, Paris.

Franzén, S., Alm, H. and Nilsson, L. (Eds) 1991, 'PROMETHEUS MMI Checklist, version 2.1', PROMETHEUS Office, Stuttgart.

24

Towards a system architecture of driver's warning assistant

G. Geiser and G. Nirschl

Introduction

A Driver's Warning Assistant (DWA) differs from an automatic co-pilot by only presenting warning messages to the driver, instead of acting upon the vehicle or the environment. It is an open question whether combinations of both types of assistance system are feasible, with automatic intervention if the driver has been warned without success, or if there is no chance for a warning message leading to an appropriate action by the driver in due time. In this case difficult questions of task allocation and of responsibility have to be answered. The evolution has begun and will continue with DWAs dedicated to special driving tasks, e.g. parking, lane keeping, overtaking, crossing. These DWAs have a competence limited to one of the driver's tasks, so that the driver can only expect support for this single task. In a car equipped with such a specialized DWA the driver has to learn the limits of the DWA's competence. Additionally, the case has to be considered in which a specialized DWA has limited competence for its task. This means that the support of a driver by the DWA is limited to a subset of all possible situations (e.g. to situations with good sight conditions). Under certain circumstances (e.g. hidden obstacles) the driver might not be supported. It has to be clarified whether this can be learned and will be accepted by the driver.

A crucial topic in the development of a DWA is the timing of warning messages for the driver. The moment of warning on the one hand has to be chosen early enough to enable the driver to react appropriately. This means that the inevitable reaction time, which the driver needs to receive and process the DWA's message and to prepare a reaction, has to be considered. On the other hand the warning message should not be presented too early as this would annoy the driver and result in poor acceptance of the DWA. Therefore, it is important to integrate a time-budget analysis in the development phase and in the run-time version of a DWA, leading to specific strategies for the warning generation process.

System architecture of a driver's warning assistant (DWA)

Based on different research described in the literature a system architecture of a DWA is proposed as a basis for the planning of future activities aiming at DWA realization. Research activities with related goals are carried out in the US—American 'Pilot's Associate (PA)'—programme funded by the DARPA (Defense Advanced Research Projects Agency). Organized in the so-called Lockheed and McDonnell Douglas teams (Rouse *et al.*, 1990; Lizza *et al.*, 1990) pilot decision aiding systems running in real-time are developed. Besides participating in the PA-programme the Center for Human-Machine Systems Research, School of Industrial and Systems Engineering, Georgia Institute of Technology, Atlanta, USA, is working in the field of an operator's associate for the supervisory control of a complex dynamic system (Rubin *et al.*, 1988; Jones *et al.*, 1990). Rouse *et al.* (1988) developed the architecture of a general intelligent interface premised on automation as a backup and not as an objective of system design. The components are world, system and operator state, error monitor, adaptive aiding and interface manager. The elements of the operator state include activities, awareness, intentions, resources and performance. In the EUREKA-project PROMETHEUS the objective of a warning electronic co-pilot is pursued jointly by different research groups. The system concept of Kopf and Onken (1991) is based on the 'model driver', proposing actions for the assessed situation, and on the 'driver's picture', collecting information about the individual driver's real behaviour. In a 'discrepancy interpretation' action proposals and real actions are compared and if necessary warnings or messages are generated. Sandewall (1990) proposes a system block specification for a driver-support system with the aims of automating or monitoring driver activities as well. Within the European research programme DRIVE the objective of the project GIDS (Generic Intelligent Driver Support) is to design an integrated driver support system (Godthelp and op de Beek, 1991; Michon and McLoughlin, 1991). The GIDS architecture consists of the two central components 'analyst/planner' and 'dialogue controller'. The analyst/planner compares the actual behaviour of the car to an acceptable reference driving model, and initiates warnings and recommendations of the dialogue controller.

The modelling approach for traffic scenes developed by Nagel (1985) is based on the use of generic situation schemata within an interpretation cycle by utilizing sensor information on driver, vehicle and environment. The notion 'situation' includes the state of an agent and his/her admissible sequences of actions (Nagel, 1991). Taking into consideration these research activities a general system architecture of a DWA can be derived as shown in Figure 24.1. By means of sensory information about the driver, the vehicle and the environment the situation assessment is carried out. The representation of the actual situation is obtained as implementation of generic situation representations, and includes the descriptions of the states of the driver, the vehicle and of the

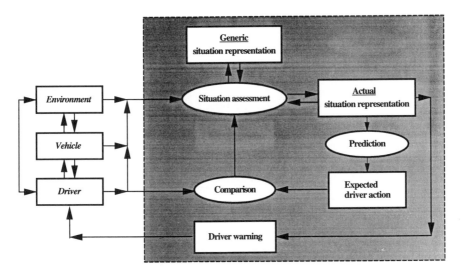

Figure 24.1   General system architecture of a driver's warning assistant.

environment and additionally the possible actions and sequences of actions. This then feeds back to the situation assessment in a situation dependent manner by controlling the input of sensory information, and by influencing the selection of generic situation representations to be considered. The state of the driver is given by his/her individual characteristics, intentions and resources being available for input, processing and output of information. The representation of the actual situation is used to derive the expected action of the driver which is compared with his/her observable actions. In case of discrepancy between expectation and observation, it has to be revised by a new situation assessment. The model has to be checked continuously not only by comparison of expected and actual driver's actions, but also by (not shown in Figure 24.1) comparison of expected and observed actions of the vehicle and the environment. The model of the actual situation is the prerequisite for the driver warning component that determines whether a warning message has to be displayed to the driver and, if necessary, which sensory modality and coding has to be chosen.

By investigating special examples of DWAs the detailed system architecture shown in Figure 24.2 was developed. The broken line encloses the components essential for driver modelling and warning message generation. The shaded regions within this area indicate the DWA's subtasks 'assessment of driver's state and intentions', 'risk analysis' and 'resource load based warning message generation'. The components in detail are knowledge bases (hatched rectangles), processing components (ellipsis) and processing results (light rectangles). Thin arrows represent the specific flow of information between single components. Thick arrows depict the compound flow of information relating to driver,

Towards a driver's warning assistant

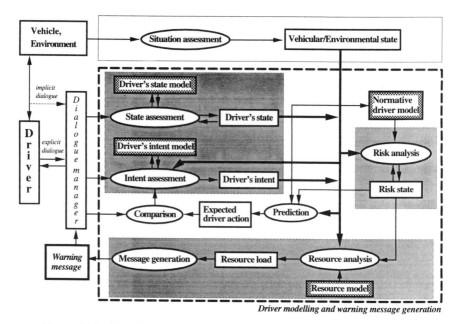

Figure 24.2   Detailed system architecture of a driver's warning assistant.

vehicular and environmental state. The assessment of the actual state of the vehicle and the environment is only roughly outlined here (thinly surrounded region on top of the drawing).

In the following, the components introduced for detailed driver modelling and their interactions are described:

The *driver's state model* represents different individual properties and states of the driver, e.g. braking and accelerating behaviour, lane and distance keeping attitude, fatigue. By evaluating the driver's actions—provided by the *dialogue manager* via explicit or implicit dialogue—in the process *(driver's) state assessment* the actual state of the driver is determined.

The *driver's intent model* describes the possible intentions of the driver, e.g. stopping, lane changing, overtaking. An intention of the driver is defined as an active goal and is recognized if the observed actions of the driver can be related to a plan for reaching this goal. Driver's actions and his/her actual state are considered, as well as the actual state of the vehicle and of the environment in order to derive the actual intention of the driver in the process *(driver's) intent assessment*.

The *normative driver model* is a knowledge base of correct behaviour of the driver when he/she has to cope with a task in a specific traffic situation, e.g. crossing an intersection. The states of the driver, of the vehicle and of the environment, together with driver's intent and with the normative driver model, are input sources for the *prediction* of driver's action. The resulting expected driver action is fed into the *comparison* process where potential differences between the actual and the expected driver action are determined. Such differences cause a revision of driver's intent assessment because the observed driver action is obviously not understood.

Based on driver's intent, the states of the driver, the vehicle and the environment, and on the normative driver model, the *risk analysis* is performed continuously. It depends on the resulting *risk state* whether the warning message generation subtask is initiated.

The *resource model* as a separate part of the representation of the driver's state describes the sensory, cognitive and motor resources of the driver available for processing of information. Together with inputs from the state descriptions of the driver, the vehicle and the environment, and from the risk state description, a *resource analysis* is carried out in order to determine the current load of the different resources. The result of this analysis influences the sensory modality, coding and timing of the message in the *message generation* process. Finally, it is the task of the *dialogue manager* to transmit the warning message to the driver.

By analyzing a simple example of a traffic situation the proposed structure is further illustrated. A driver in a vehicle equipped with a DWA is approaching an intersection where he/she has to cross a priority road regarding a stop sign (Figure 24.3). The task of the driver considered here is to come to a stop before crossing the priority road. The specialized DWA is only giving support when the driver disregards the stop sign. The description of the DWA is based on the following assumptions:

- Only the approach part of the crossing is investigated.
- No other traffic is considered.
- As no other warning functions are given the resource analysis is not included in the DWA discussion.
- The assessment of the states of the vehicle and of the environment continuously results in knowledge of the actual distance s between the vehicle and the stop line and of the velocity of the vehicle $v(s)$ as a function of the distance from the stop line with the initial values s_0 and $v(s_0) = v_0$. Additionally, the maximum deceleration for emergency braking b_{max} is assessed which is assumed to be constant.
- Driver's reaction to a warning message of the DWA is determined by a delay time due to his/her reaction time.
- Driver's normal approach behaviour is characterized by the function $v(s)$ with v velocity and s distance from the stop line. Driver's individual characteristics are given by the normal deceleration $b_{norm}(v_0)$, the normal position of onset of braking $s_{norm}(v_0)$, and the reaction time ΔT_{rea}. Furthermore the deceleration $b_{norm}(v_0)$ is assumed to be constant from onset of braking to stop.

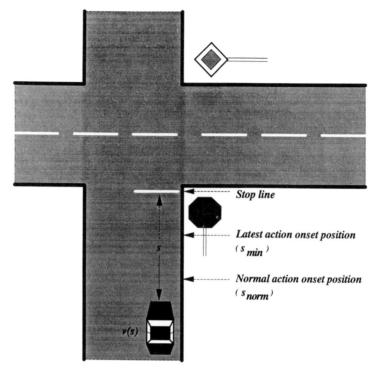

Figure 24.3   Approach to an intersection with a stop sign.

With these assumptions which have to be verified by experimental investigations (see p. 260), the DWA can be specialized in the following manner. The assessment of the situation concerning the vehicle and the environment continuously results in s and $v(s)$ with the initial values s_0 and v_0, furthermore the maximum deceleration b_{max} of the vehicle on the road is assessed. The interesting characteristics of the driver's state are his/her braking behaviour and reaction time. The parameters to be assessed are $b_{norm}(v_0)$, $s_{norm}(v_0)$ and ΔT_{rea}. The assessment of the driver's intentions has to decide between the two alternatives 'stopping' or 'not stopping' with the first one as initial value in order not to give a premature warning. The normative driver model allows to predict the expected driver action in the form $b(s)$ which is specified by the driver's individual parameters $b_{norm}(v_0)$ and $s_{norm}(v_0)$. When the position $s = s_{norm}(v_0)$ is reached, the expected driver action is braking with deceleration $b_{norm}(v_0)$. If this action occurs then there is no reason to generate a warning message as a consequence of the risk analysis. On the other hand if the expected brake action is not observed the risk of missing the stop sign has to be examined based on the time-budget analysis (see p. 257). In order to avoid premature and therefore annoying warnings on the one side, and too late warnings on the other side a window for the presentation of the warning message has to be derived which is given by the farthest and the closest warning positions. The farthest distance from the stop sign where a warning message can be presented is the position at which the driver normally begins to brake in order to come to a stop. An earlier warning position could be annoying to the driver. The closest warning distance from the stop sign has to be determined in such a way that the driver has a last chance to obey a warning message in due time. A more detailed discussion of the time-budget and an application to this example will be described below.

Time-budget analysis method for risk analysis and generation of warning message

Time-budget analysis (Figure 24.4) of driver-DWA interaction is defined as the detailed analysis of the minimum time needed by the driver and the DWA for their interaction on the one hand and of the time available on the other hand. These time requirements are due to the tasks and properties of the driver, the vehicle and the environment. As a result, a time schedule is derived based on time intervals like (Figure 24.5)

- minimum time ΔT_{min} needed for driver's (re)action(s),
- normal time ΔT_{norm} elapsing from the moment when the driver reaches the position s_{norm} of his/her normal (re)action onset until the end of his/her (re)action,
- driver's reaction time ΔT_{rea} to warning message.

By comparison of the time available and the time needed the following results are obtained. Firstly, the earliest warning moment τ_{min} is given by the moment when the driver arrives at the normal action onset position s_{norm}

$$\tau_{min} = T_{norm}.$$

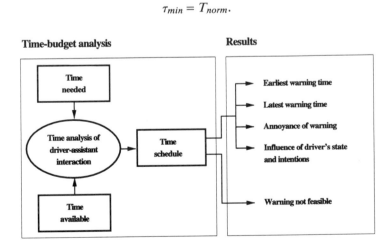

Figure 24.4 Time-budget analysis of driver-DWA interaction.

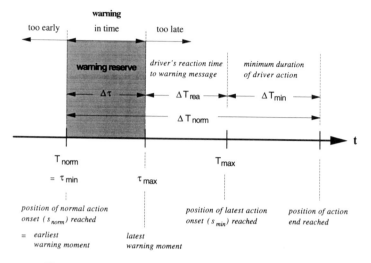

Figure 24.5   Time schedule of driver-DWA interaction.

Secondly, the latest warning moment τ_{max} is calculated from the latest possible beginning T_{max} of the driver's action at the position s_{min} by subtracting the reaction time of the driver ΔT_{rea} which he/she needs to perceive and to process the warning message and to prepare an action,

$$\tau_{max} = T_{max} - \Delta T_{rea}.$$

It is assumed that within the interval $T_{max} - T_{norm}$ the vehicle's velocity is constant $(v(s) = v_0)$ and within ΔT_{min} the maximum deceleration b_{max} is applied constantly. The time window for warnings, called *warning reserve* $\Delta \tau$, is given by the earliest and the latest warning moment

$$\Delta \tau = \tau_{max} - \tau_{min} = \Delta T_{norm} - \Delta T_{min} - \Delta T_{rea}. \qquad (1)$$

Thirdly, the annoyance of a warning message can be assessed by looking for those cases in which the warning reserve is negative, i.e.

$$T_{norm} > T_{max} - \Delta T_{rea}.$$

Here, the warning message would have to be presented before the driver normally begins his/her action. So, if he/she has the intention to act correctly such a premature warning would be useless if not annoying.

Fourthly, the influence of the state of the driver and of his/her intentions can be determined. Parameters like T_{norm}, b_{norm} and ΔT_{rea} have to be provided by the driver's state model. Furthermore, if the driver's intention is known exactly from the intent assessment premature warnings could be avoided.

As an example for time-budget analysis the simple traffic situation of approaching a stop sign is chosen again. The time intervals needed for the calculation of the warning reserve according to equation (1) can be derived as follows:

$$\Delta T_{norm} = (s_{norm} - s_{min})/v_0 + \Delta T_{min},$$
$$\Delta T_{min} = v_0/b_{max}.$$

This leads to

$$\Delta\tau = (s_{norm} - s_{min})/v_0 - \Delta T_{rea}.$$

With

$$s_{norm} = v_0^2/(2b_{norm}) \text{ and } s_{min} = v_0^2/(2b_{max})$$

the warning reserve follows as

$$\Delta\tau = ((v_0/2) \times (b_{max} - b_{norm})/(b_{max} \times b_{norm})) - \Delta T_{rea}. \qquad (2)$$

Figure 24.6 shows the warning reserve $\Delta\tau$ as a function of the normal deceleration b_{norm} for different initial velocities v_0 with the assumed maximum deceleration $b_{max} = 5$ m/s^2 and driver's reaction time $\Delta T_{rea} = 1$ s. If the driver approaches the stop sign with an initial velocity of $v_0 = 30$ km/h there is a positive warning reserve for $b_{norm} < 2.3$ m/s^2. For normal deceleration $b_{norm} > 2.3$ m/s^2 the warning message has to be presented before the driver normally begins to brake. For initial velocity $v_0 = 50$ km/h the upper limit of normal deceleration inducing positive warning reserve increases to $b_{norm} = 2.9$ m/s^2. It might be surprising at the first glance that approaching with a higher initial velocity v_0 yields a bigger warning reserve $\Delta\tau$. This is due

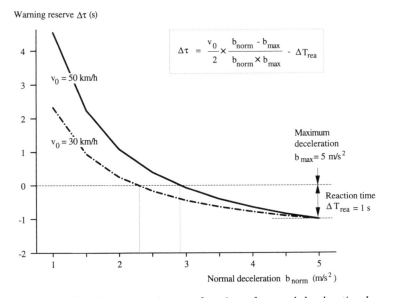

Figure 24.6   Warning reserve $\Delta\tau$ as a function of normal deceleration b_{norm}.

to the fact that assuming b_{norm} and b_{max} unchanged the increase of ΔT_{norm} exceeds the increase of ΔT_{min}. As can be seen from Figure 24.5 and equation (1), with ΔT_{rea} remaining constant this results in broadening the time window $\Delta \tau$.

In the case where the driver intends to stop the vehicle at the stop sign with normal deceleration b_{norm}, it can be expected that he/she will be annoyed by warning messages occurring before he/she normally begins to brake. It has to be shown by experimental investigations how the driver chooses his/her deceleration b_{norm}. Furthermore, if it would be possible to recognize the intention of the driver, the problem of warning annoyance could be avoided.

Experimental investigations

Experimental results from literature concerning the driver's braking behaviour when approaching a stationary object have been reviewed (Johansson and Rumar, 1971; Claffey, 1976; Van der Horst, 1989) in order to get references to the driver's state parameters 'normal deceleration' and 'reaction time to warning messages'. Furthermore, the first results from our experiments with an instrumented vehicle are available. For the example chosen above 'approach to a stop sign', driver's pedal activities have been measured together with vehicle parameters like distance from the stop sign, velocity and deceleration.

There is little in literature on 'normal braking behaviour' when stopping at an appointed position. Claffey, 1976 states a normal deceleration rate of $2.22 \, \text{m/s}^2$ when approaching a stop mark with a velocity of 24 km/h. This compares favourably with own observations! When drivers approached a stop sign with velocities in the range of 25–30 km/h deceleration rates were measured in the range of 1.5–$2.7 \, \text{m/s}^2$. Claffey, 1976 further states a normal deceleration rate of $1.94 \, \text{m/s}^2$ when approaching with a velocity of 48 km/h. In contrast to this, own experiments have shown increasing deceleration rates with increasing approach velocity. Drivers approaching the stop sign with 35–40 km/h chose deceleration rates in the range of 1.7–$3.2 \, \text{m/s}^2$. Van der Horst and Brown, 1989 report on different strategies of drivers when choosing the moment of brake onset in relation to the TTC (time-to-collision) measure. The results reveal that TTC at the onset of braking increases with speed, but less than could be expected from a constant deceleration model. The minimum TTC as reached during the approach is rather independent of speed, normal or hard braking instruction, and driver experience.

As an important result of first own experiments the normal deceleration rate seems to vary interindividually (between various drivers) to a great extent, whereas intraindividually (between various runs of the same driver) rather small variations have been observed. This can be seen in Figure 24.7 where the warning reserves derived from experimental observations of 8 drivers are depicted. When approaching a stop sign without velocity regulations the initial velocity v_0 and the (mean) deceleration b_{norm} have been measured. Assuming

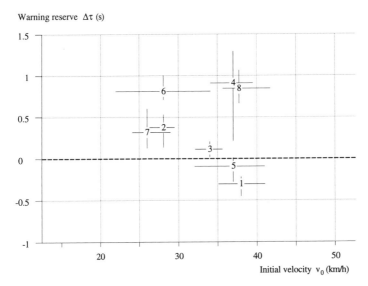

Figure 24.7   Observed warning reserve $\Delta\tau$ as a function of initial velocity v_0.

$b_{max} = 5$ m/s^2 and $\Delta T_{rea} = 1$ s the warning reserve $\Delta\tau$ has been calculated. The mean values of $\Delta\tau$ and v_0 from 3 runs of each driver are depicted in Figure 24.7 together with the ranges of variation. It can be seen that 2 drivers out of 8 are located in the area of negative warning reserve and would get premature, i.e., annoying warning messages from the DWA. With the assumption that inattentive drivers, especially left to support by a DWA, would have reaction times distinctly longer than 1 s, the separation line between positive and negative warning reserve in Figure 24.7 has to be shifted upwards. With an assumed reaction time to a warning message of $\Delta T_{rea} = 1.7$ s already 5 drivers out of 8 are located within the area of negative warning reserve, i.e. annoying warnings.

Conclusions

- The detailed system architecture of a DWA shows the knowledge bases needed for driver modelling: driver's state, driver's intent, driver's resources and normative driver. The task of setting up such knowledge bases for real traffic situations is a big challenge. A major topic of actual research work is the development of convenient knowledge representation techniques and tools (Krüger, 1991).
- Sensory information from the driver, the vehicle and the environment is needed in order to be able to assess the state, the intentions, the expected actions and the resource load of the driver, as well as the risk state of the

actual traffic situation. It has to be considered, which information can be gathered with acceptable expense.

- Time-budget analysis shows the feasibility of specified DWA-functions and sets time-limits for the presentation of warning messages which can be used by the driver in a successful manner. Furthermore, the importance and the needed extent of the knowledge of driver's state and intentions can be shown.
- As first experiments show, the warning reserve as a time window for useful warnings is strongly limited. Drivers' reaction times seem to be a major bottleneck in driver-DWA interaction.
- Further experiments are necessary in order to measure and to validate parameters of driver's state and intentions.
- In order to investigate the usability and acceptance of a DWA a prototype version will be realized within an experimental vehicle. Driving experiments will be conducted.

Acknowledgements

This work has been supported by the PROMETHEUS programme of the Bundesministerium für Forschung und Technologie and the following companies: BMW AG, Daimler-Benz AG, MAN AG, Dr.-Ing. h.c. F. Porsche AG, and Volkswagen AG. The authors thank H. Fehrenbach for his contribution to this work by performing the driving experiments.

References

Claffey, P. J., 1976, Vehicle Operating Characteristics, in Baerwald, J. E. (Ed.) *Transportation and Traffic Engineering Handbook*, Institute of Transportation Engineers, pp. 16–37. Englewood Cliffs, NJ: Prentice-Hall.

Godthelp, H. and op de Beek, F., 1991, Driving with GIDS: Behavioral Interaction with the GIDS Architecture, *Advanced Telematics in Road Transport*, Proceedings of the DRIVE Conference, Brussels, pp. 351–70, February 4–6, Amsterdam: Elsevier.

Horst, A. R. A. van der and Brown, G. R., 1989, *Time-to-collision and driver decision-making in braking*, TNO-report IZF 1989 C-23, TNO Institute for Perception, Soesterberg, The Netherlands.

Johansson, G. and Rumar, K., 1971, Drivers' Brake Reaction Times, *Human Factors*, **13** 1, 22–7.

Jones, P. M., Mitchell, C. M. and Rubin, K. S., 1990, Validation of intent inferencing by a model-based operator's associate, *International Journal of Man-Machine Studies*, **33**, 177–202.

Kopf, M. and Onken, R., 1990, *A Machine Co-Driver for Assisting Drivers on German Autobahns*, European Annual Manual, No. 9, I-ISPRA.

Krüger, W., 1991, Begriffsgraphen zur Situationsmodellierung in der Bildfolgenauswertung, Dissertation, Fakultät für Informatik, Universität Karlsruhe.

Lizza, C. S., Rouse, D. M., Small, R. L. and Zenyuh, J. P., 1990, Pilot's Associate: An Evolving Philosophy, The Human-Electronic Crew: Is the Team Maturing? 2nd Joint GAF/RAF/USAF Workshop, 25–28 September, Ingolstadt.

Michon, J. A. and McLoughlin, H., 1991, The Intelligence of GIDS, *Advanced Telematics in Road Transport*, Proceedings of the DRIVE Conference, Brussels, pp. 371–6, February 4–6, Amsterdam: Elsevier.

Nagel, H.-H., 1985, Wissensgestützte Ansätze beim maschinellen Sehen: Helfen sie in der Praxis? in Brauer, W. and Radig, B. (Hrsg.) *Proc. GI-Krongreβ Wissensbasierte Systeme, München, Informatik-Fachberichte*, **112**, pp. 180–98, Berlin: Springer-Verlag.

Nagel, H.-H., 1991, Wissensbasierte Systeme für Anwendungen im Verkehr, Entwicklungslinien in Kraftfahrzeugtechnik und Straβenverkehr, *Forschungsbilanz 1991, 14, Statusseminar des Bundesministers für Forschung und Technologie (BMFT), in Zusammenarbeit mit dem Bundesminister für Umwelt*, Naturschutz und Reaktorsicherheit (BMU) und dem Bundesminister für Verkehr (BMV), pp. 457–66, 13–15 May, Dresden.

Rouse, W. B., Geddes, N. D. and Curry, R. E., 1988, An Architecture for Intelligent Interfaces: Outline of an Approach to Supporting Operators of Complex Systems, *Human-Computer Interaction*, 1987–1988, **3**, 87–122.

Rouse, W. B., Geddes, N. D. and Hammer, J. M., 1990, Computer-aided fighter pilots, *IEEE Spectrum*, **27**, 3, 38–41.

Rubin, K. S., Jones, P. M. and Mitchell, C. M., 1988, OFMspert: Inference of Operator Intentions in Supervisory Control Using a Blackboard Architecture, *IEEE Transactions on Systems, Man, and Cybernetics*, **18**, 4, 618–37.

Sandewall, E., 1990, Proposal for a ProArt Specification Platform, Department of Computer and Information Science, Linköping University, LAIC-IDA-90-TR18, Linköping, Sweden.

25

Intelligent accelerator: an element of driver support

H. Godthelp and J. Schumann

Introduction

One of the challenges connected with the use of modern communication and information technology in road traffic, is the development of a so called 'co-driver' system. Where in airplanes the use of smart sensors, fly-by-wire systems and intelligent failure detection and identification methods is becoming common, the lack of using these modern techniques in automobile control becomes more and more visible. Given the accident and pollution figures related to road traffic one could even suggest that conventional techniques have dominated land traffic technology far too long. Recent European research programmes, such as PROMETHEUS and DRIVE, have recognized this situation and have initiated a series of projects which aim to develop knowledge and techniques that may help to regain control over road traffic in Western European countries.

The GIDS project is one of the DRIVE-projects (GIDS = Generic Intelligent Driver Support). Its main objective is to develop requirements for an intelligent co-driver system, that may help the driver to behave safely and efficiently in tomorrow's traffic. In its earliest form GIDS will involve a dialogue system which supervises and presents navigation, anti-collision and vehicle control information in a user-friendly way. One of the major subgoals of GIDS is to reduce driver workload in critical conditions. Several important research issues are included in the GIDS project:

- Development of technical sensor and communication systems which might provide adequate information to the GIDS system.
- Making available existing and newly developed knowledge about the driver's needs in critical driving situations; making the GIDS system adaptive to changes in driver needs as effected by experience, knowledge, etc.

● The design of an optimal interface which provides the driver with GIDS information in a suitable manner, i.e. spatially and temporally balanced in relation to the user's capabilities and needs.

The present study deals with the last research issue and focuses on the question whether an active gas-pedal may serve as an element in an integrated information system in a future GIDS car. The idea behind the use of active controls is to reduce driver workload by using the controls (accelerator, steering-wheel) not only as a control device, but also as an information system for the driver.

The use of an active accelerator for speed control

Many road accidents happen because one drives too fast. Entering a sharp bend, driving through a village and misjudging one's own speed, following a high speeding lead vehicle, etc., are all examples of every day driving which may result in dramatic consequences. During the last two decades most European countries have developed various kinds of speed measures. Millions of Ecu's have been spent in order to drastically change the geometrical design of city and village streets as measures to force the car driver to reduce speed. In addition, Rutley (1975) has suggested the clever use of advisory speed signs, carriageway markings and even head-up display speedometers to influence driver speed. The success of enforcement and of all these technical solutions has proved to be of limited value in each case. Tenkink (1988) analyzed the effectiveness of speed reducing measures and concluded that they only will be effective if, (1) the user population considers them as logical, and (2) the measure has verifiable negative consequences if neglected. Furthermore, a disadvantage of most of the regular speed reducing measures (signs, road design) is that they are permanent and non-adaptive to the local/current traffic situations, e.g. a 30 km/h limit may be acceptable during peak-hours near a school-area. However, at the same place a 60 km/h limit may be sufficient at 11 o'clock in the evening. Such an adaptability of elementary traffic rules seems of high importance for their acceptance (Oei and Papendrecht, 1989). Given these limitations of conventional speed reducing measures several authors recently wondered why such a small research effort has been put in the development of adaptive speed limiters or warning systems that operate inside the car. Echterhoff (1985) investigated the acceptability of such speed limitation systems. Subjects drove a car which automatically adapted its maximum speed to the local speed limit. They were quite positive about this device and argued that they would accept it 'provided that everybody would use such a system'. In a similar study Malaterre and Saad (1984) discovered that when only a limited number of cars are provided with such a speed limiter, other traffic participants may react unprepared, which may cause unwanted behaviour. However, it is not appropriate to consider this result as a negative aspect of such a speed reducing measure. When evaluating the usefulness of different types of road transport informatics the effectiveness of speed regulating devices is often estimated to be enormous (Marberger *et al.*, 1990). In its present form the gas-pedal hardly provides any systematic proprioceptive feedback to the driver about the current speed of the car. Yet the workload reduction provided by kinesthetic information from a manipulator has been demonstrated several times in aircraft-control (Merhav and Ben Ya'acov, 1976; Hosman and Van der Vaart, 1988). Figure 25.1 shows the principle of an active side-stick controller tested by Hosman and Van der Vaart (1988). The pilot force is used to control the aircraft, whereas the side stick position is artificially connected to the aircraft roll angle (or in another test condition to the roll rate). From a series of experiments it appeared that these types of position feedback may serve as a very efficient cue to the pilot.

When developing an active accelerator as an integrated element in a GIDS system, the control loop of Figure 25.2 is proposed. Given the limited position range of an accelerator it is chosen as a starting point to use the pedal-force as an information-carrier. The schematic diagram of the system indicates that several speed or headway related dimensions can be fed back as a force cue,

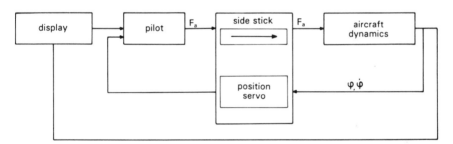

Figure 25.1   Control loop with a servo controlled side stick as an active controller (Hosman and Van der Vaart, 1988), with $\varphi, \dot{\varphi}$,—roll angle, roll rate.

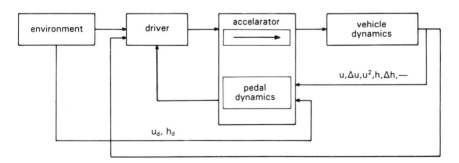

Figure 25.2   Schematic diagram of the man-vehicle control loop with an active accelerator providing kinaesthetic feedback about the error between the actual situation and some normative behavioural model.

i.e. the force level can be chosen in correspondence with the current speed error (Δu) or headway error (Δh), etc. With respect to both speed or headway several functions can be given to such an active pedal, i.e.:

- To provide kinaesthetic information about speed or headway errors as compared to some normative safety model.
- To serve as a speed or headway limiter by making the pedal force more or less infinite as soon as a speed limit is reached.
- To serve as speed regulation device providing force feedback based on a normative model which is related to fuel savings or minimizing pollution.

As a first stage in the process of specifying the feedback requirements of an active accelerator, a prototype of such a device has been developed for experiments in the TNO driving simulator. Two explorative experiments were carried out to demonstrate the potential effects of such a facility (Färber *et al.*, 1990), and in chapter 14 Janssen and Nilsson further illustrate the effectiveness of the active pedal as an element of an anti-collision system. The present study focuses on the question of which force feedback characteristics will be most appropriate for driver support in speed regulation. In the driving simulator subjects performed a speed adaptation task on a straight road approaching a speed limit area. Different pedal force feedback characteristics were compared, i.e. the force being related to the error in speed and/or accelerator position. Driver performance was evaluated in conditions with normal and with limited view on the speedometer.

Method

Driving simulator

The experiment was carried out in the fixed base driving simulator of the TNO Institute for Perception. This simulator involves a Volvo 240 mock-up, with regular steering-wheel and pedals. The steering-wheel axis is connected with a potentiometer which measures the steering-wheel angle. Steering force is generated by means of an electric torque motor (Axem MV 19), mounted on the steering axis. In a similar way the gas-pedal position is measured by way of a potentiometer connected to the pedal-rotation axis, whereas the gas-pedal force feedback can be regulated by a torque motor (Axem F12 M2) which is connected to the pedal-axis by a gearbelt drive. The perspective view of the outline of the road was electronically generated by an Evans and Sutherland PS300 system and projected in front of the mock-up with a horizontal field of view of $50°$.

Task

Subjects' task consisted of driving on a two lane rural road with a speed of

100 km/h. Drivers started in the right lane and approached a lane barricade at B (see Figure 25.3). At a predetermined location A (7 seconds before B, except pedal signal 3, see below) the speed of 100 km/h should be stable (± 5 km/h), otherwise the programme stopped and started over again. At the moment of passing location A the requested goal speed at B (100, 85, 70 or 55 km/h) was presented on the central part of the screen.

The appearance of the visual goal speed also served as a warning for the subject to change lanes, since the right lane was blocked at B (see Figure 25.3). The barricade could easily be seen from A. After B the subject should stay in the left lane and keep the goal speed for a period of 10 seconds, i.e. till point C. To control the subject's looking behaviour with respect to the speedometer an occlusion device was installed:

No occlusion is defined as a condition in which the speedometer information is available continuously until A and may be asked for shortly (0.5 s) by pressing the horn lever after A until the end of the run.

With occlusion is defined as a condition in which the speedometer is activated continuously until A and not at all during the period A to B. After B the feedback via the speedometer is available again on request (0.5 s looks, see above).

Accelerator characteristics

In the present study the conventional, 'passive' accelerator was compared with three types of 'active' pedal configurations. In each case the gas-pedal position s_g ranges from 0.0 (loose) to 1.0 (full). The following force feedback characteristics were applied:

- **passive pedal:**
  Gas-pedal force depends on pedal position $F_g = 10 + 50\, s_g\,(N)$.
- **active pedal 1:**
  Gas-pedal force depends on speed error: $F_g = 25 \pm 60\, u^2_e\,(N)$.
  Speed error u_e (m/s) is defined as the difference between the actual speed and the desired speed in the area after A.

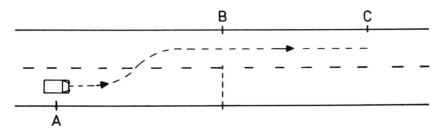

Figure 25.3   Geometry of the manoeuvre path with the start of the lane change at A and the lane barricade and speed limit at B.

Desired speed is defined as a linearly decreasing function from the approach speed at A (100 km/h) to the requested goal speed at B. Pedal force is limited at an upper and lower boundary of 250 N and 10 N respectively.

- **active pedal 2:**
  Gas-pedal force depends on pedal position, i.e. in relation to the pedal position belonging to the requested goal speed at B.

  $F_g = 25$ N with $s_g \leqslant s_{g \text{ goal speed}}$

  $F_g = 80$ N with $s_g > s_{g \text{ goal speed}}$

- **active pedal 3:**
  Gas-pedal force as in case of passive pedal with the addition of a short vibration (0.5 s, 10 Hz, force amplitude 20 N) which is given at point A. This point is now located such that releasing the pedal after A leads to the right (goal)speed at B.

Subjects

Eight male subjects (Ss) participated in the experiment, all of them had previous experience with the driving simulator. All Ss had their driving licence for at least 5 years with a driving experience of at least 10 000 km per year. Age varied between 23 and 38 years. They were paid for their services.

Procedure

Two subjects participated on one day. Each S made four sessions of runs with the four gas-pedal characteristics respectively, the order of which was randomized. A session lasted about 30 minutes with a training period to begin with. After each session Ss alternated giving each S a break of about half an hour between sessions. Before the experiment S received a written instruction which explained the background and procedure of the experiment. After a short introduction S took place in the mock-up.

The *training* for the first session contained 32 runs. It started with eight runs of the highest goal speed (100 km/h). For the first four runs the speedometer could be activated after location A on request (no occlusion condition), for the following four runs the speedometer could be triggered for a short look only after point B (with occlusion condition). After that, the next lower goal speed (85, 70, 55 km/h) was presented in the same sequence. The training for the remaining sessions consisted only of 4 runs each (goal speed in descending order; 2 runs with no occlusion condition, 2 runs with occlusion condition).

During each *experimental* session S made four blocks of eight runs. Occlusion of the speedometer alternated between runs, starting with no occlusion. During each block the four different goal speeds were randomly presented twice.

Data-analysis

Data-storage started at point A with a sampling rate of 10 Hz. The following signals were recorded:

u   = vehicle speed   m/s
obs = observations at speedometer   –
S_g = gaspedal position   –
F_g = gaspedal force   N
F_b = brake force   N

From these signals the following characteristics were derived:

u_e = algebraic and absolute speed errors at location B and C
N_{obs} = number of speedometer observations between A and B/B and C.

Differences between conditions were tested on statistical significance by way of analysis of variance (ANOVA), which contained the following factors: Subjects (S), Accelerator characteristics (ACC), Occlusion of speedometer (OCC), goal speed and replica.

Results

Figure 25.4 presents the algebraic speed errors at locations B and C. The ANOVA reveals a significant effect of accelerator characteristics ($p < 0.01$) indicating that the speed dependent configuration (1) leads to smallest speed

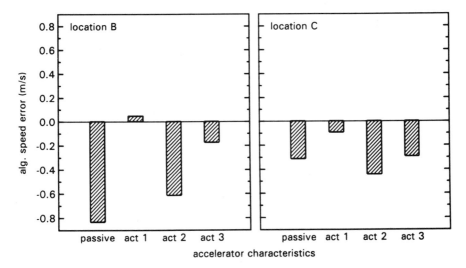

Figure 25.4   Algebraic error in driving speed at locations B and C for the different accelerator configurations.

error. This figure also shows that in most cases speed is slightly too low
compared to the goal speed. Absolute speed errors are given in Figure 25.5.
At location B absolute errors are smallest (p < 0.01) for pedal configurations
(1) and (3), which points to the efficiency of speed dependent feedback in terms
of force (conf. 1) or location (conf. 3). Keeping the correct speed between

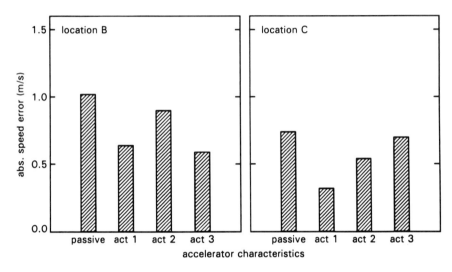

*Figure 25.5   Absolute error in driving speed at locations B and C for the different
accelerator configurations.*

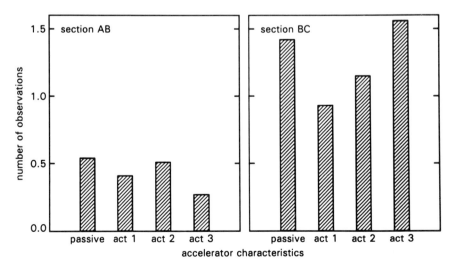

*Figure 25.6   Number of speedometer observations at the sections AB and BC for the
different accelerator configurations.*

location B and C is not supported by configuration (3), whereas configuration (1) helps the driver to reduce the speed errors in this area as well. As a consequence the speed dependent feedback of configuration (1) leads to the smallest speed errors at location C.

Results on driver glance behaviour at the speedometer are shown in Figure 25.6. The left part of this figure shows the number of speedometer observations between location A and B for the 'no-occlusion' condition. Configuration (3) informs the driver about where to release the gas-pedal, in which case hardly any speedometer observations are needed. Here also the supportive characteristics of configuration (1) are most efficient in section BC. At this section configuration (3) requires most observations whereas the permanent speed error feedback of configuration (1) allows the driver to keep the number of speedometer observations low.

Discussion and conclusion

The results of this explorative study show that 'intelligent' force feedback from the accelerator may serve as a useful way to reduce speed errors. Comparing the different acceleration configurations considered in this study it appears that force feedback related to some speed error criterion (configuration 1) leads to the lowest speed errors. Results in driver speedometer observations furthermore indicate that this type of control support may also effect glance behaviour. This finding confirms the idea that force feedback from the accelerator may improve the quality of the driver-car interface from a workload point of view.

In a parallel study the authors analysed driver performance as effected by force feedback given by the steering-wheel in case of an anticipated lane exceedance. Although the result of this study should also be considered as preliminary, the findings again suggest that active control information may serve as an efficient information system. Additional simulator and field trials are needed to verify these results for a broad range of users, i.e. young and elderly, experienced and novice drivers. Special attention should be given to questions regarding driver reactions to the combined presentation of accelerator and steering-wheel force feedback (Schumann *et al.*, 1992). Providing steering-wheel force feedback in case of lane exceedance may serve as an appropriate warning in semi-automated lane-keeping systems as presently under development in the PATH-program (Chira-Chavala and Zhang, 1992).

In any case it seems most important that active controls form a part of a well balanced, integrated driver support system which involves all support aspects, i.e. navigation, speed, headway, lane-keeping, etc. and takes account of the acceptable workload levels of a particular driver. Janssen and Nilsson (1991) illustrated the potential benefits of accelerator force feedback as information carrier for a collision-avoidance-system. In case of an intelligent cruise control system such force feedback may inform the driver about locally or temporarily changing speed limits. This leads to the question whether the intelligent accelerator should serve as a warning or advising system or as a pure, be it intelligent, speed limiter. Field trials with instrumented cars and small fleets should be carried out to answer this question. In a paper on the characteristics of future 'self-explaining' roads Theeuwes and Godthelp (1992) argue that speed limitation is required on urban and low category rural roads, whereas speed-keeping support might be appropriate on high category rural roads and motorways.

References

Chira-Chavala, T. and Zhang, W. B., 1992, *Phased implementation of lateral guidance systems in HOV-lanes*, Paper 920844, 71st Annual Meeting of the Transportation Research Board, Washington DC, USA.

Echterhoff, W., 1985, Umgang mit einer Steuereinrichtung zum Vermeiden unabsichtlicher Geschwindigkeitsüberschreitungen, *Zeitschrift für Verkehrssicherheit*, **31**, 22–3.

Färber, B., Färber, Br., Godthelp, J. and Schumann, J., 1990, *State of the art and recommendations for characteristics of speed and steering support systems*, Deliverable Report DRIVE V1041/GIDS-CONO1, Traffic Research Centre, University of Groningen.

Godthelp, J., 1984, Studies on human vehicle control, Dissertation, TNO Institute for Perception, Soesterberg, The Netherlands.

Hosman, R. J. A. W. and Van der Vaart, J. C., 1988, Active and passive side stick controllers: tracking task performance and pilot control behaviour, *Proceedings AGARD conference on Man-machine interface in tactical aircraft design and combat automation*, AGARD-C8-424, 26-1/26-11.

Janssen, W. H. and Nilsson, L., 1991, An experimental evaluation of in-vehicle collision avoidance systems, *Proceedings of the 24th ISATA Conference*, Florence, Italy.

Malaterre, G. and Saad, F., 1984, Contribution a l'analyse de la vitesse par le conducteur: evaluation de deux limiteurs, *Cahiers d'Etudes de l'Organisme National de Securité Routière*, nr. 62.

Marburger, E. A., Klöckner, J. H. and Stöcker, U. W., 1990, Eine erste Abschätzung der möglichen unfallreduzierenden Wirkungen ausgewahlter Massnahmen aus dem Prometheus-Project, *Zeitschrift für Verkehrssicherheit* **36**, 43–6.

Merhav, S. J. and Ben Ya'acov, O., 1976, Control augmentation and workload reduction by kinesthetic information from the manipulator, *IEEE Transactions on systems, man and cybernetics*, Vol SMC-6, 12, 825–34.

Oei Hway-Liem and Papendrecht, J. H., 1989, Hebben snelheidsborden effect op verkeersveiligheid? (in Dutch), *Verkeerskunde* **40**, 4, 179–83.

Rutley, K. S., 1975, Control of drivers speed by means other than enforcement, *Ergonomics*, **18**, 1, 89–100.

Schumann, J., Godthelp J. and Hoekstra, W. H, 1992, *An exploratory simulator study on the use of active controls in car driving*, Report IZF 1992 B-2, TNO Institute for Perception, Soesterberg, The Netherlands.

Tenkink, E., 1988, *Determinants of driver speed* (in Dutch), Report IZF 1988 C-3, TNO Institute for Perception, Soesterberg, The Netherlands.

Theeuwes, J. and Godthelp, J., 1992, *Self-explaining roads*, Report IZF 1992 C-8, TNO Institute for Perception, Soesterberg, The Netherlands.

26

Driver-vehicle interface: dialogue design for voice input

R. Leiser

Introduction

The majority of the work reported here was carried out with part-funding from the Commission for the European Communities under DRIVE Project V1041: Generic Intelligent Driver Support System (GIDS). An aim of the project was to identify techniques for supporting driver interaction with a range of in-car electronic devices. The role of BAeSEMA in this project was to focus on the use of speech as an input and output medium. To this end, a number of empirical studies were carried out, literature was reviewed and specialized interfacing techniques were developed and demonstrated. This paper draws heavily on the main results of the DRIVE work to present issues and design recommendations for the use of voice as a medium for driver interaction with in-car devices.

The need for voice input and output in in-car user-system interfaces

Controlling the vehicle safely must be regarded as the primary task of the driver. This task has heavy visual and manual requirements, but makes negligible demands on the driver's hearing and speech. These channels may be regarded as suitable candidates for use in tasks to be carried out concurrently with the primary driving task. The driver's restricted movement and the limited space available for additional input/output devices provides further motivation to use speech input and output. This medium requires little movement by the user. The input and output devices take up little space and need not be close to the user. As well as using a perceptual channel that is underused in driving, speech output provides explicit outputs (i.e. words) rather than the symbolic output that devices like indicator lamps use.

Problems in using speech input and output

Speech output has drawbacks. Visual output can be displayed for as long as necessary, but speech is transient, so if the user does not hear the message the first time, it has to be repeated. This requires action from the user. Speech information cannot be scanned for the bits that are relevant in the way that text can. This means that the user may be forced to listen to a long message just to hear the small part that is relevant. Speech recognition devices can misrecognize spoken input or mistake other sounds for spoken input. Two simple strategies for dealing with both types of error are;

- assume that no errors have occurred, and leave it to the user to recover from any situation that arises as a result,
- check all recognitions by asking the user to confirm that the interpretation is correct.

Both have obvious drawbacks; the first will force users to spend much of their interaction time recovering from errors. The second technique is time-consuming and irritating, particularly with users and operating environments where few errors occur. Specialized techniques are required to deal with this problem in a way that is convenient to the user.

Special user interfacing problems of in-car applications

Background noise

Background noise is an obvious problem in speech input and speech output in the in-car environment. Speech recognition problems arising from background noise impact system design in two ways. Firstly, design of speech processing equipment has to take account of the level and spectral character of background noise. Secondly, user-system dialogue design has to take account of misrecognition problems that may arise. Speech processing engineers have long been trying to find ways of cancelling background noise. The in-car application is less problematic than others because the levels and characters of the noise are predictable (compared say to the background noise in a railway station or busy office). Some progress has been made in this area (e.g. Ruehl *et al.*, 1991). The complex signal processing issues will not be discussed further here. It is enough to say that speech recognition is possible in moving cars, though dialogue design should still take into account the possibility of errors.

A less obvious problem with background noise is the effect it has on the driver's speech. Anyone who has spoken to someone who is listening to music on headphones knows that speakers naturally adapt their voice to the perceived level of background noise. The adaptation is complex, involving changes in pitch, pitch variability, stress pattern and timbre as well as overall energy level. This is an unconscious and context-dependent adaptation, so it is different for every different background noise. There are two possible approaches to speakers' intuitive adaptation to background noise: stop the user from behaving in this way, or modify the system to be able to cope with it. Stopping them from doing it is against the best principles of user interface design: if a particular behaviour is natural, it is preferable to find a way for the system to adapt to it. One method of enabling the system to adapt to the user's response to background noise is to use recognition templates collected in a variety of conditions of background noise.

The effect of background noise on audibility of spoken system outputs is more controllable because the output can be modified by the system. Approaches to this problem lie in the domain of signal processing and are not discussed further here. In summary, background noise causes some problems, but they can be dealt with by appropriate signal processing and interface design techniques.

Stress and cognitive load effects on recognition

A more psychological problem is the effect of stress and cognitive load on speech recognition performance. There are many scientific and anecdotal reports that, although a speaker or listener may be unaware of the changes, a speaker's voice will change under stress or cognitive load so that it causes difficulties for a speech recognition device that uses templates collected under normal non-stress, low-load conditions. (See Hecker, Stevens and von Bismarck, 1968; Armstrong and Poock, 1981).

An anecdote illustrates the point. The leader of a project developing user-system dialogues was using speech recognition to give a demonstration to a senior member of the company staff. He rehearsed the entire procedure the day before. The rehearsal ran smoothly, the points of interest were presented, and there were no problems in speech recognition. The following day, when the presentation took place, recognition performance was so bad that it was impossible to complete the demonstration. The user's speech was affected by the stress of the situation. Other people can get successful recognition performance under a wide range of stress and cognitive loads. Stress of various types can arise in the in-car environment. Special methods are required to overcome these effects.

User-initiated and system-initiated interaction

An unusual feature of user interfaces to in-car devices is that there will be a combination of user-initiated and system-intitiated interaction. For example, interaction with a car stereo will be largely user-initiated. A car telephone will demand a roughly equal mixture of user-initiated and system-initiated inter-action, user-initiated when the driver wants to make a call, and system-initiated when an incoming call arrives. Interaction between a driver and an engine monitoring system will be largely system-initiated. The driver rarely needs to ask the system for output; it is presented only when necessary (e.g. when a fault is detected). A mixture of system-initiated and user-initiated dialogue types is unusual, and requires special techniques in dialogue design.

Varying levels of urgency and authority in system outputs

The outputs from an in-car system will vary much more than those in conventional interfaces in terms of such things as urgency and authority. For example, outputs may include messages from:

- a CD player, saying that the disk is finished, asking the driver whether or not to play it again,
- a telephone, indicating an incoming call, asking the driver if it should be taken or held,
- a road traffic information service, notifying the driver of black ice and strong cross-winds in the area and advising extreme caution and reduced speed,
- the engine monitoring system, indicating that the engine is overheating dangerously and advising an immediate stop to let it cool.

Each of these is different in terms of how urgently the driver should act and what legal authority the instruction has. The messages must be correctly prioritized by the presentation system and presented in a way which is appropriate to their content, so that the driver can judge how best to react.

Need for driver to interrupt interaction to respond to other events

Another unusual feature of an in-car user-system dialogue arises because the primary task is not interaction with the system: the driver must be able to suspend interaction to respond to events in the driving environment. However irrational it may seem, a driver will feel pressure to complete a trivial interaction with an in-car system even when this requires some risk in the driving task. We have all continued to fiddle with the radio when we should have been using our hands to steer past an obstacle or switch on a direction indicator. This is partly due to a fear of causing difficulties in future interaction. The dialogue must therefore, over continued interaction, provide the driver with implicit reassurance that it is robust against sudden interruptions. A less obvious cause of this behaviour is the need for a Gestalt: successful completion of one task before another is taken up. Again the answer to this difficulty is for the system to provide reassurance to the user that an abandoned task can be successfully taken up again later.

Anecdotal evidence suggests a third pressure against the driver interrupting interaction—social pressure. There is evidence that although they are clearly machines, users are not immune to social effects arising from interaction, particularly when the medium is speech (Leiser, 1992). One such pressure commonly encountered in human-human dialogue is the need to suspend a dialogue gracefully, rather than simply leave it hanging. Solutions to this could be of two types: reduce the social pressure exerted by the system (perhaps by using synthesized rather than digitized speech to emphasize the machine-like aspects of the system); or immunize the driver against feeling such pressure through continued demonstration that no negative effects arise—the system *will* talk to the user again despite the abrupt way she ended the last interaction.

Integration of interaction through a single interface

A driver may need to interact with a wide range of different devices. It is desirable to integrate user interaction with these through a single interface. This allows consistency to be imposed on interaction conventions, so that each user action has a parallel effect in all of the devices. Similarly, all outputs are encoded in the same way, so that when it is necessary to use arbitrary or non-explicit coding of output, a single consistent coding convention is all the user needs to understand. One danger in this approach is that the user may become confused about which application is in use. When a variety of devices is used, without an integrated interface, each has its own look-and-feel and gives the user cues as to what actions are appropriate and inappropriate at any stage in the dialogue. Most people have had the experience of being unable to remember the command to carry out a particular action in a software package until they see the screen with the package running. Any single integrated interface must balance the value of consistency across applications against this danger of lack of context marking.

General recommendations for solutions

Our approach to design of dialogues was heavily influenced by the notion of the electronic co-driver. Just as studies of human expert behaviour have been invaluable in developing systems with intelligence, and other interface design techniques have arisen from studies of human-human communication, many ideas came from consideration of the cooperative behaviour between a driver and a human co-driver. This provided a good model of cooperative dialogue. Many ideas came from the social psychology literature. This identifies many strategies used by human conversants that are extremely effective in regulating dialogue, and can be adapted for use in human-computer dialogue. This flavour of our work emerges in some of the techniques described below.

Experimental user interface studies

We also carried out some experimental studies to address specific issues in the design of user interfaces for in-car use. Two of these are described briefly below.

Detecting and adapting to stress

Background

The effects of stress and cognitive load on speech recognition are described above. We carried out a study to see whether current speech recognition technology could be made more robust against these effects. A more ambitious aim of this study was to see whether it would be possible to detect different types of stress through their effects on the speaker's voice. Then the dialogue could be made adaptive to other loads on the driver.

Experiment

The question we were asking in this study was: 'If we collect recognition templates under various stress conditions, can these be made available subsequently to improve recognition performance and to identify the current stress type the speaker is experiencing?' We hypothesized that:

(a) the availability of templates recorded under a range of stress conditions would reduce the number of misrecognitions,
(b) when a speaker provided an utterance under a particular condition of stress, the best match would be provided by the template for that utterance that had been recorded under the same stress condition.

The subject's task was to speak numbers in the range 1 to 10 under four conditions:

- In the *normal* condition, speakers were under no artificial stress. They merely had to speak the answer to simple sums presented aurally (e.g. 6−4).
- In the *concurrent* condition, the same task had to be carried out at the same time as a manual spatial tracking task. This was analogous to driving.
- In the *frustration* condition the software introduced apparent recognition failures so that the user would often have to repeat the answer two or three times (in response to 'I couldn't get that, please say it again').
- In the *urgent* condition, the subjects' responses were provided in the context of a game. Ten numbered vertical lines appeared on a computer screen. Apparently at random, one would start reducing in length. The subjects' task was to say the number of the line as quickly as possible. On accurate recognition of the line number, the line would stop reducing. Subjects' scores were dependent on how quickly they named the line.

These four conditions were designed to simulate realistic stress conditions that might occur during driving. The concurrent condition represented the situation where the driver is speaking to the interface while carrying out a difficult part of the driving task. The frustration condition represents the situation where the driver is having difficulty being recognized by the system. The urgent condition represents situations where the driver needs to achieve the desired input urgently. Each subject took part in two phases, a template collection phase and a template testing phase. In the collection phase, when subjects thought they were responding to the machine their utterances were recorded as recognition templates. In this way each subject provided 4 different templates for each of the numbers from 1 to 10. In the testing phase, all of the previously collected templates were made available for recognition and subjects carried out the tasks again. The template providing the best match to each response was recorded.

Results

Results indicated that the technique did provide a better-than-chance means of identifying a speaker's state, but not a perfectly reliable one. The strength of association varied amongst speakers, but only for one was the observed correct identification of state less than would be expected had the system been guessing at random. A more detailed description of the study is available in Hillary (1989).

Application

The simplest application is to collect templates from a variety of stress situations and, as in the experiment, make all versions available during recognition. These would improve recognition because they cover a wider range of likely user states than a 'normal' set. A more advanced application would use this effect for adapting the dialogue to take account of the perceived stress state of the user. Thus if the user's speech is providing best matches with templates collected under urgent conditions, an abbreviated version of the dialogue could be used, with briefer and faster spoken messages to meet the user's current need for fast interaction. This kind of adaptation can be observed in human-human dialogue where it allows conversants to mutually tune their speech to the other's needs (Giles *et al.*, 1987). The critical issue in any application of the experimental result is the balance between the value of the facility as perceived by the user, and the negative effects of the inevitable occasional errors in state identification. Empirical assessment of each application would be necessary to determine its value.

Encoding message meaning through phrasing and intonation

Background

Intonation is used constantly in human speech to add refinement of meaning.

A simple sentence can be given a variety of meanings by changing intonation alone:

John made fish soup on Friday—means it was John and no-one else who made the soup.
John *made* fish soup on Friday—means he made it and did not buy it.
John made *fish* soup on Friday—means it was fish soup and no other kind.

Stressing 'soup', 'on' or 'Friday' will give further different shades of meaning. This has two implications for the use of speech for in-car announcements:

- Outputs which have no intonation, or inappropriate intonation, may not have the desired impact on the user. For example, an urgent warning message that has flat intonation may be ignored or not treated as urgent, despite the content of the message. In this respect it would be less effective than visually presented messages, because when visual messages are interpreted, more emphasis is placed on the content and less on the style. To get the desired listener reaction, spoken output must sound urgent.
- When intonation is used it must be carefully designed to produce the desired reaction.

Experiment

We carried out a study to address this issue. It was not our intention to unravel the theoretical issues of intonation, but to investigate the possibility of identifying some very simple rules. These would allow design of spoken output messages that get the desired reaction from listeners. Our study followed the simple rationale that since humans are good at encoding and interpreting meaning as intonation in speech, we could learn a lot from observing their encoding and interpreting behaviour. We devised a paradigm where intonational cues of specific meanings could be recorded and their effects on listeners could be monitored. Of particular interest in the in-car context were encoding of urgency, reliability, importance and authority:

- *Urgency* means how soon the driver has to react. If a tyre blows out, an immediate reaction is required. If the road is blocked 10 miles further along and there are many exits before that, the situation is less urgent.
- *Reliability* means how likely it is that the information presented is true. A notification from the road authority that there are road works is more reliable than a phone message from a member of the public to the police saying that a herd of elephants has escaped from the zoo and is heading across the motorway.
- *Importance* means how serious the consequences will be if the driver does not react. Black ice on the road is more important than a queue of traffic, because black ice is much more likely to cause accidents and injury.

- *Authority* refers to the legal status of a message. A long-standing speed limit has complete authority, whereas an instruction to drive carefully has low authority because it is difficult to define and so difficult to enforce.

In a series of studies, subjects were asked to speak predefined texts as they would be spoken in various situations and to write texts reporting situations of various degrees of urgency, reliability, importance and authority. Other subjects were asked to interpret these. The recordings and texts that were most reliably identified as having the intended meanings would be considered the best models for encoding those meanings.

Results

Although the spoken texts sounded effective at the time of recording (when the experimenters were aware of the intended meaning), subjects showed a high degree of non-random confusion between the intended categories. In the written text study, texts were not shown directly to subjects for assessment. Instead they were used to extract a commonly-used pattern of phrasing used for the intended meanings. This was used to generate new messages which were then presented to subjects, who, as in the spoken text study, had to indicate which of the possible intended meanings they thought each conveyed. Responses conformed very closely to the predictions.

Summarized conclusions from the studies are:

- Contrary to our expectations and common intuitions, intonation alone is not sufficient to encode various levels of different variables.
- Although intonation is not always sufficient to indicate what meaning *is* intended, subjects showed strong agreement as to what meanings were *not* intended (correct and incorrect responses fell into only a few categories, rather than being spread randomly across all of the available categories). Therefore, although intonation is not sufficient to encode a meaning explicitly, it should at least be consistent with the intended meaning, otherwise listeners will not consider it a possible interpretation.
- Phrasing is more effective for encoding meaning than intonation. The simple scheme developed in the study was as follows;

To encode *reliability*, the phrasing of the *situation* component was modified:

Low reliability: There could be {situation} + {action}
e.g. There could be an accident ahead. Slow down.
Medium reliability: There seems to be {situation} + {action}
e.g. There seems to be an accident ahead. Slow down.
High reliability: There is {situation} + {action}
e.g. There is an accident ahead. Slow down.

Authority was encoded by varying the *action* component;

  Low authority: {situation} + maybe you should {action}
  e.g. There is an accident ahead. Maybe you should slow down.
  Medium authority: {situation} + you should {action}
  e.g. There is an accident ahead. You should slow down.
  High authority: {situation} + {action} immediately.
  e.g. There is an accident ahead. Slow down immediately.

The levels of urgency were encoded in a slightly different way. As well as adding wording, the order of the situation and action components in the medium and high urgency conditions were reversed (because in urgent situations the considerate strategy of explaining the situation before demanding action has to be abandoned to ensure immediate action by stating the required action immediately). The situation component in the high urgency condition was also truncated because this seemed to evoke a stronger sense of urgency;

  Low urgency: {situation} + get ready to {action}
  e.g. There is an accident ahead. Get ready to slow down.
  Medium urgency: you should {action} + {situation}
  e.g. You should slow down. There is an accident ahead.
  High urgency: {action} + {truncated situation}
  e.g Slow down. There is an accident.

- Rules for combining levels of more than one of these variables in a single message (e.g. high urgency and low reliability) need to be established and tested, because some of the coding strategies conflict with each other.
- Subjects' assessments of messages in the laboratory may not be a reliable predictor of how they would react to them in practice. Further studies of these methods in real applications need to be carried out over a long period of use. Subjects' actual responses (in terms of whether or not they react to the messages in the appropriate way) would need to be recorded, rather than their subjective assessments of the messages' intended meanings.
- For maximum effect in spoken warning messages, appropriate phrasing must be used, perhaps in line with the successful scheme described above, in combination with intonation which is consistent with the intended meaning

Application

The coding and intonation strategies described above were implemented in a laboratory demonstrator of a user interface to a range of electronic in-car devices. Although no formal evaluation was carried out, users reported that the phrasing and intonation of the messages seemed consistent with the intended meanings and, furthermore, that the urgent messages used did indeed have the intended alerting effect. A more ambitious application would be to the generation of output messages using synthesized speech. In preliminary work in our laboratory simple manipulations of pitch, pitch variability and speech rate produced credible affective impact. Some more development in this area would allow a system like the one shown in Figure 26.1 to be realized.

The output message is optimized by the system as follows:

1. The application producing the output to the user presents the necessary parameters to the dialogue controller. This decides first whether any message should be relayed to the user, given its analysis of the output and its knowledge of the current environment and conflicting demands of other applications.
2. If a message is to be produced, the dialogue controller assigns the necessary urgency, reliability, authority and importance variables. A salience variable (with values low, medium or high) is also assigned, which reflects how

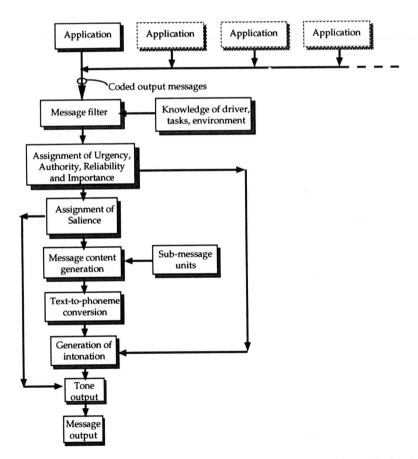

Figure 26.1 A system for filtering and assigning appropriate intonation and phrasing to synthesized voice output messages.

important it is for the driver to pay attention to the message. This is generated by combining, with weightings, the four affective variables.

3. The dialogue controller pulls together the necessary sub-message units from storage to generate the necessary text.
4. The text is converted to phonemes.
5. The phoneme list is converted to speech, with intonation generated on the basis of the urgency, reliability, authority and importance variables.
6. The message is output, preceded by a tone reflecting the overall salience rating of the message.

The feasibility of steps 1 and 2 has been demonstrated in DRIVE Project V1016: Infosafe, where a rule-based system was developed to perform this function with road safety-related messages. Step 3 is a standard procedure in state-of-the-art speech input/output interface design. A wide range of successful systems have been developed which could carry out Step 4. Step 6 is straightforward. Step 5, as indicated above, is the only one which requires further work. Because there is only a limited range of affective variables, each with only three possible values (high, medium, low), and because the aim is only to be consistent with, not to unambiguously encode the required meaning, and given the positive indications of our preliminary work in this area, this should not be a major problem.

Guidelines for in-car user interface design

The various problems in designing user interfaces to in-car applications identified above were largely resolved by reference to relevant literature, by experimentation and by explorations with prototype interface modules. Below are some guidelines which summarize the results of our studies. These are followed by a sample dialogue between a driver and an integrated in-car user interface.

Do not force interaction

The user interface must not demand input from the user. Its role should be on a purely 'take-it-or-leave-it' basis: the user is made aware of what features and activities are available and is allowed to choose whether or not to take advantage of these.

Support interruption and recovery

Some of the applications will need to initiate interaction with the driver themselves (e.g. in response to events in real time like an incoming phone call or the driver getting too close to the car in front). If the situation is critical and the driver is already interacting with another application it will be necessary for the on-going dialogue to be interrupted. The user will find this unacceptable unless the criticality of the situation is clear. This means that:

- A dialogue should only be interrupted in truly critical situations,
- If the criticality is not immediately obvious, the system should provide some justification.

These guidelines may sound familiar. Most of us learn them at a young age in the form of manners. So again we find a parallel between guidelines for acceptable human-computer interaction and human-human interaction. Humans know they should only interrupt another conversation if the situation is critical, and when the situation is not obviously critical, they usually make sure they make it clear: 'I'm sorry to interrupt, and I know it sounds trivial, but what is your home phone number?'.

Mark interaction context

Users of screen-based computer systems quickly learn to associate the visual appearance of the screen with the legal inputs for the package they are currently using. Visual information provides valuable cognitive cuing as to what is available, allowable and not allowable at any time. In the absence of supporting visual information, as may occur when the dialogue is mediated solely with speech or when the driver's eyes are elsewhere and unable to receive these visual cues, interaction errors may result. As a substitute for visual cues, auditory cues are required to provide context markings. Possibilities in the auditory mode include use of characteristic equalization (enhancing particular frequency bands within the speech signal) or other effects such as reverberation. We opted to use a coding system which mirrors the real world—different speaker's voices deal with different functionalities. Each application in the system has output messages recorded by a different speaker, so that not only do applications become associated with voices, but switches between applications are emphasized by the change in voice.

Balance consistency with adaptivity

Consistency has been a golden rule in user interface design for many years; but it can be over emphasized in design to the extent where an easier or faster way of doing something is not implemented because it would be inconsistent with the way things are done elsewhere. This is a particular problem in an in-car dialogue, where the user will typically interact with the system in many long sessions over extensive periods. Where possible, dialogues should be reduced to a more truncated form as the user becomes familiar with them through regular use.

Sample dialogue

Below is a sample dialogue that illustrates some of the design features
discussed above. The notes explain what the system is doing and how this is
done. The reader should remember that this is a time-compressed series of
interactions: In normal driving the system would be idle most of the time.

Dialogue (U = User, S = System)	Notes
S (Supervisor): ⟨click⟩⟨click⟩⟨click⟩	In standby mode, the system emits a regular ticking to indicate that it is available. The volume and periodicity are adjusted so that the sound is audible but not obtrusive.
U: Attention	
S (Supervisor): What service?	The system only responds to certain commands in the standby state, so that it will not be woken up by conversation, radio, traffic noise etc. Here 'attention' brings the system out of standby mode.
U: Phone	The car phone is controlled by voice. The interface translates spoken commands and sends them to the phone control circuitry until a call is successfully initiated. It then redirects spoken input directly to the phone. When the call is terminated it goes back into input recognition mode.
S (Phone): What phone service?	Each different facility uses a different digitized voice. This provides context marking in the absence of visual information. The experienced user can simply state the required command.
U: ⟨no response⟩	Inexperienced users may forget which commands are available.

Dialogue (U = User, S = System)	Notes
S (Phone): Phone services available are call and messages	Receiving no response, the system tells the user which commands are available.
U: Call	The user wants to make a call.
S (Phone): Who do you want to call?	Here the user can say 'call ⟨name⟩', but because no name was heard, the system prompt for one.
U: Office	The name is matched against templates set up previously by the user for people who are called regularly. The appropriate number is stored along with the name. When the name is recognized, the appropriate number is called.
S (Phone): Office is busy. Do you want to keep trying?	On detecting the engaged tone, the system offers the user a facility to keep trying.
U: Yes	The user elects to keep trying
. . .	The system waits some time before trying again.
S (Phone): Office is still busy, do you want to keep trying?	The number is still engaged after a number of attempts, so the user is asked whether it should keep trying.
U: No.	
S (Phone): Do you want to use another phone service?	It is an interface convention that before leaving any application, the user is offered the chance to use another facility in that application.
U: No.	

Dialogue (U = User, S = System)	Notes
S (Phone): Returning to supervisor.	Transitions between applications are always announced explicitly to support the user's awareness of context.
S (Supervisor): What service?	
U: Standby S (Supervisor): Did you say 'status'? U: No. S (Supervisor): Did you say 'standby'? U: Yes.	The user tries to send the system back into standby mode. The recognition system detects a doubtful recognition, and so requests confirmation of its best match with input. This match is wrong, so the system presents its next guess. This is accepted by the user. The correct input is identified without the need for the user to repeat the input.
S (Supervisor): ⟨click⟩⟨click⟩⟨click⟩	
. . .	Neither the system nor the user wish to interact for some time.
S (Navigator): ⟨Low salience tone⟩. Message from navigator. Turn left at the next traffic light. Then turn right after 40 metres. Did you get that?	The system needs to initiate an interaction because an important navigation action needs to be executed. To attract attention and give the user time to break off any conversation with passengers, a tone is played before the message begins. The salience of the tone is designed to convey the necessary sense of urgency. A low salience tone attracts the user s attention while indicating that it is not essential to listen to the following message. Thus, even in system-initiated interaction, the system gives the user the necessary information to allow a decision whether to interact or not. The user is offered the chance to hear the message again. Rather than force the user to learn another command, the prompt is phrased so that a yes or no response is sufficient.

Dialogue (U = User, S = System)	Notes
U: No.	
S (Navigator): Turn left at the next traffic light. Then turn right after . . .	The message is repeated at the user's request.
S (Collision avoidance): ⟨High salience tone⟩ You are too close to the car in front. Please pull back. . . . . . . You are still too close. You must pull back. . . . OK, now keep that safe distance. Returning to navigator.	The collision avoidance system identifies a critical situation. According to the system's prioritization scheme, its message to the driver is assigned a priority higher than the message being played from the navigator. It is therefore allowed to interrupt it. A high salience tone is used so that the user knows it is important to listen to the following message. The phrasing of the basic message from the collision avoidance system, 'pull back' is modified according to the user's response. First, a polite message is used, the basic message is preceded by 'please'. When the user fails to respond appropriately, 'please' is replaced by 'you must'. These versions are generated from prerecorded message segments according to a pre-defined grammar for generation of messages with different urgency, authority etc.
S (Navigator): Turn left at the next traffic light Then turn right after 40 metres. Did you get that?	
U: Yes	
S (Phone): Returning to supervisor. S (Supervisor): What service? U: Standby S (Supervisor): ⟨click⟩⟨click⟩⟨click⟩	

All of the facilities illustrated in the sample dialogue are feasible and have been demonstrated in desk prototypes in our own laboratory. This is a first step towards full implementation. A number of outstanding issues need to be resolved before full in-car implementation is possible.

Outstanding issues

Cultural adaptation

It is well-known that national temperaments differ, and more specifically, attitudes to driving differ even between countries within a small continent like Europe. Experience in other projects has indicated the need to adapt the design of speech interfaces to the national characteristics of the target user group. The same applies to the design of speech interfaces for interaction with in-car devices. Adaptation for use in other countries would require more than simple translation of the output messages. The balance between politeness and brevity would need to be set differently in different countries. Languages such as German and French have distinct familiar and polite forms of personal pronouns and verbs. The issue of whether a German-speaking interface and its user should 'duzen' or migrate towards this familiar form over time, can only be resolved by German-speaking users and designers who are familiar with the language and its use in user interfaces.

Long-term evaluation

The short timescales of the project work in the DRIVE programme made it impossible to assess user interface designs in long-term real-life use. Despite the availability of many techniques for building-in usability at the design stage and assessing usability thereafter in early system use, there are often interface design problems that only emerge after users have had the system available for use in their daily lives for extended periods. Systems often fall into disuse after a period because the perceived value of the system is not enough to justify the perceived necessary effort of interaction. A proper long-term evaluation of an in-car integrated user interface would require many users to be given the equipment, and their use of it to be monitored unobtrusively. Only then can the practical usability of the interface be established objectively. A related issue is the need for adaptation over long periods of use. The long-term evaluation proposed above may reveal what appear to be necessary design changes, but these may be the result of the user having adapted to using the system. If the changes were made under these circumstances, it might be the case that new users cannot use the system successfully. The need in such cases is for long-term adaptation of the system.

A final word of caution: the driving task itself does not change as a result of advances in user interface design. However well we design user interfaces to in-car devices, we are still adding an additional load to the driver's task space. Although this may to a large extent be simply taking up spare capacity, this is not limitless. There must come a point when the driver cannot deal with any more non-driving activity before the driving task suffers. Any perceived success in producing easy to use, adaptive user interfaces should not allow us to overlook this.

References

Armstrong, J. W. and Poock, G. K., 1981, *Effect of operator mental loading on voice recognition system performance*, Technical Report NPS55-81-0160, Monterey, CA: Naval Postgraduate School.

Giles, H., Mulac, A., Bradac, J. J. and Johnson, P., 1987, Speech Accommodation Theory: The First Decade and Beyond, in McLaughlin, M. L., (Ed.), *Communication Yearbook*, **10**, pp. 13–48, Beverly Hills, CA: Sage.

Hecker, M. H. L., Stevens, K. N., von Bismarck, G. and Williams, C. E., 1968, Manifestations of task-induced stress in the acoustic speech signal, *Journal of the Acoustical Society of America*, **44**, 993–1001.

Hillary, C. R., 1989, Are the effects of situational variables on a speaker's voice consistent enough to allow identification by speech recognition device: An empirical investigation, MA Honours thesis, University of Glasgow, Department of Psychology.

Leiser, R. G., 1992, The Presence Phenomenon and other problems of applying Mental Models in User interface design and evaluation, in Rogers, Y., Rutherford, A. and Bibby, P. (Eds) *Models in the Mind: Perspectives, Theory and Application*, London: Academic Press.

Ruehl, H. W., Dobler, S., Weith, J., Meyer, P., Noll, A., Hamer, H. H. and Piotrowski, H., 1991, Speech recognition in the noisy car environment, *Speech Communication*, **10**, 1.

27

Interface design for navigation and guidance

M. C. Ashby and A. M. Parkes

Introduction

An old joke goes as follows:

> James was driving around some quiet hinterland on a hot and dusty summer's afternoon. All the roads looked similar, as did the junctions. It was useless. After he had driven for the third time past an old gentleman leaning on a five-bar gate and chewing a straw, he decided to swallow his pride and ask for a bit of help. Reversing back to the elderly local, he asked for the quickest way to the town. 'Now then,' he began, 'you go up that way, take the first left and head north, then the second right. Er ... no, hang on.' More silence. 'No,—you go back the other way, take the first right after the pond and then head east until you go over a cross-roads and ... no, wrong again.' James looked at him. 'No, sorry,' said the old man with an air of resigned indifference, 'if I were you, young man, I wouldn't start your journey from here.'

It can be argued that the world falls into three categories.

1. Those who laugh at the joke.
2. Those who do not laugh at the joke.
3. Those who failed to see it was a joke, noting instead a possible opening for an appropriate road traffic informatics system.

Although mindful of the fact that readers are unlikely to fall into the first category, it is postulated, with tongue firmly in cheek, that the third category may have had a sizeable following in years gone by. It represents the technology-led approach which has generated many and varied solutions to problems encountered by drivers. It represents a way of identifying the problem without, possibly, actually knowing the full extent of the problem—the human problem.

Looking at recent DRIVE estimates (DRIVE Programme, 1989) shows how much wasted effort results from traffic congestion and inefficient route choice. Five-hundred billion ECU are lost in Europe each year in a transport system that has an abundance of maps, landmarks, signposts and people to point the way. The financial cost to Europe points to an individual cost to the driver, as navigating in an unfamiliar environment is assumed to be a demanding and stressful task, increasing the probability of an accident. So, surely, giving the driver some extra navigational information is the best answer? Several route information systems are already being tested or available in Europe (Bosch TRAVELPILOT, Autoguide/ALI SCOUT, TrafficMaster, CARIN and more). There are other systems in the United States and Japan, and it is likely that the marketplace will see more activity and new arrivals as time and technology progresses.

This assumes that the decision has been made that route information systems are a good thing and provide positive cost benefits to both individuals and society. Finding exact numbers to indicate those benefits seems less easy, especially as the calculations require the systems to work efficiently and to affect the whole traffic system positively. If a route information system turns out to influence other parts of the traffic system negatively, then the picture is quite different. The UK Department of Transport estimates the cost to society of a fatal accident at well over £0.5M (Department of Transport, 1989). If system failure rate is high, then the potential fuel, time and environmental cost savings could easily be eroded as a result of increased accident rates. The appropriate use of a human factors knowledge/research base, and the realization that the individual navigation system impinges on the total traffic system, are important starting points in the design of an interface.

What is route information?

The term route information is used to describe the kind of navigation information a driver might receive at any given point in his or her journey. This can be subdivided into three categories:

(i)   route planning,
(ii) route navigation and
(iii) route guidance.

These make a logical distinction in the functions provided. Some systems may, of course, exhibit more than one of these functions.

Route planning

This task is usually performed not only before the journey begins, but also outside of the vehicle. Some sort of route planning activity might result from:

● needing to travel to an unfamiliar destination,
● needing to reach several destinations within the same journey,
● needing to find alternative routes around congestion or road works.

For drivers to plan the best route(s) they not only need to know the most logical route, but also information that affects this choice and is variable with time. Weather conditions, traffic flow, road accidents, diversions all fall into this category of fluctuating information. The routes some drivers take may also depend on the facilities along the way, so planning for accommodation or recreation needs attention.

Route navigation

Route navigation systems provide the driver with information which they can use to facilitate the navigation task through the road network. No specific route following information is generally available, but the information from which to make navigation decisions is provided, usually in the form of a map. The in-vehicle system generally consists of a display for the driver, a map database, a position location device and a computer. Information displayed is therefore basically geographical in nature—where am I, where is my destination, what are the surroundings, how far have I travelled? The drivers are left in control of the situation, and have the flexibility to decide his or her own route strategy based on the information provided on the display. This also means that more junction-specific decisions such as lane changes are left to the driver to decide.

Route guidance

Route guidance refers to real-time route information that is dynamic and may be part of a traffic management system. A route guidance system calculates the most appropriate route in the circumstances, given a start point and destination, and then feeds the driver with the route following instructions needed at any given time. It too, uses a map database, a position location device and a computer, but also has further information coming to it from outside of the vehicle. This can be more positional or traffic information. The system is therefore fulfilling the navigation task en route, and presents the drivers with redundancy-free information only when they need it. The drivers are not necessarily presented with a view of the road network, and may therefore lack real understanding of their precise location.

Human factors guidance for navigational guidance

The paragraphs above have discussed the nature of the systems that might be used in a navigational task. There are advantages and disadvantages associated with all the system types as described. The difficulty lies in designing a system, for use by many different kinds of user, with differing characteristics and diverging needs, which is to be placed in an environment that is as information-competitive as a vehicle on today's roads. Any expected technical and societal impact or benefit, as mentioned by Schraagen (1993) elsewhere in this volume, is bound to be negated if the system does not fit harmoniously into the current driving task. Simply because a navigation system is seen to work well in a marine or aeronautical setting is not a valid enough reason to employ the same sort of system in the earth-bound automobile setting. The primary task involved is different, as are the users, as are the reasons for having a navigation system in the vehicle in the first place.

So, what is there to do? Firstly, the needs of the users of navigational guidance must be considered and investigated. Without the knowledge of drivers' needs, capabilities and expectations it would be difficult to understand what sort of information drivers would find appropriate at a given point in a journey. It follows that this will have a bearing on the kind of output medium selected to convey the information. For instance, auditory presentation would be the preferred medium for information which is simple, short or deals with events in time. Visual presentation is more suited to information which might be referred to later, deals with a location in space or is being presented in a noisy environment (McCormick and Sanders, 1983). It is then important to consider the driving tasks required for competent control of the vehicle, and how these requirements on the driver impinge on the design of the functionality and, therefore, interface of the system. The intention is to approach interface design in a driver-centred way.

Who are the users?

There are a number of ways to describe the user. The field of human-computer interaction uses the following four terms to do so: naive, novice, skilled and expert. In driving terms these could translate into learner, recent licence holder, commuter and business driver. Drivers are separated along many dimensions, and the result is different needs and abilities. For example:

Younger......................Older
Experienced......................Inexperienced
Frequent......................Infrequent
Social......................Business
Special needs......................Typical needs
Prefers motorways......................Prefers B-roads

It is obvious that the older, infrequent, social, B-road-preferring driver will have a wholly different set of requirements to a younger, experienced, frequent business driver with special needs. It therefore follows that to design a system that relies on the reaction times, visual accommodation and information processing abilities of a younger, frequent driver all but excludes an older, more occasional driver from even being able to use it. This demonstrates both the importance of knowing the target groups a system may be aimed at and the importance of designing a system to take into account the range of likely users. The nature of the journey taken and the information required for the individual journey is a further variable, independent of the characteristics of the user.

What information do users need?

The starting point for deciding on the functional specification of a route information device should be an investigation of the information needs of the proposed user market. Different driver needs, capabilities and expectations should dictate different information provision, which in turn should dictate different media and styles of information transfer.

Southall and Twiss (1988) identified three main areas where drivers require information: journey planning, system status and route guidance. Journey planning may require much varied information for the driver. Obviously information is needed about road numbers, street and town names, but also much more specific types of information might be required. For instance, a driver may wish to know where the most suitable place to stop for refreshment is, or where places of interest or natural beauty are along the route. Other information may be needed on hotels, railway and bus stations, restrictions on certain roads, and petrol stations. Drivers with special needs may require specific information regarding parking and wheelchair access. The driver should be able to retrieve this information on demand rather than have it displayed continually, so that only the information required at a particular moment is presented. Drivers also need information on what the system is doing at any given time, i.e. system status. The driver might receive feedback on what function has been selected, and an indicator if the system is busy, e.g. when it is searching for a route. Too often systems leave the user mystified about which stage of interaction has been reached. The third area of information requirement is the information the driver needs to navigate from starting point to destination. Current systems differ in the 'level' of information they present. In addition to location information, guidance information might concern time, distance and direction to destination, as well as fuel consumption.

How effective a new route information system will be, and how acceptable to the individual, will depend to a large degree on the method previously used for such purposes. At present there is a great variety of strategies for route planning and guidance by the general public. Routes may be chosen for various reasons; speed, distance, traffic volume, complexity, beauty of surroundings, familiarity, superstition, economy or on-route facilities. Similarly, there is a great variety in how individuals choose to attempt to follow the planned route once underway. Some people will use maps, though preference will be given to a variety of scales and styles of representation; while others prefer to make detailed lists incorporating road names or road numbers; yet others may place more importance on landmarks, compass headings or the presence of local experts. It is important to recognize that drivers vary widely in their ability to use navigation aids, particularly maps.

The results of Parkes and Martell (1990) from a survey of drivers who used a PC based route planning product, provide a clear indication of factors to be taken into account in the design of future route planning and guidance systems. Currently most people find their way on novel journeys with the use of notes taken from maps, and confirm their progress by reference to local landmarks. Few people in younger or older age groups ($< 15\%$ under 25, $< 19\%$ over 55) usually follow a road map whilst driving to a novel destination. Around 70 per cent of users, across the age groups had either a slight or definite preference for an automated text-based route guidance system, over their previous method of route following. When making recommendations for improvements to a demonstration system, over 50 per cent suggested the addition of landmark information and over 13 per cent wanted the inclusion of real-time traffic information. It is important to note that drivers information needs are not homogeneous. Different drivers have varying requirements from a system e.g. a courier requires the quickest route, whereas a holidaymaker may want to follow the most scenic roads. An expert user of a system may require much less detail than a novice user. The needs of an individual driver may also vary over time. For example, whilst travelling on a motorway the driver requires little information, but when entering a city centre, or nearing the final destination, much more detail is required. Similarly, detail is needed to reach the motorway in the first place!

Figure 27.1 tries to indicate how information complexity changes over time for a journey to an unfamiliar destination. The comparison is made between a route planning system, a navigation system and a preferred guidance system. It shows how a PC-based route planning system provides the driver with information for the heart of the journey, i.e. in terms of the roads to be taken and facilities to be visited away from the start and destination areas or towns, but unable to provide the kind of information at a detailed level to aid the driver to reach the major roads in the first place. The route navigation system provides global information to the driver all the time. The driver must pick out the appropriate items at any given time throughout the journey, including finding parking at the destination. The preferred system gives guidance at pertinent points of the journey (e.g., leaving parking slot to reach the right road out of town, at select junctions on the journey to the destination, then entering the destination area and finding a car park within easy distance of the

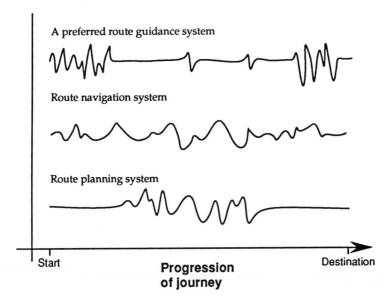

A preferred route guidance system

Route navigation system

Route planning system

Start | **Progression of journey** | Destination

Figure 27.1 Information interchange activity during a journey (after Höök and Karlgren, 1991).

meeting). It limits itself to proving guidance when it is needed, so fitting in to the driving task and allowing the driver to concentrate on controlling the vehicle.

When introducing in-vehicle systems questions arise about the amount of knowledge that can be assumed within the user population and what training is required for efficient and safe use of the system. Training may take the form of a manual or set of tutorials or even after sales tutoring. Whatever training is required should be based on an in-depth analysis of the knowledge of potential users and pre-set safety standards for user performance.

How should the information be presented?

Once the system designer has decided on the information content of a new device, the next step should be a decision on how to present that information to the driver. Various media can potentially be used by in-vehicle systems to give information, each with their relative merits. These include text, symbols, maps and speech. Streeter (1985) concluded that speech output is superior to maps in terms of errors and distraction, and systems should be voice-orientated. However, no comparison was made to alphanumeric or symbolic displays. Her conclusions were based on the premise that voice displays are less visually distracting, but this does not necessarily mean they are less demanding on attention or perception. Indeed, processing of verbal instructions such as 'left' or 'straight on' may be more demanding and time consuming than symbols such as arrows or text instructions. An obvious problem with speech is the possibility that instructions will not be heard in the noisy vehicle environment. If the message is heard correctly, there is still the problem that the transient nature of voice messages might mean that the instruction could be forgotten by the time the desired action is required. Also restriction will be placed on the driver and passenger conversation, and on the use of music and radio systems. Dingus *et al.* (1988) compared a number of navigation methods (memorized route, paper map, electronic navigator) and found that the electronic navigator could be used 'effectively' by the driver. However, visual scan patterns and visual sampling processes were quite different from those used for either paper maps or memorized routes. For example, some tasks such as determining the name of a road made heavy demands on the driver's visual capacities. However, Dingus *et al.* found little evidence for the intrusion of the navigation method on indices of driver performance such as lane exceedences and brake actuations. These authors suggest that any additional time-sharing demands of the navigator drew largely upon spare driver resources.

In a similar fashion Labiale (1989) investigated the effects of different electronic map displays as part of in-vehicle navigation systems. He varied the level of guidance information (map alone, map plus auditory message and map associated with written message) and the complexity of the map display, using memory recall techniques as a primary source of data. It was concluded that drivers preferred simple to complex map displays, that drivers reduced speed when consulting the system, and that the complexity of the display had significant effects on memory recall.

In a laboratory based study, Parkes and Coleman (1990) compared the effectiveness of directional symbols, printed text, and voice simulation as methods of presenting drivers with the directional information required for route guidance. Task completion times, error scores, eye movements, blink rates and subjective measures were the criteria used for evaluation. Voice simulation was clearly considered the easiest method of route guidance to follow, and task completion times were significantly faster in this condition. Directional symbols consistently ranked higher than printed text. Up to 25 per cent of visual scanning time was spent looking at visually presented information, instead of on the primary task (a figure supported in other research e.g. HUSAT, 1990). Although fewer errors overall were made when information was presented by voice, it is worth noting that the number of 'uncorrected' errors increased, i.e. those where the subject made an assumption about the instruction that was incorrect, due to misinterpretation, and went on to make the wrong manoeuvre.

Given the variety of ways information can be presented and the variety of driver information needs and capabilities that exist, it seems clear that systems should be flexible and adaptable to the characteristics of the individual rather than generic.

What are the consequences of interface design on the driver?

The driver has only a limited information processing capacity and is presented with an enormous amount of information at one time whilst driving. It is clear that selection of information is necessary for successful negotiation through the driving environment. Every piece of information competes with every other for the attention of the driver; as Rumar (1988) highlighted, some will win and some will lose. The information selected by the driver is determined largely unconsciously, but is based on his experience, expectation, and motivation. Rumar (1988) suggests that the the most selected information is that which the driver judges to reduce his uncertainty in a situation. Thus the argument suggests the most important information is that which is of relevance to the road. The next level of information relates to the interaction with other road users. The lowest importance would be given to navigation information. In the situation of overload, information is missed in reversed order.

The role of memory in the driving situation is a key consideration. Memory is involved in such fundamental processes as recognition and identification of stimuli. Moreover, its role is of greater significance in particular circumstances: when a stimulus is only available for a limited time; when there is a delay between recognition and recall; or when information load is high and attention is divided with another task. It is important to note that:

- other inputs impair recall making irrelevant material harmful,
- very similar inputs impair recall,
- immediate recall of details of complex images is poor,
- recall is better for text plus symbol than either individually,
- people remember in the short term (< 30 seconds) by scanning back along input.

Sutcliffe (1987) points out that; 'the fact that we are basically sequential machines should be apparent from our poor ability to do two or more mental tasks concurrently'. Research into short term memory (STM) has indicated that in situations where the individual has to store information from many sources, capacity is soon reached. Miller (1956) gave the number of items as seven, plus or minus two, and highlighted the importance of chunking information into logical units for successful recall. More recently Hitch (1987) has suggested that there are two subsystems in STM, one language based, the other visual-spatial. Experimentation into the whole, termed working memory, has shown that distraction causes the forgetting of recently learned material. Even a small number of simple chunks are lost within twenty seconds, if there is distraction during input. These limitations have major implications for the design of route information systems.

Complex displays which require the driver to locate and interpret relational information, i.e. maps, will necessitate continual rechecking of the display to update progress. This is a distraction from the primary task of observing the road ahead, and likely to be inefficient and confusing. Map based information is not amenable to verbal support, and thus an important medium of transfer is ruled out. In short, it is clear that map based navigation systems are difficult (at least for a large subset of the population) to use whilst driving. Anyone who has tried to use an 'A to Z' type road map whilst driving in a strange town, can attest to the innate problems of simultaneously trying to orientate oneself on the map, find roadside place names, and fully attend to the movements of other traffic in the system. Attention is obviously selective, therefore there exists a problem of users switching primary focus between vehicle control and route information device. How does the driver cope with such overload? One means of coping is to predict, further and further ahead, what will occur in the environment. A driver may turn to look at a navigation system if he glances at the road ahead and perceives no risk. This assumption that whilst the driver has looked away from the road ahead, no event will occur to disrupt the safe progression of the vehicle, may of course lead to accidents. A vehicle travelling at 50 kph is covering the ground at a rate of nearly 14 metres per second. Systems which require long glance durations for drivers to assimilate the desired information from a display are dangerous. These systems obviously run the very real risk of distracting the driver from the primary task of monitoring the road ahead, and the likelihood of some event that would require compensatory action occurring and being missed is increased. It is conceivable that drivers might misinterpret the potential benefits of an information system.

Take, for example, a driver who has often experienced a high level of stress whilst attempting to navigate in an unfamiliar environment. This driver might feel that his normal performance in such a situation is so poor that he has a high likelihood of being involved in an accident. Providing such a driver with an information device that takes away a large part of the uncertainty associated with the task, may markedly reduce his stress level, making him feel he is coping more effectively and safely with the traffic system. This perceived increase in safety may be purely illusory, if what in fact happens is that the driver spends long periods looking away from the road ahead, and has an increased potential for being involved in a collision. Indeed Zimmer (1990) argues that: 'In order to achieve safer driving behaviour and thereby less accidents, the information provided should counteract the illusory security in traffic situations, and instead should raise the perceived risk, which most drivers tend to underestimate'.

According to Sabey and Staughton (1980), human factors are a component in 95 per cent of all accidents. Treat *et al.* (1977) highlighted improper lookout, excessive speed and inattention, as the leading causes of accidents. Smiley and Brookhuis (1987) argued that attention and perception errors predominate over simple response errors in accident causation mechanisms. These findings, though not trivial, are not unexpected. They do, however, indicate the potential danger of introducing an element into the total traffic system that demands that the visual attention of the driver is taken away from the roadway ahead, for even the briefest of periods. For the purposes of discussion, we can accept the hierarchical classification (Michon, 1989) of driver behaviour being:

● Strategic   → PLANNING
● Tactical   → MANOEUVRING
● Operational → CONTROL

It is clear that behaviour such as 'improper lookout' or 'inattention' are most highly associated with the tactical level of behaviour, where manoeuvres are enacted. Route guidance devices are clearly designed to help a driver make decisions at the tactical/manoeuvring level of behaviour. They do this by providing the driver with information that will directly influence the control movements of the vehicle. To receive this information a certain amount of attention must be directed away from what is regarded as the primary task, which is immediate safe control and progression of the vehicle. Indeed, there are a number of potentially negative effects of new driving aids, from well understood mechanisms such as distraction or overload, to promoting over-reliance on the system by the driver, or encouraging misuse, or engendering a change in behaviour based on a real or perceived shift in the locus of control within the vehicle.

The way in which information is presented to the driver can directly influence the ability of the driver to work as an efficient element of the total traffic system. A prescription for information systems in vehicles would include these elements:

● The information should be timely,
● The information should be relevant to the current task,
● The information should be easily discerned without error,
● The information should be easily understood,
● The information should match the expectations and experience of the recipient,
● The information should not distract the receiver, nor compete directly for limited attention resources.

If systems are to be effective they must meet certain criteria. The following discussion is intended as an outline of what designers of in-vehicle information systems should attempt to achieve and the issues they should consider. Systems should provide the optimum display for identification, perception and interpretation of information so that the driver is distracted no more than is necessary from the driving task. Any in-vehicle system should require a minimum amount of prior knowledge and its usage should be easily learned. Consider, for example, the complex in-car entertainment systems which possess numerous small buttons and controls and provide a multitude of complex functions. A driver should not have to struggle to operate systems when his attention is obviously needed for the driving task itself.

The psychological capabilities, such as attention and information processing, of the driver can be overloaded by poorly designed displays and inappropriate media. Displays should be easily read and understood, so issues such as glare, flicker, contrast and illumination are important physical considerations. Other issues concern the use of colour, design of graphics, symbols and text, layout of information and the amount of information on the screen at one time. Physiological capabilities are also important to consider in design. Perhaps the most obvious is the use of visual and auditory senses to take in information. The amount of detail presented on a screen or the length of instructions must be tailored to meet human capabilities. The most important information should be the most prevalent. Also special consideration should be given to the elderly as regards visual functions, e.g. greater susceptibility to glare and a longer time for accommodation.

At the heart of the design of a good route information system would be a full consideration of the nature of information transfer itself. Before any decisions are reached concerning hardware, the designer should be able to answer the following task-specific questions:

- which of the possible channels should be used for a specific task?
- what kind of information should be chosen (natural or symbolic)?
- how much information is necessary to guarantee simplicity, robustness, and meaningfulness of the communication?
- is the available information directly related to concrete actions?
- is the amount of support to be given by the device sufficient for the information needs of the driver?

The road ahead

The preceding discussion has outlined the problems associated with introducing a new information system into an already complex multi-task environment. The most important consideration must be the safety effects of any development. Systems that do not take this into account will sooner or later disappear from the marketplace, even if they reach other goals, such as efficiency or environmental protection (Alm, 1990). What becomes clear is that a navigation system on its own is not the preferred avenue for development from an interface design perspective. The visual cost of a purely map-based display becomes too high to bear when driving in unfamiliar road systems, which, it can be argued, are the reasons navigational guidance systems were invented. Map displays may still, however, be of use for stationary route planning activities. What is needed is some form of dynamic guidance system, though the problem still remains that such systems make limited road/traffic-contextual information available.

The problem of context can be overcome by incorporating some form of traffic messaging device, able to either pre-warn of roadside information en-route (such as signposts, crossings or junctions), or to provide road reports and justifications for rerouting the driver in mid-journey. Whilst this offers opportunities for very flexible voice displays, work must still be done to investigate the optimum lengths and timings of voice-displayed guidance. Current systems have been found to raise driver workload due to inappropriate message timing (Ashby *et al.*, 1991). For Europe-wide acceptance of navigational guidance systems, it may be necessary to assess exactly how complex junctions and other items new to the in-vehicle display are to be presented to drivers, and how this should, or should not, vary between countries.

How should designers and legislators proceed? Unfortunately, it is not possible to consult a fully formulated set of design guidelines or standards that are relevant to new information devices, such as navigation systems in vehicles. The research findings are simply not yet available that can validate any performance standards that might be proposed. Does this mean that all development work should be halted until the framework for performance standards exists? Although there have been strong voices in favour of this view (e.g. Zwahlen, 1988), it is more realistic to attempt to support designers in their current tasks. The first stage must be to provide a detailed checklist that can be used in a product-independent manner, and which is useful at both the product specification and evaluation stages of system design. Various initiatives are being taken within both the DRIVE and PROMETHEUS European programmes. A useful starting point has been made by Alm (1990), who has produced a behavioural checklist that details the following design considerations:

- impact on drivers abilities to perceive information outside the car,
- impact on drivers ability to identify important information,
- impact on drivers ability to attend to important information,
- impact on drivers ability to implement judgements into action,
- impact on drivers judgements of ability to drive the car,
- impact on drivers motivation to drive safely,
- changes in the drivers stress level,
- possible system misuse.

There is a growing body of literature on the application of human factors principles to the software and hardware design process. Findings show (Eason, 1989; Harker *et al.*, 1990) that of the nominal stages of design—specification, prototyping, build, evaluation and implementation—it is prototyping and evaluation which typically come under pressure when cuts need to be made, for whatever reason. These are the very stages where real usability trials need to be conducted to ensure acceptable safety performance, but they tend to be expensive to conduct, and the software design community in general seem unconvinced that user-centred design necessarily leads to 'right first time' solutions. However, the development of safety checklists may be the first step towards providing marketing and product managers with an easily used tool, that will help them make design decisions appropriate to the safety goals required. Once a system reaches a certain level of market penetration, it becomes increasingly difficult to make major revisions, either functional, or aesthetic. It is more common to simply offer updates, or 'new improved' versions, than to effect a fundamental revision of the interface principles employed. It is, therefore, of paramount importance for safety and usability criteria to influence design at the product specification stage, rather than be seen as an optional extra at stages after the critical system decisions have already been made.

Designers will also need to recognize the constraints and opportunities presented by other activities in the DRIVE and PROMETHEUS programmes. The heavy demand for dashboard real-estate will push the visual interface for these information systems in two directions. Firstly towards shared multi-function dashboard displays, where, for example, a single screen may be home to a number of functions such as air conditioning, in-car entertainment and route information. This could be organized in either shared time or concurrent overlapping window systems analogous to many office systems products. An alternative would be to move the visual display away from the dashboard. The availability of Head Up Display (HUD) technology is being pushed heavily in DRIVE and PROMETHEUS, and has obvious potential in the areas of vision enhancement, collision avoidance, and co-operation driving. Route information text or symbols can quite readily be presented in a head up position for the driver, with the potential advantage of reducing the need for moving focal vision away from the road ahead, and by focusing the information at infinity, visual accommodation times for the driver can be reduced. Whether these potential advantages are outweighed by other influences on driving behaviour need to be the subject of extensive research.

Another technological advance that will greatly influence the interface design will be the improvement in speech recognition technology in the vehicle. Displays can be simplified by leaving supporting information off unless demanded specifically by the driver. A default display might only include very basic guidance information. Extra information, such as presence of petrol stations, time to destination, distance to destination, local toll road fees etc. may be presented only at the request of the driver, using a natural language vocabulary. Speech recognition could allow detailed interrogation of a system, whilst satisfying handsfree and eyesfree requirements.

Information presented in HUD format should, however, be strictly limited to avoid visual clutter, and in particular distraction from, or masking of, important aspects of the road scene. It is likely that future systems will require a limit to the time of presentation for certain types of information. Text messages that are neither part of alarms or warnings will 'time out'. However, it is also likely that there will need to be complementary displays in the head down (dashboard) position that will maintain the message beyond the duration of the HUD to allow driver confirmation and consultation if needed later. This may be particularly important if the driver requested information that provides choices (e.g. options for park and ride facilities) that need consideration before a response is made.

This paper has shown that route information systems are entering the market rapidly, and already considerable thought has been devoted to MMI issues. The generation of systems currently being advanced in DRIVE and PROMETHEUS will be more sophisticated and incorporate many ergonomic and human factors recommendations. The systems to be used in vehicles by the year 2000 will need to be yet more sophisticated and take even greater account of the presence of other competing and complementary systems within the vehicle, and the variety in requirements by the user population. Route information devices will merely form a subset of driver information systems, and the problems facing the MMI designer will not merely be those surrounding the choice of 'visual or auditory' guidance information, but how to develop fully integrated MMI solutions for the whole driving task; which occasionally has the component of route finding and following as important elements.

References

Alm, H., 1990, *Behavioural Checklist*, Swedish Road and Traffic Research Institute, Ref:441.

Ashby, M. C., Fairclough, S. H. and Parkes, A. M., 1991, *A comparison of two route information systems in an urban environment*, DRIVE V1017 Report 49, HUSAT Research Institute.

Ashby, M. C., Parkes, A. M., Fairclough, S. H. and Lorenz, K., 1991, A comparison of route navigation and route guidance systems in an urban environment, *Proceedings of 24th ISATA Conference*, Florence, Italy.

Department of Transport, 1989, *Road Accidents Great Britain 1988: The Casualty Report*, Transport Statistics Publications.

Dingus, T. A., Antin, J. F., Hulse, M. C. and Wierwille, W. W., 1988, Human factors issues associated with in-car navigation system usage, *Proceedings of the Human Factors Society*, 32nd Annual Meeting, 1988.

DRIVE, 1989, *The DRIVE Programme in 1989*, Doc X111/F/DR0789 Brussels.

Hitch, G. J., 1987, Working memory, in Christie, B. and Gardiner, M. M. (Eds) *Applying Cognitive Psychology to User Interface Design*, New York: John Wiley.

Höök, K. and Karlgren, J., 1991, *Some principles for route descriptions derived from human advisers*, SICS Research Report R91:06, Swedish Institute of Computer Sciences, Kista, Sweden.

Jeffrey, D. J., 1986, *The potential benefits of route guidance*, TRRL Report 997, Crowthorne, England.

Labiale, G., 1989, *Influence of in Car Navigation Map Displays on Driver Performance*, SAE Technical Paper 891683, Future Transportation Technology Conference and Exposition.

McCormick, E. J. and Sanders, M. S., 1983, *Human Factors in Engineering and Design*, New York: McGraw-Hill.

Miller, G. A., 1956, The magical number seven, plus or minus two: some limits on our capacity for processing information, *Psychological Review*, **63**, 81–97.

Michon, J. A., 1989, Explanatory pitfalls and rule-based driver models, *Accident Analysis and Prevention*, **21**, 341–53.

Parkes, A. and Coleman, N., 1990, Route guidance systems: a comparison of methods of presenting directional information to the driver, in Lovesey, E. J. (Ed.), *Contemporary Ergonomics; Proceedings of the Ergonomics Society 1990 Annual Conference*, London: Taylor & Francis.

Parkes, A. and Martell, A., 1990, A usability analysis of a PC based route planning system, in Lovesey, E. J. (Ed.), *Contemporary Ergonomics; Proceedings of the Ergonomics Society 1990 Annual Conference*, London: Taylor & Francis.

Rumar, K., 1988, In-vehicle information systems, *International Journal of Vehicle Design*, vol. 9, nos 4/5.

Sabey, B. E. and Staughton, G. C., 1980, *The Drinking Road User in Great Britain*, TRRL Supplementary Report No. 616, Crowthorne, England.

Schraagen, J. M. C., 1993, Information presentation in in-car navigation systems, in Parkes, A. M. and Franzén, S. (Eds) *Driving Future Vehicles*, London: Taylor and Francis.

Smiley, A. and Brookhuis, K. A., 1987, Drugs and traffic safety, in Rothengatter, J. A. and de Bruin, R. A. (Eds) *Road Users and Traffic Safety*, Van Gorcum: Assen.

Southall, D. and Twiss, M. K., 1988, Human factors considerations in the design of in-car navigation systems, *Proceedings of 18th ISATA Conference*, Florence, Italy.

Streeter, L. A., Vitello, D. and Wonsiewicz, S. A., 1985, How to tell people where to go: comparing navigational aids, *International Journal of Man-Machine Studies*, **22**, 549–62.

Sutcliffe, A. G., 1988, *Human-Computer Interface Design*, Basingstoke: Macmillan Education Ltd.

Zimmer, A., 1990, *Driver Information Needs*, DRIVE Project V1017 (BERTIE), Report No. 25.

Zwahlen, H. T., Adams, C. C. and Schwartz, P. J., 1988, Safety aspects of cellular telephones in automobiles, *Proceedings of 18th ISATA Conference*, Florence, Italy.

28

Effective design: ensuring human factors in design procedures

I. Howarth

Introduction

In the past twenty years or so, our understanding of the nature of the design process has been much improved. This understanding makes design less dependent upon the inspiration of individuals and has enabled many industries to set up very effective design teams. These insights, which are now used so extensively, have come from many different, but complementary sources. Some come from work on the psychology of design and problem solving (Simon, 1969, 1985; Jones, 1972; Howarth, 1988). Some come from work which is more specifically focused on developing computer systems (Keen, 1981; Kidder, 1981); or computer programmes (Booch, 1990); or human computer interfaces (Hartson, 1985; Norman and Draper, 1986); or industrial production systems (Riggs, 1981); or national economic systems (Beer, 1981); or transport systems (Salter, 1976; Mannheim, 1979; Howarth, 1990); or on the use of predictive models (Sivak, 1987; Roberts *et al.*, 1983). From this extensive literature, we can extract common factors which are now used, prescriptively, to guide the activities of designers and design teams.

Elements of the design process

The following stages and elements are common to most descriptions of the design process. They will also be found in most descriptions of problem solving and planning.

Stage 1. The definition of the problem

This will involve the following two elements.

(a) An analysis of the structure of the system in which the problem exists. This may be called a task analysis, or a systems analysis, or any one of several other names. For any complex system the structure will have an underlying hierarchical and modular structure which must be clearly revealed by the analysis.

(b) A clarification of, and agreement about, the goals and purposes which the system is failing to achieve. This will include a comparison of the relative importance of different goals, such as safety, traffic flow, cost and convenience. Herbert Simon has proved that it is impossible to optimize on all such goals, so we should follow his advice and set 'good enough' targets for them. Some goals are hierarchically related. For example, cost savings may enable us to implement a larger system and so improve safety. Alternatively, cost savings may also result in less reliable components and so reduce safety. At this stage of the analysis, we should also be able to identify impediments to the implementation, functioning and evaluation of a new system.

These two elements, between them, define the nature of the problem to be solved.

Stage 2. The conceptual design stage

This will also involve two rather different elements.

(c) An analysis of the consequences of changing the structure of the system in various ways. This will involve modelling the working of the system and using it to predict the effects of making changes to it. Initially the model will be no more than a description, in words of what is likely to happen. But the more precise and quantitative the model becomes, the more useful it will be. The model will usually be validated, initially, against the functioning of the existing systems under various conditions. It is at this stage that creative ideas develop, and are given their first, conceptual, evaluation.

(d) The development of performance indicators to guide the design process and to facilitate self-monitoring and evaluation. An important function of modelling is to identify key stages in the working of the system. These will obviously include the final output of the system, but should also include intermediate stages, which critically determine what happens at later stages. Measures should be devised which assess the performance of the system at these key stages. Such 'performance indicators' should measure the extent to which the desired goals are achieved, and should be built into the new design in such a way that the new system becomes self-regulating. The speed and accuracy with which performance indicators control the working of the system is often the chief determiner of the success of a new design.

The modelling techniques, which are used in elements (c) and (d), may involve no more than the careful use of ordinary language, or they may use mathematical equations, or computer simulations of the working of the system. In whatever form the model is constructed, it will be used to facilitate and clarify discussions about the problem and about possible design solutions. At this stage, many ideas can be tried relatively quickly, and the most likely ones kept for more rigorous development and evaluation. It is very important to delay creative thinking until the problem has been adequately clarified in elements (a) and (b), and until a way of thinking about the problem quickly and accurately has been developed at elements (c) and (d). An enormous amount of effort is otherwise wasted in solving non-existent problems, or in the detailed development of designs which, with clearer thinking, would have been eliminated very quickly.

Stage 3. The concrete design stage

There are also two elements in this stage.

(e) A survey of existing resources and existing designs. There is obviously no point in reinventing the wheel. It is equally shortsighted to ignore the existence of 'user friendly' information systems, which can be adapted to use in road transport informatics.

(f) The design of an implementable system which has a realistic chance of solving our problem. The clear thinking which is facilitated by stages 1 and 2 will itself facilitate detailed design work. In addition to the conceptual work in steps (a) to (d), designers must make use of existing techniques when possible, and commission the development of new technologies only as a last resort.

Stage 4. Preliminary evaluations

Careful evaluations should be carried out on the new proposals throughout the design process. These evaluations tend to become progressively more realistic and expensive as the design becomes more detailed and more settled, but at all stages they will be guided by the objectives determined in stages 1 and 2 above.

Four fairly distinct elements can usually be detected in the evaluation stage.

(g) Conceptual evaluation based on the ideas developed in elements (c) and (d). Computer simulations or mathematical modelling are commonly used at this stage.

(h) Demonstrator systems may then be produced which resemble the final product, but which are not yet incorporated into a working system. These are usually evaluated in a largely subjective way.

(i) Field trials can then be developed in which a pilot version of the design is tried in a part of the total system into which it will eventually be placed. At this point the system may be worked realistically and objectively evaluated, but the results may not be fully representative of the way the new design will work when implemented as a total system. However, if the conceptual modelling has been sufficiently realistic, it will predict which types of field trial are likely to be representative of the working of the complete system.

Stage 5. Implementation and evaluations of the system in action

Final evaluations have to be delayed until the total system is implemented, so it is essential that the possibility of evaluation and of consequent changes should be part of the overall design specification. Far too often, designs are difficult to modify and this leads to a reluctance to evaluate. If, as a result, the system fails to meet its purposes at this late stage, the consequences can be disastrous for all involved. The disaster can be minimized if failure can be detected quickly and corrected efficiently.

If Stages (1), (2), (3) and (4) of the design process have been done properly, failure should be unlikely, but any expensive project should have the possibility of damage limitation if the early stages of the design process have failed to predict exactly the performance of the system, as it is finally implemented. For this reason, the design specification should include the requirement that the system should be easily modified to correct any weaknesses in performance and to accommodate any changes in circumstances and demand.

The characteristics of self-organizing systems are beginning to be relatively well understood. As Simon (1985) has indicated, a modular structure greatly facilitates the modification of a system because it enables it to make changes in one or a small number of modules, which, because they are relatively simple, are easy to modify and to evaluate, and which, because of the modular structure, can be modified without affecting the functioning of other parts of the system. Von Neuman (1966) showed that self-organization, self-repair and self-reproduction, all require a high degree of complexity in the system. There is a 'complexity barrier' below which self-organization is impossible. At the moment, it is usually uneconomic to try to automate these processes. The complexity needed can more easily be provided by people in the system. But even so, their task can be enormously simplified by an appropriately modular design. With a hierarchically organized system of feedback and evaluation, errors can be detected early and quickly attributed to one part of the system. These is a large literature on fault finding, which demonstrates how difficult this can be in badly designed systems, but which also shows how good designs can facilitate fault detection and correction.

Using standards and objectives in the design process

All complex designs should have a modular structure, each element of which can itself be a major design problem. The parts must then be integrated into a working whole. This integration can be facilitated by the adoption of common objectives and standards. Objectives are determined during the analyses of the problem at Stages 1 and 2 of the design process. Standards are derived from these objectives. They act as performance indicators and as such can be used in the evaluations of Stages 4 and 5. However they can also be used as guides in Stage 3, the concrete design stage. Standards are of three kinds:

1. *Product standards* require an agreement (or a directive) to use identical and compatible products in all parts of the system. For example, a common voltage may be adopted, or compatible hardware systems. These are familiar and obvious examples. Slightly less familiar is the need to help human users of information technology by adopting familiar and consistent information and control systems. These latter standards are mostly derived from research on human factors and ergonomics.
2. *Performance standards.* These specify the required performance of elements in design to ensure that the disparate elements will work effectively together. They should also specify the required performance of the component or total system in order to achieve its design objectives. Performance standards are perhaps the most important kind of performance indicators, since they must be chosen to measure the most essential aspects of the functioning of the new design.
3. *Procedural standards* specify the appropriate sequences of the different procedures which make up the design process. Stages 1, 2, 3, 4, and 5 in the design process are themselves one kind of procedural standard.

Figure 28.1 shows a related procedural standard which specifies, by means of a flow diagram, the sequence of activities involved in efficient design, and the way product and performance standards can be used to set objectives at the appropriate stage in the sequence. It is, of course, a great simplification of the sequence of activities involved in the design process. The cycle, assessment, design, prototype, assessment, may be traversed several times to correct initial misconceptions. It will also be traversed fairly independently, in relation to each of the elements in a modular design, before the cycle is finally completed in relation to the total system. Not surprisingly the design process also has a modular structure.

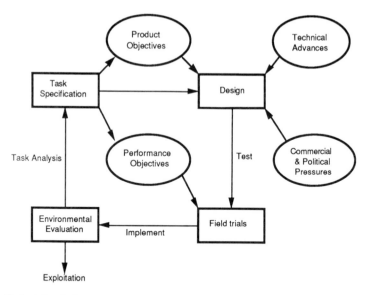

Figure 28.1   The design cycle (after Jones, 1970). This diagram shows that Stages 1 to 5 of the design process are not related in a simple linear manner, but that they are part of a continuing cycle. When the system is implemented it changes the environment in which it is placed, so that the original analyses on which the design was based may be invalidated by the process of implementing it. For this reason, the design process is never finished, but must be brought to a temporary halt when a good enough solution of a problem has been achieved. The diagram also shows how product and performance objectives enter into the design process.

Human factors standards

Standards and objectives are usually applied to the physical and engineering aspects of design. As a result, many new systems work well at a mechanical level, but cause unnecessary difficulties for their human users. The reason for this seems to be that, in relation to relatively simple technologies, one could rely on human flexibility and intelligence to cope with any of the alternative designs being considered. Hence the human factor could safely be ignored in making decisions about the final form of the design. As technologies have become more complex, the demands they make on human intelligence has become more critical, so that we can no longer safely ignore the human factor during the design process.

One reaction to the increasing complexity of modern technology has been to insist on ever increasing ability in the human operators of the technology. This can be achieved by ever more stringent selection and by ever more complex training programmes. It should be recognized that neither of these would be necessary if equipment and methods of working were designed to be more 'user friendly'. Accidents blamed on 'human error' are also a consequence of poor design, in the sense that good design should make human error less likely. Similarly, the increasing difficulty in finding employment for the less intelligent and less well educated segments of our population can also be seen as a failure of the design process to take full account of the human factor.

How can we make sure that these common errors of design are less common in future? The most obvious method is to insist that the human factor should be taken into account at all stages in the design process. This can be facilitated by setting 'human factors standards' as well as engineering standards. We have already referred to the use of ergonomics in setting objectives of ease of understanding and ease of operation when deciding on product standards. There is a readily accessible literature on this, but it is still ignored more often than it should be (Boff and Lincoln, 1988; Card *et al.*, 1983). But there is a limit to what can be achieved by taking the human factor into account when setting product standards. Such standards make it more likely, but do not ensure, that the final system will work efficiently when operated by a variety of human beings. The adequacy of the end product can be made much more likely by setting human performance standards at all stages in the design cycle. These have the same function as those physical performance standards which specify such things as the speed, reliability, accuracy and robustness of the physical components in the design in any system which involves humans.

It is foolish to specify the accuracy of the physical components, while ignoring the accuracy with which people can use those components. The latter will be determined more critically by the accuracy with which people can read the dials or other indicators, and by the accuracy with which they can operate the controls. We commonly measure the variability in the performance of a machine on a test bed, but do not measure variability in the performance of the machine when it is being controlled by a person. It is foolish to assume that the former will be a good indicator of the reliability of the machine in operation. If it is so difficult to operate that it is tiring to use then variability will be more critically determined by variability in the behaviour of the operator. It is useless to set a standard which requires a machine to have a response time of less than a millisecond, if the people operating it then impose a delay of several seconds, while they figure out what to do. It is much better to give the machine ten or even one hundred milliseconds to provide the humans with a signal to which they can respond reliably within half a second.

This type of problem can be overcome by setting performance standards for the system including the human operator. These can be called human factors performance standards. They can be applied to elements of the system in a modular design and usually provide the best predictors of the adequacy of the performance of the total system. A common error is to assume that such human factors performance standards should only be applied to the total system when it is implemented. As a result, corrections for human factor design errors are often detected too late and are very expensive to correct. The reason why this error is so common seems to be because designers tend to have engineering training and are unaware of the regularities and predictability of human performance, or of the ease with which the human factor can be accommodated at all stages of the design cycle. It is becoming increasingly obvious that human factors experts should be involved in all stages of the design cycle (Howarth, 1988).

Examples of successful design

Unfortunately there are no examples of the successful application of these design principles to transport. It is far easier to give examples of bad design. Perhaps the simplest of all these is the positioning of the horn on most cars. This ensures that we can sound the horn reliably in order to encourage our spouse to stop chatting and get in the car or to complain about the behaviour of another driver, but in an emergency, when the horn should be used for its principal purpose we are more likely to spray the windscreen with water. This is because human factors standards were not set or used in evaluation during the design of the cars controls. An appropriate performance standard would have required the designer to check in a simulator, whether the proposed position of the horn enabled a representative sample of drivers to take appropriate action in a simulated emergency. Even easier would be to set a product standard, since it is well known to psychologists that under stress people forget newly learned skills and regress to older patterns of behaviour. Thus, in an emergency, we are likely to try to sound the horn by taking the action which was appropriate in an earlier car, perhaps the one in which we learned to drive, or the one which we kept for a long time and in which we became very well practised. Hence the appropriate product standard should forbid any variation in the position of the horn or of the action which is required to sound it. It would not matter very much which position or action were chosen so long as it could be standardized.

There is nothing very revolutionary about this idea since it has been applied successfully for more than seventy years to the positions of the clutch, brake and accelerator pedals, and to the actions required to operate them. What is extraordinary is that this application of a psychological principle (or of common sense, since the principle is so obvious) is ignored by almost all designers of the controls of new cars. It would not, therefore, be too surprising if, in the design of new types of 'Transport Informatics', the same type of error were made.

When the human factor is taken into account during the design of advanced technology, the benefits, both human and financial, can be very great. There is, at the time of writing, a very effective advertisement for Apple Macintosh computers in which it is claimed that 'the most cost effective personal computers are those which people like to use'. This paper is being typed on an Apple Macintosh and I can vouch for the validity of this claim. When I bought this machine after trying without success to work with several other makes, I looked at the manual for about five minutes, started to play with the machine, and have never returned to the manual. The machine is so easy to use and such a pleasure to use that I have never needed to look at the manual again. It was, therefore, no surprise to discover that the human factor was considered from the very earliest stage in the design of the Macintosh and of its most popular software. The psychological expertise of the design team was much greater than in the teams which designed competing systems and machines. Moreover, there was much more careful evaluation of human factors performance standards during development, than had been the case for any previous system. The growth of the Apple Computer Corporation is one vindication of this approach. In relation to transport systems, proper attention to the human factor could also be commercially successful, but even more importantly, it could save many lives.

Safety implications of the human factor

If improved technology is to improve safety as well as efficiency, the design problem becomes more complex and challenging. We cannot predict improved safety directly from an increase in efficiency, because people adapt to some kinds of improved efficiency by taking greater risks. This has been called 'behavioural adaptation' by Evans (1990), but the phenomenon has also been called 'risk compensation' and 'risk homeostasis'. Evans' terminology is preferable because the adaptation may be to features of the situation other than risk, while homeostasis implies near perfect adaptation or compensation. We now know that behavioural adaptation can sometimes decrease risk rather than increase it, i.e. that the feedback may be positive as often as it is negative. The designers task is to identify those features of the situation which lead to adaptations which decrease rather than increase risks.

Howarth (1989) and Evans (1991) have summarized what is known about the factors which affect the direction of the behavioural adaptation. Risk is reduced by safety measures which are not readily perceived as reducing risk, and by measures which give the driver clear and timely warnings about situations and actions which increase risk. Measures which appear to reduce risk more than they do in fact, or which provide danger signals which are not precisely related to the real risks, may, because of the inappropriate behavioural adaptation which they induce, actually increase the danger of an accident rather than reduce it. An example of the latter is the use of speed warning signs which are so excessively cautious that they are ignored and lead motorists to ignore warning signs which may be real indications of danger.

For the designer, behavioural adaptation presents an unusually difficult problem, because the adaptive behaviour may not be revealed in preliminary evaluations in simulators or field trials. The best that can be done is to make use of everything we know about behavioural adaptation, attempt the most realistic of trials during the design cycle, but, most importantly, we must make it as easy as possible to modify the final design in response to the behaviour of drivers who are using it as part of their normal driving experience. Modular structure, and the provision for easy modification of key parameters in the system, are probably the most important design requirements for this purpose.

References

Beer, S., 1972, *Brain of the Firm*, Chichester: John Wiley.

Boff, K. R. and Lincoln, J. E., 1988, *Engineering data compendium: human perception and performance, Vols 1, 2 and 3*, Aerospace Medical Research Laboratory, Wright Patterson Air Force Base, Ohio.

Booch, G., 1991, *Object Oriented Design with applications*, Redwood City, CA: Benjamin/Cummings.

Card, S. K., Moran, T. P. and Newell, A., 1983, *The Psychology of Human-Computer Interaction*, Hillsdale NJ: Lawrence Erlbaum.

Evans, L., 1991, *Traffic Safety and the Driver*, New York: Van Nostrand Reinhard.

Hartson, H. (Ed.), 1985, *Advances in Human-Computer Interaction Vol. 1*, Norwood NJ: Ablex.

Howarth, I., 1988, 'Psychology and information technology', in Blackler, F. and Oborn, D. (Eds), *Information Technology and People*, Leicester: British Psychological Society.

Howarth, C. I., 1989, Perceived risk and behavioural feedback, *Work and Stress*, **1**, 61–65.

Howarth, I., 1990, Making residential areas safer for children', in Trench, S. and Oc, T. *Current Issues in Planning*, Aldershot: Gower Press.

Jones, J. C., 1970, *Design Methods*, Chichester: John Wiley.

Keen, J., 1981, *Managing Systems Development*, Chichester: John Wiley.

Kidder, T., 1981, *The Soul of a New Machine*, Harmondsworth: Penguin.

Mannheim, M. L., 1979, *Fundamentals of Transportation Systems Analysis Vol. 1*, Cambridge MA: MIT Press.

Norman, D. A. and Draper, S. W., 1986, *User Centered System Design*, Hillsdale, NJ: Lawrence Erlbaum.

Riggs, J. L., 1976, *Production Systems: Planning, Analysis and Control*, Chichester: John Wiley.

Roberts, N., Anderson, D., Deal, R., Garet, M. and Shaffer, W., 1983, *Introduction to Computer Simulation*, Reading MA: Addison-Wesley.

Salter, R. J., 1974, *Highway Traffic Analysis and Design*, London: Macmillan.

Simon, H. A., 1969, 2nd edition 1985, *The sciences of the artificial*, Cambridge MA: MIT Press.

Sivak, M., 1987, A 1975 forecast of the 1985 traffic safety situation: what did we learn from an inaccurate forecast? in Rothengatter, J. A. and de Bruin, R. A. (Eds) *Road Users and Traffic Safety*, Assen/Maastricht: Van Gorcum.

Von Neuman, J., 1966, *Theory of Self reproducing Automata*, Urbana and Chicago: University of Illinois Press.

29

Adaptive interfaces and support system in future vehicles

E. H. Piersma

Introduction

Drivers are not all the same, neither are the situations in which they find themselves from one moment to the next, nor are drivers the same as they were last year—they have learned things on the one hand, but may have lost some abilities on the other. The question is, then, under what circumstances will support systems in cars have to be adaptive and to what degree. As this involves evaluating possible support systems, criteria for evaluation are required. Three come to mind: pollution reduction, traffic system efficiency and safety. Safety will be the focus of the present discussion.

To increase traffic safety with regard to vehicles there appear to be multiple options that correspond to separate components of the driver-car-traffic environment regarded as a system. This paper will focus on the driver-car side of the system. It will be argued that enhanced safety must not be expected from improving cars beyond current state of the art, but by improving driver performance. As long as the driver is in control—and for various obvious reasons this will be the case for several decades—the driver's ability to perceive and anticipate the driving situation, to decide upon and to implement the required actions, and the development of the driver skills will determine traffic safety. The aim to improve safety by improving driver performance may be achievable by providing the driver with the necessary information by adding 'support' systems to the driver-car system. There are various ways of reducing the difference between required and actual performance, including, for example, tutoring, anticipatory warnings, warnings of immediate hazards, control support, interventive support, enhanced perception.

Each support function producing a message or display, however, adds a subtask to the driving task. Without care this will lead to reduced safety, because the driver will have to attend to many types of messages, many of which will appear relevant or attract attention to themselves as the traffic

.tuation gets more critical. Therefore, support should be made dependent on the driver's *current* limitations and abilities with regard to processing information and performing the driving task. Systems able to accomplish this will have to incorporate various driver or user models: a general model of the driving task for producing objective safety based support; a model of the driver's processing limitations, to adapt format and timing of messages to; a model of the driver's driving style and deficiencies to allow for effective personal tutoring; and a model of the driver's preferences regarding types of support and types of presentation of information. Each type of model can be devised at various levels of sophistication and be more or less adaptive with regard to various aspects: the driver's abilities, state, preferences, driving history, current driving subtasks and secondary tasks. The integration and adaptivity of various types of support were the focus of the GIDS project (Drive V1041, Generic Intelligent Driver Support Systems) and are the focus of the ARIADNE (Drive V2004) project. This paper will discuss the design of the GIDS I prototype system developed in the GIDS project and its likely extensions in the future as an example adaptive interface and support system, along the dimensions identified.

Improving cars? (nutshell psychology)

To evaluate the effects of changing cars on safety, that is also determined by the people driving them, a little understanding of the driver's psychology is needed. A simple psychological theory considers the human being as a difference engine (Newell and Ernst, 1965): if the current situation differs from the desired state of affairs the agent takes action to reduce the difference. A plausible extension allows the anticipation of situations that the current situation may develop into, and the determination whether they will differ from the desired state of affairs. This conception suffices for present purposes: imagine a driver as maintaining a set of invariants, propositions that are to hold for the current situation, for example 'traffic laws not violated', 'safe distance to lead vehicle', 'well within lane'. If the driver detects, or anticipates, a violation of one of these invariants, he will take action to restore the invariant affected, for example reduce speed to increase distance to lead vehicle, or, in the case of anticipation, take action to prevent violating a constraint. It seems safe to assume that the driver has some ability to recognize (not necessarily consciously) deviations from the desired driving performance, and an ability of limited anticipation for situations in the near future that the current situation might develop into. (e.g. for lane keeping; Godthelp, 1984). And finally, a human being is a learning system: most obviously, novel situations and recognized errors provide opportunities to learn; less obviously, performance tends to keep changing, as a result of practising available behaviour patterns.

The determinants of a driver's performance are therefore the quality of perceiving the current situation and anticipating situations it might develop into, the quality of detecting deviations from the acceptable ways of driving, the ability to select appropriate actions for coping with the current situation and, finally, the ability to successfully implement the selected actions. The driver's future performance depends on the ability to evaluate current performance to translate perceived errors into different behaviour in the future, and the development of habits through practice. Having discussed the driver, it can be judged whether cars can be changed to improve driving safety in the light of the theoretical considerations we just made. There are two obvious reasons why it is hardly, if at all, feasible to improve on the safety of cars with respect to driving performance, because so called improvements either will allow more dangerous behaviour and invite the driver to engage in that behaviour, or make one or more of the processes of perception, anticipation, decision, action and/or evaluation more complex. Two illustrative examples will show this.

A friend of mine had loaded the front baggage compartment of his Volkswagen Beetle with 90 kg of stones. This allowed him, as he boasted, to negotiate 90 degree curves and sideroads at 90 km/h without skidding. The obvious problem, however, is that most 90 degree curves cannot be sufficiently inspected at 90 km/h before entering them, and any obstacle or road user around the corner may be easily crashed into (my friend was lucky). Any driver's reaction would be too slow. In general, cars that allow taking curves at high speeds should not be expected to increase traffic safety.

Another example concerns ABS (anti-blocking system) on the car brakes. A possible explanation of the fact that ABS is most likely ineffective for enhancing traffic safety (Biehl *et al.*, 1987) can be easily formulated. In terms of the theory discussed both anticipation, decision and action are affected: anticipating the car's behaviour is more complex; deciding required actions is more difficult because the system will support alternative courses of action (e.g. reducing speed early, brake forcefully later); and some action is affected, e.g. blocking the wheels (as required in the situation of absent friction where the safe course is straight on) is impossible without switching the contact off (as taught in slip courses in The Netherlands). This is supported by Biehl's finding that 10 per cent of taxi-drivers interviewed thought ABS justified smaller safety margins and higher speeds in curves. An alternative theoretical approach can be found in Wilde's risk homeostasis theory (Wilde, 1986), discussion of which is beyond the scope of this paper.

One important point has now been made: a safe car is not a car with superior performance to today's (1992) ordinary cars, but a car the behaviour of which can be easily predicted by the driver under most, if not all, circumstances. Given reliable predictable cars of today, one should expect the already amazing performance of current car drivers. If changing car performance will not enhance traffic safety, what might? If our aim is to improve safety, one must focus on improving the driver's performance.

Therefore, the next section will consider ways of improving driver performance by adding support systems to the car, without changing the car behaviour.

Improving drivers! (support options)

If changing cars is unlikely to improve driver's safe performance, changing aspects of the driver's behaviour must be considered. The following determinants of driver performance were identified in the previous section: perception, anticipation, decision, action and evaluation. The usual analysis of the driving task distinguishes three hierarchical levels: navigation (strategic level), manoeuvring (tactical level) and control (operational level). Table 29.1

Table 29.1   Examples of each process for each task level.

Navigation
> Perception
>> Where am I?
> Anticipation
>> What are the next landmark and decision points to be encountered
> Decision
>> Take the rural way to avoid traffic jam announced on radio
> Action
>> Turn right at intersection (implemented by next level)
> Evaluation
>> Late for work; take other route next time

Manoeuvring
> Perception
>> Distance and relative speed of lead vehicle
> Anticipation
>> Lead vehicle expected to slow down for curve
> Decision
>> Overtake lead vehicle now
> Action
>> Take overtaking course (implemented by next level)
> Evaluation
>> Aborted manoeuvre because curve ahead; check before overtake next time.

Control
> Perception
>> Speed and course relative to desired course by manoeuvre
> Anticipation
>> Move foot to brake as gas pedal is released
> Decision
>> Brake now
> Action
>> Press brake pedal
> Evaluation
>> Too big, unpleasant deceleration; brake less but earlier next time

considers the outerproduct of these levels and the theoretical processes proposed, and provides an example for each cell, to which support functions can be added. Perception can be supported in principle by adding sensors. The problem is, however, that information from sensors needs to be passed on to the driver, by having the driver perceive this information in some form, absorbing some capacity to perceive. Anticipation can be supported by warning the driver in advance. Decisions can be supported by telling the driver what to do next. Actions can be supported by having the system take over, or by suggesting actions using feedback on the controls. Evaluation can be supported by providing the driver with appropriate feedback about current performance or by providing the evaluation itself.

The meaning of adaptive

A useful definition of adaptive appears to be the following: a system is adaptive if its performance is both sensitive to its environment and changed in ways to improve the quality (on average) of the system's performance. A system is adaptive if it changes its behaviour mostly for the better dependent on the momentary circumstances.

Driver support can be adaptive in various ways. First, and apparently trivially (but see the discussion on context sensitivity p. 330), it can adapt to the driver's preferences regarding types of support and the modalities used for presentation, e.g. switch lane keeping support on/off and present as a pulse on the steering wheel or alternatively by a spoken warning. Second, driver support can adapt to the secondary tasks (non-driving tasks) the driver is engaged in, e.g. while on the phone, spoken messages are presented visually as text or, if possible, are presented only after the end of the conversation. Third, systems can adapt to the currently available information processing capacity or to workload (its complement), for example if the driver is approaching a complex intersection, no visual information is presented to prevent distracting the driver and overloading the available visual processing resources, or the rate of presentation of some information is self-paced. Fourth, driver support can adapt to the traffic situation, e.g. the telephone never rings just before and during driving across an intersection. Fifth, driver support can adapt to the driving tasks that are currently being performed, e.g. no navigation instruction during overtaking. Sixth, systems can adapt to the history of the individual driver, either with regard to the environment or specific driving tasks. For example if the driver is in an unfamiliar area, navigation messages may need to be adapted and workload should be expected to be higher; if a driver has not encountered roundabouts for years since obtaining a driver's license, anticipatory warning and explanation may increase the likelihood of correct behaviour. It should be noted that these are not independent aspects, e.g. workload is most likely estimated best, considering both individual history with respect to the specific area, driving task and current secondary tasks (if known).

Design decisions for GIDS I prototype

This section will discuss some issues in driving support system design and the options taken in the design of the GIDS I prototype (Piersma, 1990).

The first issue concerns task allocation. In principle different and potentially varying allocations of tasks between driver and system are possible. In GIDS I the following fixed allocation of tasks was chosen: navigation is performed by the system; manoeuvring and control by the driver, for multiple reasons. First, adequate manoeuvring depends on detailed perception of all relevant objects in the vicinity of the car. Though part of this can be achieved using radar sensors and the like, only a traffic system where each relevant static (e.g. a traffic light) and dynamic (e.g. vehicle, pedestrian) transmits its position and state to other vehicles accurately could support this completely. If ever, it may take decades to achieve this situation, and still then, people may drive from these high infrastructure areas to low infrastructure areas, which takes us to the second reason. People should not become dependent on the presence of driving support as it would disable them to drive successfully in unsupported areas. Therefore, the driver should be in full charge and fully responsible for driving and only non-interventive support should be used to attempt improving performance. In practice this means that well performing drivers will only receive navigation instructions, until they make an error with regard to maintaining appropriate safety margins (see p. 32).

The second issue concerns the way the support system addresses the driver. The GIDS concept was born from the horrification of the idea of future vehicles potentially being equipped with various support systems and applications with separate displays, warning lights and sounds etc. that might attract attention to themselves at times unrelated to the driver's state or each other. Currently available examples are detailed maps on monitors absorbing visual processing capacity; cars informing the driver that oil needs to be refreshed during the next 3000 km (disregarding, for example, whether the driver is busy aborting an overtaking manoeuvre to avoid a collision); hand-held car telephones; and dash-board mounted faxes. GIDS was proposed to integrate applications by 'integrating, prioritizing, filtering and presenting' information provided by applications (Smiley and Michon, 1989).

Imagine a front end sensor detecting a lead vehicle and telling the driver to slow down while a rear end sensor would suggest to speed up to avoid collision with a tailgater. This will be further clarified when the driving model is described (see pp. 327–328). A side issue, concerning dynamic task allocation, is the problem that drivers can only be expected to behave correctly if the allocation of tasks between driver and system is clear to them at all times. A positive exception is cruise control where the driver can detect the state directly from the position of a limb: the foot is not on the gas pedal.

The modalities used for presenting information to the driver were highly constrained during design, simplifying the presentation problem somewhat, as the system need not determine presentation modalities dynamically. People tend to process information quicker the more familiar the format and the more predictable the timing of the message (Verwey, 1991a). This suggests making only limited, if any, use of dynamically alternating between modes of presentation. Driving being a primarily visual-motor task, the auditory-verbal modalities seem most apt for interaction with the system. Therefore, voice command was included for commanding the system as an alternative to buttons, and spoken messages were considered as the primary means of addressing the driver. For control support, however, reaction times should be minimal and research (e.g. Janssen and Nilsson, 1990) has shown tactile feedback on the gas pedal to be free of undesired compensation effects caused by other forms of warning (e.g. excessively large headways or driving in the left lane of a motorway just to prevent warnings from ever happening). Therefore, active controls are used: a gas pedal force to suggest speed changes by increasing or diminishing counter force; a pulse on the steering wheel to suggest moving back to centre lane. A final issue concerns the phrasing of verbal warnings. Research has shown (Färber and Färber, 1984) that straight forward spoken warnings tend to have blaming properties especially in the presence of fellow passengers. 'Slow down to increase distance to car in front' implies the driver either did not detect the lead vehicle or did not react to it properly. 'Check distance to car in front' and 'Check rear view mirror' are non-blaming ways of conveying basically the same type of information.

Models and adaptivity

To support the driver, the GIDS I prototype design comprises three models: the driving model, the workload model and the history model, and allows the setting of some user preferences. These models will now be discussed along with possible extensions and the desired adaptivity of the models or the support they provide.

The driving model

The GIDS driving model (van Winsum, 1990) consists of a set of situations: possible road segments and vehicle configurations on those segment types determine the task the driver is currently engaged in (e.g. a lead vehicle implies car following). For each of these tasks a set of rules determines acceptable ranges for speed and course. The rules take into account driver reaction time and maximal G-force acceptable when decelerating or in curves, collision courses and traffic laws. The GIDS I prototype supports a limited set of situations that are possible in the 'small world' defined to limit the scope to match the available effort in the project, consisting of two lane roads, inter-sections, roundabout and curves, where only cars appear as traffic. This driving model describes acceptable driving at the manoeuvring level and allows detecting the situations that require support. Usually, support is provided at the control level by a counter force on the gas pedal or a pulse on the steering wheel, but verbal warnings are possible. The model itself could be extended to be tuned to the individual by setting the reaction time and comfortable G-force parameters or even adapt them to momentary driver state. Other forms of adaptation do not appear reasonable, however, as 'acceptable driving' does not differ among drivers beyond catering for their reaction times. Both traffic laws and collision courses and time-to-collisions do not depend on personal characteristics. The determination of the driving situation and task the driver is engaged in and is anticipated to be engaged in the next few moments is used to support traffic situation and driving task adaptivity in the rest of the system. While the model does not adapt, limited adaptation of the support it provides is possible. User preferences may determine what modalities are used for warnings and what type of support is given (e.g. lane keeping support may be switched on and off and be either via steering wheel feedback or by spoken warning).

As control warnings on the active controls tend to be time critical and safety related only minimal adaptation to workload and secondary tasks is allowed, but the converse would be a possible extension to the GIDS I prototype. When on the phone, the telephone audio may be switched off to present an urgent warning without interference with the person speaking on the other side. Though not found in the prototype a model of the specific user's driving style might be used profitably for early detection of deviations from usual driving, but this may be of limited use for driving support for at least two reasons: first, the acceptable driving model already prevents entering dangerous situations without warning; second, if the driver deliberately changes his/her driving style within acceptable bounds, 'support' towards old habits appears both annoying and useless.

To complete the picture of this side of the GIDS I prototype system, note that it also supports, apart from the manoeuvring and control levels already discussed, the navigation level. As the driving support module anticipates driving situations, including decision points in the road network, it is the most appropriate module to time navigation support messages. Navigation support consists of a spoken message (e.g. 'Turn left at traffic lights') presented simultaneously along with a simple pictogram on the GIDS screen as a reminder. This is the only use of the GIDS screen while driving.

The workload model

The workload model should allow us to determine the expected ability of the driver to process support messages in the next few moments, to suppress messages when they are likely to overload information processing capacity and to time them either earlier or later in time when workload (the complement of spare information processing capacity) allows. Inspired by Wickens' (1984) multiple resource theory, in the GIDS I prototype, the workload model consists of single resources for each relevant input and output modality (hands, feet, voice, and vision, audition, tactile hands and tactile feet), and one for central processing (cognition), that are modelled as either in use or not (Piersma, 1991). For each modality and each time granule in the time window considered two parameters specify the constraints used for scheduling messages: the maximum time a resource may be claimed by a support message and the minimal interval between two claims. Whenever a support message is generated in the GIDS I system (by the driving support module, the tutoring module or the secondary task support module, e.g. the ringing of the telephone) it is passed on to the 'scheduler' with the relevant parameters for scheduling, consisting of: expected load durations per modality, importance of the message, relevance interval and preferred time of presentation. If work-load constraints do not allow scheduling the message at the preferred time of presentation, the scheduler attempts scheduling at another time instant in the relevance interval. If scheduling in the relevance interval is not possible, the message is suppressed because current workload does not allow the driver to process it. Only very high importance messages are allowed to overrule this and can then be presented disregarding estimated workload. Figure 29.1 illustrates the workload based message scheduling in the GIDS prototype. The 'scheduler' times or suppresses messages to the driver from various support modules depending on workload constraints obtained from the 'workload estimator', based on anticipated tasks and situations from task models. For example a request to have the telephone ring from the secondary task support to the scheduler when driving up to a complex intersection will cause it to be timed just after the intersection is passed, as the 'negotiate intersection task' anticipated by the driving model causes the workload estimator to constrain severely the duration and number of loads on modalities in the workload model, and the driver is expected to be both cognitively and visually loaded by scanning the intersection to decide whether to yield to other traffic etc.

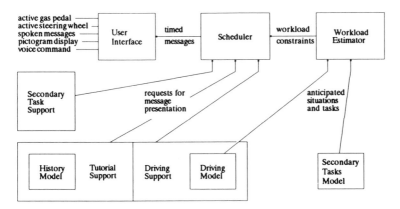

Figure 29.1   Workload based message timing in the GIDS I prototype (see text for explanation).

The workload model in the GIDS I prototype is severely limited as workload constraints were derived from averaging over experience groups (Verwey, 1991b). Obviously the model, and hence the scheduling by the system, can be expected to perform better if individual characteristics are taken into account. Adaptation to secondary tasks has already been incorporated as far as possible, e.g. while on the phone, the workload estimator is informed of that fact (but, for example, the driver counting dogs on the sidewalk, a visually and cognitively loading task, cannot be detected directly of course).

Traffic situation and driving tasks are already taken into account, but here there is room for useful extension as well. Workload estimates currently specify fixed worst case estimates for entire manoeuvres (e.g. during overtaking certain messages will be blocked throughout the manoeuvre). More detailed workload estimates, based on extending the driving model's task analysis to the control level and the level of psychological processes, may allow more accurate prediction of fluctuations of workload within manoeuvres, preventing missing opportunities to present information to the driver as may occur in the GIDS I prototype.

Other useful extensions can be expected in making use of the driver's personal history. On the one hand the familiarity with a particular area, on the other hand the frequency and average success of performing certain driving tasks can be expected to allow more accurate workload estimation. These extensions will be considered in the ARIADNE project (DRIVE II V2004), in the design and implementation of the GIDS II system prototype. Another unused possibility is that of self-paced message presentation as studied by van Wolffelaar *et al.* (1990), who showed that the adverse effect on elderly drivers having to divide attention between support system messages and driving was smaller when the driver controlled the timing and rate of information. Another possible extension is with regard to context sensitive user preferences: allow the user to specify situations where support messages are undesired, either off-line or on-line, immediately after the event and have the system change its behaviour accordingly.

The history model

Many instances have been mentioned so far of the desirability to adapt to the individual history of the driver. This also applies to tutoring, as it would be very annoying if occasional errors are commented upon, but tasks hardly ever or hardly ever correctly performed by the driver would appear to require support. The feasibility of a simple version of such history will now be shown by describing the design of the tutoring support module designed for the GIDS I prototype (Groeger, 1990). The computing through manoeuvres by the driving support module allows routine collection of the frequency of manoeuvres performed and what types of error occurred (because support was required). A simple database, consisting of a table relating manoeuvres to their frequency of occurrence and to the frequency of occurrence of associated errors allows for tutorial messages explaining the 'proper way to drive' with respect to a particular aspect of a particular manoeuvre, when a set criterion of unacceptable performance is reached. Ideally, this table would be updated throughout a person's driving career. This possible implementation of a personal driving history, used only for tutorial information, can be easily extended to support workload estimation. Another useful extension would be to include a representation of the familiarity of specific roads.

User preferences

The GIDS I prototype allows for the setting of a few user preferences; basically to switch certain types of support (navigation, speed support, lane keeping support) on and off, and to switch between the use of active controls versus spoken warnings for control support. One could extend user preferences to adapt to the situation. One could allow the user to view a menu consisting of a 'tree' of possible driving situations and indicate the desirability of certain types of support in certain situations. For example, automatically switch off navigation in familiar areas, never present anything in curves etc. Alternatively, one could allow the user to react to support messages with 'undesired', 'too late' and 'too early', to have the system adapt workload estimation and default timing of messages to the preference of the user. Of course safety related messages can seldom be given later than as currently implemented in the GIDS I prototype, but in practice, navigation messages and incoming phone calls may well be more frequent for well performing drivers than manoeuvring and control support, and will therefore be the ones most susceptible to causing annoyance, if in disagreement with the users' preferences.

Conclusion

If future driving is as current driving, except for a safety net catching us just before we tend to fall, then traffic may become safer if drivers like to consider it a sport to drive from origin to destination with the least possible number of support messages along the way, instead of with maximum speed and entertainment by engaging in risky manoeuvres. In the same spirit, the possible long term adaptation of people to support systems will be a positive development only if it does not mean people become dependent on the presence of support, and does not make them lose their available skills. Support systems may even support better learning, as consistent feedback may be available more than in ordinary traffic, that tends to be ultimately forgiving for erroneous manoeuvres as other traffic avoids collisions at all costs (Brown *et al.*, 1987). Finally support systems may be most effective when they adapt to user preferences, secondary tasks (e.g. being on the phone), familiarity of the route, driving situation and tasks, and individual driving history. Most aspects discussed still require extensive behavioural research to determine in detail both the short term and long term effects on driver skills and traffic safety.

References

Biehl, B., Aschenbrenner, M. and Wurm, G., 1987, Einfluss der Risikokompensation auf die Wirkung von Verkehrssicherheitsmassnahmen am Beispiel ABS, *Unfall- und Sicherheitsforschung Strassenverkehr* Heft 63, Bergisch Gladbach, Germany: Bundesanstalt für Strassenwesen, Bereich Unfallforschung.

Brown, I. D., Groeger, J. A. and Biehl, B., 1987, Is driver training contributing enough towards road safety? in Rothengatter, J. A. and Bruin, R. A. (Eds), *Road users and traffic safety*, Assen, NL: Van Gorcum.

Färber, B. and Färber, Br. 1984, Grundlagen und Möglichkeiten der Nutzung sprachlicher Informationssysteme im Kraftfahrzeug, FAT-Bericht Nr. 39.

Godthelp, J. 1984, Studies on human vehicle control. Ph.D. Dissertation. Soesterberg, NL: TNO Institute for Perception.

Groeger, J. A., Kuiken, M. J., Grande, G., Miltenburg, P., Brown, I. D. and Rothengatter, J. A., 1990, *Preliminary design specifications for appropriate feedback provision to drivers with differing levels of traffic experience*, Deliverable report DRIVE V1041/GIDS-ADA01, Haren, The Netherlands: Traffic Research Centre, University of Groningen.

Janssen, W. H. and Nilsson, L., 1990, *An experimental evaluation of in-vehicle collision avoidance systems*, Deliverable Report DRIVE V1041/GIDS-MAN02, Haren, The Netherlands: Traffic Research Centre, University of Groningen.

Newell, A. and Ernst, G. W., 1965, The search for generality. *Proceedings of the IFIPS Congress 1965*, Vol I. pp. 17–24, Washington, DC: Spartan Books.

Piersma, E. H., 1990, *The first GIDS prototype: module definitions and communication specifications*, Deliverable Report DRIVE/V1041/GIDS-DIA1A, Haren, The Netherlands: Traffic Research Centre, University of Groningen.

Piersma, E. H., 1991, Real time modelling of user workload, in Quéinnec, Y. and Daniellou, F. (Eds) *Designing for everyone*. London: Taylor & Francis.

Smiley, A. and Michon, J. A., 1989, *Conceptual Framework for Generic Intelligent Driver Support*, Deliverable Report DRIVE V1041/GIDS-GEN01, Haren, The Netherlands: Traffic Research Centre, University of Groningen.

Verwey, W. B., 1991a, *Adaptable driver-car interaction in the GIDS system: guidelines and a preliminary design from a human factors point of view*, Report IZF 1990 B-12, Soesterberg, NL: TNO Institute for Perception.

Verwey, W. B., 1991b, Adaptive interfaces based on driver resource demands, in Quéinnec, Y. and Daniellou, F. (Eds) *Designing for everyone*, London: Taylor & Francis.

Wickens, C. D., 1984, Processing resources in attention, in Parasuraman, R. and Davies, D. R. (Eds) *Varieties of Attention*, New York, Academic Press.

Wilde, G. J. S., 1986, Beyond the concept of risk homeostasis: suggestions for research and applications towards the prevention of accidents and lifestyle-related disease. *Accident Analysis and Prevention*, **18**, 377–401.

Winsum, van, W., 1990, *Cognitive and normative models of car driving*, Deliverable Report DRIVE V1041/GIDS-DIA03, Haren, The Netherlands: Traffic Research Centre, University of Groningen.

Wolffelaar, van, P. C., Brouwer, W. H. and Rothengatter, J. A., 1990, *Divided attention in RTI-tasks for elderly drivers*, Deliverable Report DRIVE V1006/TRC1. Haren, The Netherlands: Traffic Research Centre, University of Groningen.

30

A human view of dialogue and dialogue management in the automobile

C. G. Brown and K. Höök

Introduction

In this chapter we present some human needs in dialogue with electronic systems in the coming generation of automobile. We then discuss how such needs can be met. In particular we discuss dialogue management, the management of competing dialogues with different systems and human cognitive needs in dialogue with one system. In this introduction we shall first take up the role of dialogue in communication. The work described in this contribution took place in the context of the development of a dynamic replanning navigation system (Brown, 1991). We therefore use the navigation system and the needs of the user in navigation as a basis of discussion.

Dialogue

Dialogue in its broadest sense is the exchange of information between two agents. Successful communication depends upon cooperation and mutual knowledge as well as upon the contents of the communication and the intentions behind the communication. In dealing with dialogue between a human and a complex system we must realize that (for the present and for a considerable time in the future) the principles that guide human to human dialogue do not all hold.

The type of complex system we refer to is found in the automobile and will become increasingly common. One such system is the in-car telephone. This system can deliver information to the driver (the phone rings) and the driver can deliver information to the system (picking up the phone, dialling a number). The car radio is another example. The most common is simply the system which monitors the several functions for the motor and reports to the driver such things as engine coolant temperature, fuel pressure and oil pressure. This system delivers information to the driver through dials and lighted symbols. These are relatively simple systems which are already common. Even today we have more complex systems of two-way communication where textual information or graphics (for example maps) are displayed in the automobile or truck. These systems require attention and input from the user as an integral part of the dialogue.

Yet more complex systems are on the way. Navigation systems which provide information (where you are, the route to follow, detours to take, when you will arrive) and require information (starting point, destination, stopping off points en route) as part of dialogue are currently under development for the marketplace. Information and warning systems will advise the driver of pedestrians, approaching vehicles or whether the driver is too close to the vehicle in front. These systems too are under development. The impact of these systems will be to provide a great deal of information to the driver and to require input at times from the driver. Inevitably problems will arise. Problems of comprehension and problems of attention. The information that is provided to the driver must be useful and quickly comprehensible. The information from one system must not interfere with the information from another system.

Human needs in dialogue and communication

Human needs in dialogue with the advanced systems we refer to are system dependent. In addition, there is one overwhelming practical human need—the safe control of the automobile.

One need is time. The in-car telephone sends simple messages to the driver; there is a call, the line is busy, the other person has hung up. A navigation system sends much more complex messages to the driver; follow Oxford Street, turn left at the next corner, follow route E24. In both systems, indeed in any communication, the receiving person needs time to attend to the message (to notice that some information is incoming), to comprehend the message (to interpret or understand the information), to decide upon a course of action (if any) and possibly to act. The need for time has obvious consequences for safety. If too much time is required the safety of the primary task (controlling the automobile) is compromised.

Appropriate and useful information is needed. If the navigation system delivers information which is inappropriate to the particular driver the consequences are severe; the driver may reject the information; the driver may not understand the information; the driver may take a long time to process the information. As a consequence the usefulness of the system is degraded. The safety of the primary task is compromised Acceptance of the system is reduced. We discuss appropriate information needs in *Navigation messages* (pp. 342–44). The user models we have developed are embedded in the navigation system as a consequence of our approach to user-centred design of the system.

The complex systems will inevitably compete with each other for the driver's attention. Competing messages will take more time to process cognitively. This is simply a matter of one cognitive task competing with another. The result is a degradation of performance in the primary task. As we have explained above, the consequences of this are reduced safety. Studies of aircraft systems have shown that even well trained pilots have difficulty in responding to competing messages. We take the approach that competing dialogues must not interfere with each other or the driver's comprehension of the dialogues. We will introduce a system of priorities and a dialogue management process to deal with competing messages.

Cognitive needs in navigation

The knowledge communication group at SICS and other researchers in PROMETHEUS and DRIVE have examined some of the cognitive and HMI factors we know have an effect on the drivers' understanding of the navigation and route planning system messages. They are:

1. Driver workload is a factor in the attention a driver pays to messages from any system. If the driver workload is too high, message content may be lost or ignored. As driver workload increases, increasingly dramatic measures are needed to get his or her attention. (Very loud or annoying sounds must be used, for example.) This, of course, applies to all systems, not just navigation. Competing messages contribute to driver workload and are thus to be avoided.
2. Drivers need time to attend to messages, to understand them, to make a decision as to whether action is required, and then to actually carry out the action. A time of fifteen to twenty seconds has been proposed by research directed particularly towards driving tasks such as turning corners. A competing message has at least the effect of increasing the time needed, and may result in failure to correctly interpret the message.
3. The audible channel is most open. That is, we have not yet loaded the sound input to the driver very much. This means that navigation information can be presented to the driver audibly. We do not yet know whether this is acceptable to all drivers, or even to most drivers. This has been confirmed by recent work at HUSAT (Parkes and Coleman, 1990). Note, however, that the length of a sentence or utterance directed at the driver has an impact on the time needed to understand it.
4. It has been demonstrated that presenting a map to drivers while the car is moving will result in drivers taking their eyes off the road for unacceptable lengths of time. This result was obtained under very specific conditions and may not apply universally. However, we believe that this interim result demonstrates the desirability of verbal rather than graphical presentation of navigation messages, from the point of view of safety.

5. It has also been demonstrated that some drivers do not understand maps at all. Some drivers have trouble understanding the spatial relationships presented on maps. This may be a matter of training. In any case, it is of enough concern that in our work we assume that verbal presentation of messages is more likely to be understandable than graphical presentation of maps or diagrams for navigation.

6. There are good and bad positions to put displays and instruments. In some positions the driver can read displays more quickly and more accurately than in others. Generally the rule is directly in front of the driver and slightly below the sight line. The farther below the line of sight, the longer it takes the driver to attend to and interpret the information.

7. A HUD is effective if the point of focus is about four metres in front of the car.

8. It is believed that symbols are effective ways to communicate information. This seems to be a folk homily. We question whether this has been clearly established in circumstances of competing messages.

9. Some messages must have higher priority than others. We make the assumption that a message warning of immediate danger must have higher priority than all others. Priorities are discussed below.

10. Sequences of messages are a critical problem. This normally occurs at or near difficult and complex intersections, traffic circles or highway interchanges. In these cases, because the driver must make a sequence of manoeuvres in a brief period of time, it is necessary to communicate the complete sequence of instructions to the driver before the manoeuvres are started.

Some questions affecting integration arise from the principles given above. How do we structure these priorities? How many degrees of priority are there? What happens to priorities when a new system is added to a car? One might assume that the driver workload plays a role in the delivery of messages but not in the priorities of messages. Is this a valid assumption? At this time, we are aware of no models of driver workload that are computable from indirect evidence (the input stream of message and sensors related to deliver controls). It should not be surprising to realize that these same factors apply to the HMI aspects of other systems in a car. The aircraft industry has known for some time that too much information can overwhelm the pilot and actually cause a pilot to turn off or ignore information; even time-critical warnings!

Dialogues and messages

A dialogue message is the term we shall use for any information passing between the driver and vehicle systems in either direction, in any format. Thus, the messages embody the commands, queries and information that are interchanges. (In several tasks, the order of messages to the driver will be important, as will the location en-route where messages are delivered.) We repeat here the previously stated concern about the ability of the driver to accept messages under certain circumstances. This is the driver-workload factor. Messages then:

- move between a diversity of systems and the driver,
- originate from a diversity of controls, systems and interface devices,
- may be presented in a variety of ways to the driver,
- contribute to the driver-workload,
- must be ordered in time and in space,
- have priorities.

The tasks of dialogue management will be influenced by these considerations. They have contributed to the conceptual architecture for the dialogue management system presented below.

Message attributes

The parametrization of messages is a reflection of the division of competence of the functionality of the systems. The representation of message content, one of the parameters, is important for specification of the functionality of the subsystems. The discussion that follows deals exclusively with HMI messages. Similar parametrization is necessary for messages between other communicating systems.

Messages are passed between subsystems. These subsystems should be viewed as *concurrently communicating processes* representing co-operating agents in the conceptual architecture. Messages have basic parameters. Individual messages are distinguished by the values of the parameters. Some parameters have already been suggested. What follows is the list of parameters found necessary to date.

Destination-process	The process to which the message is directed. The driver is included in the set of possible destinations by the inclusion of a presentation management process in the HMI-Mgt model given below.
Source-process	The process from which the message comes.
Type	Type corresponds roughly to dominating and supporting communication acts in dialogue. Some example values of type are: request, confirm, command, warn, and inform. Used by the models of dialogue.
Priority	Priority is an indication of the relative importance of a message. Values of priority are on a scale, ranking from least-important to an indication-of-emergency.
Timestamp	The point in time when the message originated.

Workload Factor	The contribution of the message to workload (relative to other messages from the *source-process*). The value of the workload the message imposes.
Workload Weighting	A weight given to all messages from *source process*. The weight is used in the determination of the message's contribution to total driver workload.
Duration	The minimum time over which the message should be delivered. Note that a driver has a minimum time requirement to become aware of the presence of a message and to interpret it. If such a duration is not available, the message will not be conveyed successfully, even if presented. This parameter is of use when messages are intended for presentation to the driver. It may be useful in other areas (however, see the *expiry* parameter below).
Persistence	A binary value indicating whether or not the message has the property of persistence. A message with persistence is, for example, coolant temperature.
Expiry	The time before which a message must be delivered. An indication of the lifetime of a message, the time over which a message is valid. For example, a message about an intersection must be delivered before the intersection is passed. Notice that this requirement is expressed in terms of time at the dialogue management level.
Medium	A partially ordered list of the media over which a message may be delivered; from most preferred to least preferred. This parameter interacts with *content* in the composed message which may involve more than one medium.
Content	A representation of the contents of a message in a form that can be interpreted by other processes. A semantic representation.
Opacity	A message may be transparent or opaque.

Some information must be pipelined to the presentation. The classic example of this is the speed (usually displayed on an analog speedometer). Pipelined information is treated in the same way as messages. In this case the pipeline itself has the parameters specified above. (Note that this is a standard computer science technique for communication protocols.)

Message content is to be represented as suggested above in a semantic representation common to all systems. This has some obvious advantages. It allows diverse systems to interpret a message, it allows messages to be interpreted in different ways by different systems, and it allows message contents to be compositional. Compositionality is the ability to represent the composition of sub-messages into larger messages. An example illustrates this. The message content sequence (turn(direction(left) move to(lane,right))) is a composition of two messages intended to be presented in sequence. Such compositionality has been found to be important in the presentation of explanatory diagrams.

Generality is achieved if all sub-systems can make use of the semantic representation when communicating. The last example might be interpreted as the utterance [Turn left, then move to the right lane] by a natural language generation subsystem or the presentation process, or interpreted as a diagram (graphics display) indicating the instruction by a diagram generation subsystem. Messages can be interpreted in different ways by different subsystems. For example, in the HMI-Mgt schema it is left to the presentation process to interpret the semantic content and format the message for the specified medium. (Recall that the specification of *medium* was a partially ordered list.) This allows the use of an alternative-medium, the next in the list, for presentation when the preferred-medium is currently used by a message of equal or higher priority.

Competing messages and priorities

Messages compete when the time interval of delivery of a message overlaps the time interval of delivery of another. One way to resolve this situation is to give every message an attribute of priority. The priority of a message is an indication of the importance or urgency of the message relative to other messages. We shall use a basic system of priorities with three levels: emergency, high and low. It is, of course, possible to refine these basic levels into a many levelled system through the definition of more specific constraints on message delivery. The three level system described below allows us to discuss the important points. Messages with emergency priority are those which convey information about a situation or state of affairs which may result in injury and thus must be acted upon immediately. High priority is given to messages which call for relatively quick action. Failure to act in time may result in inconvenience or in damage to the automobile. For example, failure to act upon a navigation instruction may result in the inconvenience of retracing part of a route. Failure to act upon a low-oil-pressure message may result in damage to the car's engine. Thus high priority messages have a time envelope of valid delivery. Low priority messages may be delivered at any time.

Dialogue management

Message parameters and the HMI-Mgt function having been discussed above, it is now time to turn to the specific question of how a dialogue management process (see Figure 30.1) can accomplish the goals of dealing with priorities and timing of messages. This lies at the heart of the HMI-Mgt system.

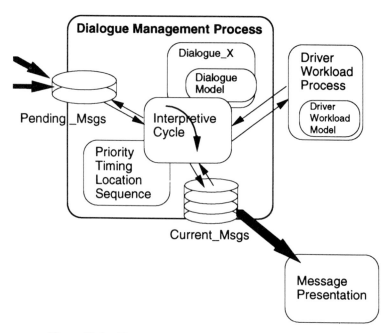

Figure 30.1   The message sets and dialogue management.

Messages are put into sets in the dialogue management process. These sets are ordered and manipulated by the process.

The dialogue management process must manage, at any given time, a set of messages to be transmitted: The *pending–msg–set*. In addition, the process must be aware of the current set of messages that have been transmitted and whose lifetimes have not expired: The *current–msg–set*. For the purpose of explanations, a history of messages may arrive to swell the pending–msg–set, messages may be transmitted which reduce the pending–message–set and swell both the current–msg–set and the history–set. Finally, messages may expire due to location in time and space which reduces the current–msg–set (and potentially the pending–msg–set). The current–msg–set is actually a set of partially ordered queues of messages. One queue for each message presentation channel. The queues are queues of unexpired messages delivered to the message presentation process. These messages are presented as soon as they are put in the current–msg–set.

The current–msg–set contains two conceptually different types of messages; *transparent* messages and *opaque* messages. Transparent messages are those which can be displayed continuously and to which a driver pays occasional attention. Typically these messages contain non-critical information which do not require immediate attention by the driver. Some examples of transparent messages (in vehicles currently on the market) are engine temperature, speed, fuel level, oil pressure, tuning frequency of car radio.

There are times when messages require attention from the driver. In such cases, the messages should be delivered as opaque messages. For example, a driver must be informed that a turn is required at a certain location. In general, navigation related messages will all be opaque. A fuel level message, however, should only be opaque when the fuel reaches a low level. The delivery of opaque messages depends upon the media available, and upon the priority of the message. Some guiding principles are:

- voice delivery is more opaque than dashboard display,
- heads up display is more opaque than dashboard display,
- a flashing display is more opaque than a steady display.

An interpretative cycle is a convenient way to handle the management of the message sets. The parameters of messages must be used in determining the sequencing in the partially ordered current–msg–set queues. This allows the embedding of a reasoning system within the dialogue management. In this case the reasoning is about time, and about the dialogue. Reasoning is provided by functional models called during the interpretative cycle. Done this way, dialogue management becomes a question of planning. One possible interpretative cycle is described intuitively below. It assumes a particular division of competence between dialogue management and sub-systems.

1. Examine message sets
   Remove all expired messages from the PENDING–MSG–SET and inform the originating process of expired/fails messages. Similarly, remove all successfully delivered messages from CURRENT–MSG–SET and inform originating process.
2. Reason
   Collect messages from PENDING–MSG–SET and CURRENT–MSG –SET. Order the messages for each presentation medium with respect to time, duration and priority. Possibly specify alternative media. Propose a CURRENT–MSG–SET.
3. Consult driver workload model
   Test the proposed CURRENT–MSG–SET on the driver workload model. Accept or reject the proposal.
4. Update message sets
   Update the proposed CURRENT–MSG–SET and PENDING–MSG–SET with proposed CURRENT–MSG–SET and remainder of messages.

Clearly there are implementation requirements. The message sets must be buffered by the dialogue management process. A flag must be set when the set is being updated by an external process. Finally, the reasoning system will have time constraints in order to provide true reactive behaviour.

Navigation messages

Human cognitive needs in dialogue with a system dictate that messages be adapted to the specific user. Stereotype user models in the system permit a system to generate messages appropriate to the individual user. We use the example of navigation messages in a route guidance system. Navigation messages illustrate how route chunking can be used to adapt the system to the user.

We have sorted out three hypotheses, derived from the RG area and cognitive science. The hypotheses were: that verbal route descriptions are feasible (although not the only solution); human route choice needs to be investigated and considered; and finally that user modelling of some kind is needed (possibly stereotypical modelling is enough). With these hypotheses in mind, we have then conducted a study with very experienced driver-navigators, from which we have deduced principles as to how route descriptions are constructed and expressed by humans.

User models and navigator stereotypes

A number of authors, (Michon, 1985; Alm, 1989; Parkes *et al.*, 1990) have pointed out how important it is that human factors, especially from a cognitive science point of view, are already taken into account at the design phase of a route planning and navigational system. One way of doing this is to incorporate a user model into the system that is allowed to influence both presentation and possibly the route planning, and the overall functioning of the system. The problem then becomes how to design that user model.

What we want is to minimize the amount of work that the driver must undertake during the trip, since any extra workload might threaten safety. We have identified the need for different descriptions for different kinds of drivers (Alm, 1989). As a first rough division we can differentiate between; drivers who are unfamiliar with a city, who will be called *tourist navigators*; drivers who live or work in the city, *resident navigators*; and drivers normally travelling back and forth between a few destinations, *commuter navigators*. Of course, there is no clear cut line between these three groups. A person might even belong to the first two groups in one city, for instance, if she knew one part of the city well and the other not so well.

In Waern *et al.* (1990) user modelling is divided into three different levels. The first is to incorporate some general knowledge of human cognitive characteristics, in the sense of the system knowing what a human can and cannot do, but not implying that the system itself should be able to stimulate human capabilities. A second level allows for some differentiation between users: using stereotypes. Users can be different on the scales of being expert-novice, being in favour-against, previous domain knowledge, learning style etc. The third level is to assess the characteristics of each individual user during the interaction with the user. One way is to assume that the user does not have the relevant knowledge when she answers incorrectly, and that she has it when she answers correctly. Choosing between these levels for our application, we can observe that an individual user model is really only possible when the interaction with the user is quite rich, preferably in natural language. Only then is it possible to diagnose each recipient individually.

In our application, the interaction with the driver is very limited; we can only observe her reactions, i.e. whether she is following or not following our proposed route-plan. Building an individual user model from that is probably impossible. Using a normative model of the driver requires that we have a normative model of the driver,which we do not have (at this time). The two first levels, on the other hand, seem almost necessary for this application. The first level will be realized through designing the interface to the driver in such a way that it takes mental load, human perceptual capabilities, short term memory limitations etc., into account. The second level can be realized through having different functionalities available that are adapted for different user groups needs, like elderly drivers, tourists and residents, commuters etc. This adapts to what Kass (1991) describes as an explicit user model acquisition technique using stereotypes.

Route descriptions for human cognitive needs

In Höök and Karlgren (1991) we turned to the problem of describing routes to more experienced drivers, residents and commuters. In all existing systems (CARMINAT, LISB, TRAVELPILOT, etc.) the tourist navigator is addressed. Through making a study of the difference between route descriptions aimed at tourists and residents, we were able to extract some interesting principles of how to describe routes to residents. These principles have been implemented and preliminarily evaluated in Gothenburg (Höök, 1991).

In the comparative study between route descriptions aimed at residents and tourists, a number of interesting results were found. Firstly, there are two fundamentally different kinds of route descriptions, the *declarative* and *procedural* route descriptions. They originate from two completely different ways of viewing a route; the birds-eye view, and the imaginary view. The first is where somebody views a city from above, and not as a map, but as a number of possible routes or parts of routes. The imaginary view, is where we take another step after having chosen the route, and try and put ourselves in the route, driving it mentally in order to give a full description of it. The grammar, describing the two kinds of descriptions, was presented together with some heuristic rules of how they should apply. For instance, in the resident declarative description, roads are left out in the middle of the trip, the start of the trip and the end of the trip. Those gaps can be expressed in some heuristic rules:

- The first 1–3 roads before a road on a higher hierarchical level (i.e. bigger roads) can be taken away from the declarative description.

- If the goal is on a high hierarchical level, the last 1–3 roads can be skipped.
- In the middle of the trip, only roads that help excluding other alternative routes are mentioned, plus some roads that help making the route description 'complete' (these roads are usually on a high hierarchical level and it would be odd not to mention them).

We name these gaps *chunks*, since a single label (a road name or landmark, etc.) is used to describe a whole chunk of roads and intersections. Apart from hierarchical level of road, some other factors (like the length of road) will help us to determine whether the road should be part of the declarative route description, or if it can be omitted. In fact, these properties were available in the map databases of cities currently being developed in Europe. It was therefore possible to implement a route chunking programme able to identify which part of a route should be more carefully described, and which parts can be chunked and only named by a label.

- Tourist routes are chosen by subjects in order to be easily explained. A route is easy to explain when it is on a high hierarchical level, contains few turns, and goes by important landmarks.
- Furthermore, we argued (Höök, 1991) for the need for explanations, or other means of convincing drivers that the RG system route choice is the best. The question is not only a matter of producing good route choices, although that is a part of it. It is also a matter of understanding how humans perceive a city, and why that perception sometimes makes them reluctant to accept route choices which violate their view of the city. We believe that studies for finding a balance between choosing the shortest (quickest) route and adapting to human preferences should be made.

Concluding remarks

We can broadly cite two motivations for the integration of human machine interface functions in a dialogue management process. The first is safety. The second is engineering.

Consider the safety aspect. We are all aware that car systems are becoming increasingly complex and increasingly numerous. Some PROMETHEUS functions represent systems of such complexity that they will supersede the ability of the driver to make decisions in a complex domain. With respect to the increase in the number of in-car systems, we only have to look in the modern 'cockpit' to confirm our suspicion. Currently we can have a primitive navigation system, a multimedia stereo sound system including cassette, CD, DAT and choice of broadcast receivers on AM or FM bands, a telephone; in taxis we also have a dispatch system, various measurement systems relating to the soundness of the car's engines, whether there is fuel, and if there is, how far we can drive on it. What is more, all of the information is delivered through the visual and audible channels primarily with the aid of meters (digital and analog), flashing or continuous lights, lighted symbols, non-lighted symbols, knobs and levers, and (increasingly) with sound, CRT display, and drivers attention. In fact, it competes for the driver's attention with the situation outside the car and with the driving task.

The second motivation is an engineering and cost effectiveness consideration. When complex systems are designed, a system view of the functionality is taken. Integration of parts of the design with other parts is sought. We consider modular design and extendibility of systems to be essential. This is especially true in the fields of systems development and software engineering which are, in fact, the fields involved in the developments of dynamic replanning IRG systems and of most PROMETHEUS functions.

Benefits of modularity and integration of modules include: maintainability (subsystems can be maintained without worrying about the effects on the rest of the systems), ease of documentation, ease of enhancement (subsystems can be enhanced individually, without effecting the rest of the system and without the need to alter the rest of the system), ease of development (which includes ease of debugging, understand ability and efficiency), and verifiability (the communications protocols between modules are verifiable and the operational semantics of modules is verifiable). These benefits are gained by the development phase of a complex system and are quantifiable as both development-time and development-cost benefits.

An additional benefit is gained in the production (and hence the cost) of a complex system. If we consider the in-car electronics and control systems it is easy to see that a system with fewer modules (physical modules and/or software modules) is less expensive to produce. A specific example is the media which communicate with the driver. The driver can only attend to one display area at once. Therefore, there is no need (today) to have separate analog instruments, CRT and other displays. All information presented visually to the driver can be displayed on one device, let us say a high resolution flat screen (upon which analogy dials, maps, text messages, symbols and even video images can all be presented). In doing so a great deal of money can be saved on hardware (gauges, lighted symbols viz idiot lights wiring and control panels). A similar argument pertains to other channels of communication to the driver, e.g. sound. The driver can attend to only a few sounds at a time. To reduce the instrumentation and control messages to a single screen and a single sound source requires control of the information stream to the driver. This is an essential part of what we have called the dialogue management.

The point must be made that the place to start is with the principles and methods of integration of the HMI function of the multitude of systems. Clearly the systems in a car today are, if taken as a whole, exactly the kind of complex system mentioned above. One can see the start of the programme of system integration in the car industry, but little sign of that integration at the HMI level. Many of the points made above simply repeat the accumulated and accepted wisdom of today. It is important to realize that these points also support very strongly the argument that the integration of the HMI functionality is the crucial problem.

In summary, both safety and cost-effectiveness are supported by full development of HMI integration. Specifically with respect to HMI, the following points are argued for in this paper.

● Effective system development is aided with standard interface formats.
● A consistent treatment of time and temporal reasoning is possible with standardized message parameters.
● Alternative sensory media may make use of the same internal message format.
● Possible benefits exist here in the developments of fault-tolerant message systems which might, in the event of faults in one sensory I/O, make use of alternative senses.
● The development of a driver workload model based on the uniform message format and other factors is thought to be possible.

References

Alm, H., 1989, *Show me the way to go home—or drive me home old chip!*, available from Swedish Road Traffic Research Institute (VTI), Sweden.

Bonsall, P. W. and Joint, M., 1991, *Evidence on drivers' reaction to in-vehicle route guidance advice, Proceedings 24th ISATA Conference*, pp. 391–401, Italy.

Brown, C., Höök, K., Lindevall, P. and Waern, A., 1991, A system architecture for an interactive route guidance system, Final report on Interactive Route Guidance 1988–1991, *SICS Technical Report*, T91, 18.

Höök, K., 1991, An approach to a route guidance interface. Licentiate thesis, DSV, Stockholm University, ISSN 1101-8526, No. 91-019-DSV.

Höök, K. and Karlgren, J., 1991, Some principles for route descriptions derived from human advisors, *Proceedings of the Thirteenth Annual Conference of the Cognitive Science Society*, Chicago.

Kass, R., 1991, Building a user model implicitly from a cooperative advisory dialog, *User Modelling and User Adapted Interaction*, 1, 203-58.

Michon, J. A. 1985, A critical view of driver behaviour models: what do we know, what should we do? in Evans, L. and Schwing, R. C. (Eds) *Human Behaviour and Trafic Safety*, pp. 485–520, New York: Plenum Press.

Parkes, A. M. and Coleman, N., 1990, Route guidance systems: A comparison of methods of presenting directional information to the driver, in Lovesey, E. J. (Ed.) *Proceedings of the Ergonomics Society 1990 Conference*, London: Taylor & Francis.

Waern, Y., Hägglund, S., Löwgren, J., Rankin, I., Sokolnicki, T. and Steinemann, A., 1990, Communication Knowledge for Knowledge Communication, LiTH-IDA-R-90-70, Linköping University, Sweden.

31

Creating new standards: the issues

T. Ross

Introduction

The primary responsibility of drivers is to use the information around them to control their vehicles in terms of speed and position. With future RTI (Road Traffic Informatics) systems comes the potential for increasing the amount of information presented to the driver. There is the possibility that drivers could be overloaded with information or presented with information they either do not see, or do not understand. This could be due to the information being presented in an unclear way, or being lost amongst other redundant information.

If the performance of a technically competent system is reduced because of inattention to usability issues, then success in the marketplace will be adversely affected, as will driver efficiency and safety. Eventually, good design might evolve through competition in the marketplace, but this will only be after a considerable time lag. In any system this delay will be expensive, and particularly so in large systems such as those being considered by the DRIVE and PROMETHEUS programmes. The safety implications of any new device used in a vehicle becomes particularly important if we look at past accident statistics. Of the errors which cause road traffic accidents, over 95 per cent (Sabey and Staughton, 1975) have a human component (as opposed to purely mechanical or technical). There is clearly a need for findings from human factors research to be integrated into the design process. The most formal way of accomplishing this (and probably the most useful way) is to develop standards to guide the designers of such RTI systems, in order that driver efficiency and safety are not compromised. To this end the four main aims of standards should be considered. These aims (ISO, 1989) are:

(i) mutual understanding,
(ii) health, safety and the protection of the environment,
(iii) interface and interchangeability,
(iv) fitness for purpose.

Current MMI standards

Standards for man machine interaction (MMI) already exist to deal with such areas as symbol colour/design, arm reach envelopes and recommended lighting levels. The majority relate to the office or manufacturing environment, as this is where the biggest increase in information technology has taken place to date. The in-vehicle environment is only just beginning to undergo drastic changes. So, why not simply transfer the office standards to the vehicle environment? Firstly, environmental conditions for the vehicle (such as ambient lighting) are more diverse and less controllable. Secondly, in the office, interaction with the computer is normally the primary task, in the vehicle, it is secondary to the primary task of maintaining the vehicle's position on the road. Lastly, current office standards refer to equipment which would never be used in a vehicle such as large CRT displays or complicated QWERTY keyboards.

Work within DRIVE Project V1037 STAMMI, 'Definition of Standards for In-Vehicle Man Machine Interaction' identified the scope of current MMI standards in Europe at a national and international level (TÜV Bayern, 1989). Three main areas of MMI were identified: system hardware; human computer interaction (HCI); and information presentation. Standards relevant to each area were then categorized into sub-groups under these main headings. This categorization shows that most current standards deal more with the hardware aspects of a system, such as those described by Simmonds (1983). Very few consider the software issues of HCI and information presentation and, of those that do, the majority are not directly applicable to the in-vehicle environment. Currently, the directly relevant standards fall into six main categories, identified in the STAMMI project:

(a) driver hand control reach,
(b) driver's eye-ellipses,
(c) location of hand controls,
(d) location of indicators and tell-tales,
(e) symbols for controls,
(f) symbols for indicators and tell-tales.

The standards in categories (a) and (b) will still be relevant when advanced RTI systems are introduced, allowing of course, for any changes in the characteristics of the driver population. However, categories (c) to (f) will be somewhat limited in their applicability. A whole new set of controls and symbols will be required for the functions of route navigation and guidance, for example.

Important areas of system design for which standards do not exist are the dialogue interface, the tool interface (manipulation of the system) and the universals of information presentation, such as human perception and information overload. Until now this gap in standards has been of little consequence, in-vehicle MMI consisted of straight forward controls and displays. The advent of RTI systems alters this and generates a need for such areas to be considered. Within the scope of the STAMMI project, existing MMI criteria were identified from the literature in an attempt to begin to fill this gap in standards. It was the aim of the project for this list of criteria to be used in two ways:

(i) to support the design of safe and useful in-vehicle information systems,
(ii) to contribute to a basis for discussion and consensus concerning evaluation and standardization of the MMI for such systems.

Types of standard

There are three basic types of standard relevant to man machine interaction. These are product, performance and process (or procedural) standards (Howarth, 1987).

1. Product standards specify the physical aspects of the system. An example in office standards could be that the home row of a keyboard should be no more than 5 cm above desk level.
2. Performance standards specify the user performance levels which should be achieved with a particular system. For example, drivers should be able to pick up the information they require without taking their eyes off of the road for more than a certain length of time (e.g. 1.5 seconds). Performance standards can also include specification of the methods and metrics to be used in evaluating a system against the standard. Performance standards are not concerned with technical system performance. They adopt a driver centred approach, and are more concerned with *enabling* task performance.
3. Process standards describe the programme of analysis and testing which the manufacturer must use, within the design cycle, to provide assurance of usability and safety. For example, an existing process standard for safety critical systems (MOD, 1991a) identifies procedures for system development including a mandatory project management structure. Procedures are specified for carrying out, managing and documenting hazard analysis and for subsequently apportioning levels of safety integrity to functions, according to the levels of hazard associated with them. Procedures for hazard analysis are described and a risk classification scheme is provided.

Each type of standard has advantages and disadvantages. Product standards provide the designer with concrete recommendations for product design and are thus easy for the designer to comprehend and comply with. As a result, these type of standards can be popular with manufacturers, so long as they are not mandatory. However, by their nature, they apply to specific parts of a system and conformance with them does not guarantee that the system (or set of systems) will be usable as a whole, bearing in mind the complex interactions which will take place. They are also technology dependent. For example, in the office environment, what happens to the 5 cm standard quoted above if the keyboard deviates from 'normal' design, as does the Maltron keyboard (Rowley, 1989)? As such, they will easily become outdated and redundant. Finally, such concrete physical standards have the potential to hinder innovative design by forcing designers to adhere to old solutions.

Performance standards are technology independent and can be applied equally to parts of systems and whole systems. Striving to achieve the standard allows, and actually encourages, innovative design. These type of standards are, however, more time consuming to apply as they require performance testing of the system to verify conformance. Also, to maximize the validity of the testing, it would be necessary for it to be conducted by experts who are familiar with the relevant methods and metrics.

Process standards ensure participation of all relevant specialists in the design process and the consideration of all the constraints upon it. In other words, they can act as an *aide-mémoire* for the design team. However, they are extremely difficult to implement unless a strong certification body exists.

Development of product standards

Product standards as already discussed, are the most problematic type of standard. Research carried out during the STAMMI project gathered the views of manufacturers and designers in an attempt to identify the way forward (Salway, 1991). If product standards are developed then a mechanism must be in place to update these in line with technological advances. This process would probably be difficult to achieve, bearing in mind the long delays intrinsic in reaching consensus amongst a large community such as the EC. If it is achieved, constantly changing standards may solve some problems but create others. In particular, for systems with long development times, altering the product late in the process carries a large overhead, so continually changing standards would cause delays and be financially detrimental.

Use of product standards

To avoid the imposition of design constraints in the form of mandatory standards, product standards would be most acceptable in the form of design guidelines with a similar style to *Ergonomics Standards and Guidelines for Designers* (Pheasant, 1987), but with a content aimed specifically at the driving environment. Manufacturers and designers also favour a checklist format which could be used to quickly evaluate prototype and final systems against certain design criteria. The DRIVE STAMMI project produced an MMI Criteria List which was developed along the two paths described above, i.e. design guidelines and an evaluation checklist (Eggerdinger and Peters, 1991). Elements of this Criteria List could form the basis for a set of product standards.

Development of performance standards

The development of performance based standards has two major inputs:

(i) empirical validation of safety and usability criteria,
(ii) scientific consensus.

Empirical studies

The empirical studies must be coordinated to ensure:

- Representative samples of user groups in terms of age, driving experience, visual ability, computer literacy, cognitive abilities etc. The size of user groups could vary from the whole population, to a small sub-set such as police drivers.
- A combination of real tasks and controlled conditions, for example, real road driving complemented by simulator studies. The former is less controlled, but will elicit realistic reactions. The latter is less realistic (for instance, lack of a view to the rear of the vehicle will alter visual scanning patterns), but has the advantage of being able to push the subjects to the limits of their abilities and simulate critical incidents. Such work was conducted within DRIVE Project V1017 BERTIE, 'Changes in Driver Behaviour Due to the Introduction of RTI Systems'.
- Set methods and metrics to promote consistency of results in order that databases of information can be built up as a step towards scientific consensus. A set method could cover aspects such as the number of subjects to use, the practice time required for the equipment, the instructions given to the subject or the environmental conditions required for the trial. Set metrics could include the definition of an 'error', the accuracy to which the length of an eye glance should be measured, the start and stop limits for duration of an operation, or the rating scales to use in a comfort assessment.

Scientific consensus

The scientific consensus which results from empirical studies and expert opinion should come from psychologists, human factors practitioners, specialists in driver behaviour and accident researchers. The European DRIVE and PROMETHEUS programmes are two arenas where this scientific consensus can begin. However, although the results of research are constantly increasing knowledge in the area, the approaches used are disparate and further efforts are needed to ensure easy comparison of findings towards a consensus.

Use of performance standards

The methods and metrics which were set up to develop a standard can be used in assessing prototype or final systems against the standard. Indeed, the methods and metrics should be an integral part of the standard itself.

For office based systems this approach has already been taken in draft standard ISO 9241 Parts 3 and 4; Visual Display Requirements and Keyboard Requirements respectively (Stewart, 1990). To conform to these parts of the proposed standard the 'test' device (which may fail to meet the physical requirements of the standard) is compared with a reference display (which meets the physical requirements of the standard) in terms of performance (reading speed, keying speed and errors) and comfort ratings, to prove that it has equal, or greater, usability characteristics (Mackay, 1990). Conformance is statistically determined by employing Barnard's sequential t-test (described in Davis, 1951). With this procedure, a statistic is computed after each subject run and compared with pre-defined boundary values. If it exceeds either of the boundary values a judgement may be made as to whether the device 'passes' or 'fails' the standard. If the statistic remains within the boundaries, no decision can be made and testing must continue.

Prior to this, the only user performance test method in standardization was BS 5321: 1975 'Reclosable pharmaceutical containers resistant to opening by children'. Here conformance is reached when performance reaches an absolute level—a certain percentage of children failing to open a container within a specified time limit (Mackay, 1990). This approach could have potential for the evaluation of RTI systems, but current research data is not sufficient to enable a decision as to whether a system is 'safe' or 'unsafe', and there is much debate over how this can be achieved.

Including methods and metrics in the standards will not alone ensure proper and accurate testing. There is a case for certified testing centres to be identified. These testing centres should be staffed by people experienced in the test methods to ensure validity, consistency and reliability of testing. The existence of testing centres would also be advantageous to vehicle and system manufacturers. Having a ready facility with experienced experimenters would remove the need for the manufacturers to invest heavily (in time and resources) to carry out their own testing.

Development of process standards

Process standards and standards for safety critical systems already exist. The main examples are:

- Defence Standards 00-55 and 00-56 (MOD, 1991a, 1991b),
- IEC SC65A, WG9 and WG10 Draft standards (IEC, 1989a, 1989b),
- ISO 9001 (ISO, 1987).

Although none of these are directed at RTI systems nor at man machine interaction issues, they can provide a basis for such standards. DRIVE Project V1051 'DRIVE Safely' has proposed a standard for the design and certification of safety related software for RTI systems (DRIVE Safety Task Force, 1991). This proposal is based around the IEC Safety Lifecycle Model (IEC, 1989b). The STAMMI project, developed a framework for an MMI design process standard which is closely related to that of DRIVE Safely (Carr, 1991). The objective of the STAMMI framework is driver safety. It specifies a process for designing out hazards during product development. The essential components of the framework are:

- Hazard analysis — to analyse the ways in which performance of the driving task might be disrupted.
- Hazard control — to design out all hazards by the use of design techniques, checklists, seeking advice on human factors.
- Safety management — including safety planning, validation planning, personnel competency (i.e. adequate and relevant technical knowledge, experience and ability including possession of appropriate qualifications) and organizational competency (e.g. ISO 9000 certification).
- Documentation — including a safety management plan, hazard log and design validation report.
- Compliance testing — either by self-certification or by an external certification body.

Use of process standards

If the framework described above is adopted then it could be linked into current procedures implemented by larger manufacturers to cover *system* safety, such as FMEA (Failure Modes and Effects Analysis), Hazop (Hazard and Operability) studies or Quality Assurance (in line with the ISO 9000 series). However, smaller organizations may not have personnel who are experienced in this area and employing experts would be too costly. Therefore, the standard should also include guidance on the procedures to follow.

Design process standards could have other benefits to manufacturers in addition to producing usable, safe and marketable products. Product liability cases may require proof of sound safety procedures. If the standard has been used correctly and all documentation retained, this will provide good evidence in defence of the system manufacturer.

The European standards environment

Some information in this section is taken from the DRIVE Standardization Workshop (CEC, 1990a). The aim of European standardization is the harmonization of standards, on a Europe-wide basis in order to facilitate the exchange of goods and services by eliminating barriers to trade, which might result from requirements of a technical nature, as well as the creation of tools for the competitiveness of European industry. Since a resolution of 7 May 1985 issued by the Council of the EC, the role of European Standardization has been increasing owing to its vital contribution to the preparation of the Internal Market foreseen after 1992.

European standardization activities mainly aim at:

- promoting the implementation of international standards prepared by ISO or IEC,
- the harmonization of national standards and technical documents,
- the preparation of new European standards where none exist,
- The creation and implementation of procedures for the mutual recognition of test results and certification systems,
- cooperation with CEC, EFTA, economic, scientific, technical, European and International organizations.

The European standards organizations

European IT standardization is based on a partnership between; standards organizations, CEC/EFTA, industry and users. The relevant standards organizations are CENELEC (electrotechnical), ETSI (telecommunications) and CEN (all other fields). Thus the most appropriate channel for the harmonization of roadside and in-vehicle RTI information presentation is CEN (European Committee for Standardization).

CEN Structure, documents and procedures

CEN consists of:

A General Assembly	(AG)	—	highest level, decision making, liaison with CEC and EFTA
An Administration Board	(CA)	—	major decisions
A Central Secretariat	(CS)	—	support the activities
A Technical Board	(BT)	—	decides work programme, coordinates the technical work

The Technical Board assigns new work items to relevant Technical Committees (TC). The national standards institutes delegate the same technical experts to CEN TCs and to ISO TCs. This guarantees that the technical decisions of CEN do not diverge from those of the ISO. The members from EFTA countries cooperate voluntarily with the countries of the CEC in order to foster a larger European market.

There are four CEN publications:

1. The Harmonization Document (HD) is a standard which carries the obligation to be implemented at national level, at least by public announcement of the HD number and title, and by withdrawal of any conflicting national standards. An HD is established if transposition into identical national standards is unnecessary or impracticable, and in particular if agreement is subject to the acceptance of deviations.
2. The European Standard (EN) carries the obligation to be implemented at national level, by being given the status of a national standard and by withdrawal of any conflicting national standards.
3. The European Pre-standard (ENV) is for provisional application, while conflicting national standards may be kept in force, in parallel. It was developed in order to deal very rapidly with advanced standardization, such as in the IT field, and can be based on Draft International Standards.
4. The CEN Report is authorized by the Technical Board to provide information.

There are three important procedures in the work of CEN (and CENELEC):

1. The standstill agreement—national activity on the same subject is suspended when work begins at the European level. This becomes a legal obligation if the CEN activity is the result of a mandate from the CEC or EFTA.
2. The consultation and voting procedure—when TC consensus has been reached on any document, it is first subjected to an enquiry where comments of a technical or legal nature are made. The TC then examine the comments and prepare a revised draft. Finally, a weighted vote of the members is taken.
3. Implementation—when the standard is adopted there are strict obligations that any conflicting national standards are withdrawn (EN & HD) and that the European standard is transformed into a national standard with no modifications (EN).

A CEN Technical Committee concentrating on standards for RTI systems has now been set up—CEN TC 278—and is, at the time of writing, in the process of defining its scope of work. Working Group 10 (WG10) is responsible for man-machine interfaces. The formation of this group and the technical committee augurs well for the consideration of usability and safety in the development of future systems.

The contribution of DRIVE and PROMETHEUS

The two avenues via which DRIVE and PROMETHEUS can contribute to standards development are:

(i) via DRIVE Central Office to CEN,
(ii) via the national standards organizations of the partner countries to CEN.

The basic research conducted under the auspices of these two programmes can make a major contribution to the development of MMI standards. Three particular sources of information for CEN TC278 WG10 could be: the STAMMI MMI Criteria List (Eggerdinger and Peters, 1991), the PROMETHEUS MMI Checklist (PROMETHEUS WG4/MMI, 1991) and the PROMETHEUS Traffic Safety Checklist (Pro-Gen Safety Group, 1990). Elements of all these documents could form a basis for the creation of product, performance and, to a lesser extent, process standards.

The future of in-vehicle MMI standards and legislation

There is a case for a European Directive which will enforce compliance with the standards, or at least will require manufacturers and employers (e.g. fleet operators) to prove that they have taken human factors considerations into account when designing and/or purchasing their system(s). European Directives are issued by the Council of the European Communities, and each EC member state must bring into force the laws, regulations and administrative provisions necessary to comply with the Directive.

Recently, a Directive was issued 'on the minimum safety and health requirements for work with display screen equipment' (CEC, 1990b). The Directive states that 'employers shall be obliged to perform an analysis of workstations in order to evaluate the safety and health conditions to which they give rise for their workers, particularly as regards possible risks to eyesight, physical problems and problems of mental stress'. This Directive explicitly excludes 'drivers' cabs and 'computer systems on board a means of transport', so the requirement for equivalent legislation for RTI systems remains.

There are European Directives in existence which apply to vehicles, but (certainly in the UK) the bulk of these are, in effect, standards applied through Type Approval Regulations (which must be met before the vehicle is allowed on the public road) governing only the *construction* of vehicles and non-RTI equipment. At present, each European country has its own type approval system. This still leaves barriers to trade which will be undesirable after 1992. To solve this problem, the European Commission has proposed a European Whole Vehicle Type Approval (EWVTA) scheme. Although this is good news for those concerned with the enforcement of standards, there is still no evidence that it will cover in-vehicle MMI standards.

One step which could be taken outside formal Type Approval or Directives and implemented more quickly is within the power of sponsoring bodies such as the CEC. Funding for research and development programmes such as DRIVE should be restricted to projects which have considered the safety implications of their work and taken the necessary steps to ensure best practice. The beginnings of such a policy was evident in the procedure for DRIVE 2 proposals which required project consortia to consult the DRIVE Safety Task Force Guidelines (DRIVE Safety Task Force, 1991). A further step could be an annual safety audit which must be 'passed' before the project is allowed to continue. This could be incorporated into current audit procedures.

There must be a new initiative towards standards for in-vehicle MMI, and the enforcement, particularly of performance and process standards. Large programmes like DRIVE and PROMETHEUS should work towards European Standards through CEN, whilst exchanging information with ISO in order that common standards can be achieved. True international standards would remove barriers to trade for the motor and component industry. This was seen as a desirable situation to strive towards by the industrial participants at an early STAMMI workshop on standardization. The formation of CEN TC278 is a long-awaited step towards achieving this.

The approach discussed in this chapter, i.e. the design of systems around the driver cannot be expected to be the definitive solution to safety problems. RTI systems will alter the current driving task to such an extent that there will be implications for driver training and licensing. Therefore, development of human factors standards should be accompanied by parallel changes in driver instruction to include use of RTI systems, plus more detailed assessment of driver's functional abilities, e.g. visual capabilities.

There is still much work to be done in the standards area, but it is necessary for experts from all fields to come together on this important safety issue. This includes manufacturers' representatives as well as academic researchers. The vast investment of industry and governments is resting on the success of future RTI systems. The development of standards is one step towards making these systems safe, usable and efficient, and hence contribute to their acceptance by the target population.

References

Carr, D. J., 1991, *Exploration of Design Process Standards for MMI*, Deliverable 8, DRIVE Project V1037, STAMMI, October 1991.

CEC, 1990a, Converting R&D Proposals to Standards, in *Proceedings of the DRIVE Standardization Workshop*, Brussels, September 1990.

CEC, 1990b, Council Directive of 29 May 1990 on the minimum safety and health requirements for work with display screen equipment (fifth individual Directive within the meaning of Article 16(1) of Directive 87/391/EEC), *Official Journal of the European Communities*, No L 156, 21.6.90, pp 14–8.

Davis, O. L., 1951, *The Design and Analysis of Industrial Experiments*, Edinburgh: Oliver and Boyd.

DRIVE Safety Task Force, 1991, *Guidelines on System Safety, Man-Machine Interaction and Traffic Safety*, Produced by the DRIVE Task Force of Behavioural Aspects and Traffic Safety, June 1991, DRIVE Central Office, Brussels.

Eggerdinger, C. and Peters, H., 1991, *The Development and Validation of the MMI Criteria List*, Deliverable 5, DRIVE Project V1037, STAMMI, December 1991.

Howarth, C. I., 1987, Psychology and information technology, in Blackler, F. and Oborne, D. (Eds) *Information Technology and People. Designing for the future*, Leicester: British Psychological Society, pp 1–19 (specifically pp. 10–3).

IEC, 1989a, Draft—Software for Computers in the Application of Industrial Safety-Related Systems, IEC SC65A WG9 (Technical Committee No. 65: Industrial Process Management and Control).

IEC, 1989b, Draft—Functional Safety of Programmable Electronic Systems: Generic Aspects, IEC SC65A WG10 (Technical Committee No. 65: Industrial Process Management and Control).

ISO, 1987, ISO 9001: 1987, Quality systems: model for quality assurance in design, development, production, installation and servicing, International Standards Organization.

ISO, 1989, Procedures for the technical work of ISO. IEC/ISO Directives—Part 1.

Mackay, C. J., 1990, Background to the development of ISO user performance test methods, Paper presented at the HI Club Systems Assessment and Evaluation SIG Workshop 'Scientific Issues and Implications of User Performance Testing', Loughborough, UK, February 1990.

MOD, 1991a, Requirements for the Procurement of Safety Critical Software in Defence Equipment, Interim Defence Standard 00-55 issue 2 (draft), Ministry of Defence, UK.

MOD, 1991b, Requirements for the Analysis of Safety Critical Hazards, Interim Defence Standard 00-56 issue 2 (draft), Ministry of Defence, UK.

Pheasant, S., 1987, *Ergonomics Standards and Guidelines for Designers*, British Standards Institution catalogue number PP 7317:1987.

Pro-Gen Safety Group, 1990, The Pro-Gen Safety Group Traffic Safety Checklist, PROMETHEUS Office, Stuttgart, November 1990.

PROMETHEUS WG4/MMI, 1991, MMI Checklist Version 2.1, The PROMETHEUS Thematic Working Group 4: Man-Machine Interaction, PROMETHEUS Office, Stuttgart, February 1991.

Rowley, H., 1989, A keyboard fit for the human hand, *Safety Management*, June 1989.

Sabey, B. and Staughton, E. C., 1975, Interacting roles of road, environment, vehicle and road user in accidents, *Proceedings 5th International Conference of the International Association of Accident and Traffic Medicine*, London.

Salway, A. F., 1991, *Product, Performance or Process Standards? Feedback from Designers and Manufacturers on Standards for In-Vehicle MMI*, Deliverable 7, DRIVE Project V1037, STAMMI, October 1991.

Simmonds, G. R. W., 1983, Ergonomics standards and research for cars, *Applied Ergonomics*, **14**, 2, 97–101.

Stewart, T., 1990, SIOIS—Standard interfaces or interface standards? in Diaper, D. *et al.* (Eds) *Human-Computer interaction—INTERACT 90*, pp. xxix–xxxiv, Amsterdam: Elsevier Science Publishers B.V.

TÜV Bayern, 1989, *List of European Standards*, Deliverable 2.1, DRIVE Project V1037, STAMMI, June 1989.

PART V

METHODS IN DESIGN AND EVALUATION

32

Methods in design and evaluation

A. M. Parkes

The introduction of new information systems into the vehicle is likely to have an effect on driver behaviour and performance. In a complex system where so much variability exists, the prediction of the exact nature of that effect is difficult to make. This final collection of papers throws some light onto the problem of how changes in driver behaviour could or should be studied. No generic methodology exists, different techniques are appropriate in different circumstances, and have been reviewed elsewhere (Parkes, 1991; Zaidel, 1991).

The DRIVE programme has tended to encourage the use of three categories of criteria; safety, efficiency and acceptability. The following chapters highlight the principle that successful research requires a tightly focused research question, and a clear view of which are the main criteria of interest in the design and evaluation. The choice of exactly which data to capture, or which environment to use for experimentation, is not straightforward. If the choice is made to capture empirical data, rather than perform a theoretical modelling exercise, then available sources would include the observed behavioural responses of the driver, the subjective impressions by the driver, the physiological reactions of the driver, or measures of vehicle performance. The researcher also has the choice of conducting experiments in natural real road environments, or in artificially controlled ones such as test tracks, driving simulators, or laboratories.

There are always pragmatic constraints on design and evaluation in the applied commercial environment in which much RTI system development is conducted. These constraints should influence the range of research questions that can be addressed adequately, rather than the way in which they are addressed. The papers in Part 5 offer a diverse sample of views, and some fresh insight from active researchers in the area. It is hoped the reader will find something new or provoking in these viewpoints.

Brookhuis argues that it may be possible to increase traffic system safety if we achieve the ability to predict decreases in driver performance due to such factors as fatigue and their consequent effect on attention. Physiological indices of driver performance are being correlated to unobtrusive vehicle parameters, such as steering wheel reversal rate in an attempt to identify the onset of unsafe driving characteristics. Brookhuis reports on current research, and also comments on some of the difficulties associated with such driver status monitoring systems, such as predicting acceptance by drivers, or making firm cost benefit analyses for safety predictions.

Fairclough, in Chapter 34, also reviews the applicability of physiological measures of driver response, in design and evaluation, but concentrates more on the aspect of driver overload and stress, rather than underload and inattention. He argues that such measures of mental workload and stress are valuable tools to be used in pre-market evaluations of products.

Lindh and Garder also look at stress; in this case with a view to evaluating simulated or prospective systems, but with the use of subjective data. They report on the production of a stress index, and detail an interesting, and perhaps controversial research method for using a natural roads environment as simulations of certain proposed RTI applications.

The difficulty of determining accident causation behaviour in real environments is the starting point for Nilsson (Chapter 36). Accidents are relatively rare and irregular events in the real traffic system, and so the appeal of conducting fruitful research in highly prescribed and constrained environments is obvious. It becomes possible to perform very early design evaluations in a safe and repeatable environment if one has access to a driving simulation tool that has the required ecological fidelity. Nilsson reviews the arguments for and against driving simulators, and highlights the need for continuing validation of such facilities.

Reichart emphasizes the changing role of the driver in the traffic system, arguing for driver support rather than replacement in future application development. He presents the concept of human centred automation from a reliability engineering point of view, and shows how careful modelling of reliability could and should be built into the design process.

Automatic policing systems are the focus for Harper, in Chapter 38 which critically examines the potential contribution of artificial intelligence developments. His conclusions are that such developments are not necessary for this application. This also brings into focus other current debates about the usefulness of many behavioural modelling exercises. Harper also raises the conceptual problem of behavioural adaptation after the introduction of new systems; the implication being that there is insufficient knowledge at present to enable such complexities to be modelled, and thus the need for empirical data remains.

Risser and Hyden look at real traffic behaviour. They argue that it is too late to look for system effects in changes in accident statistics, and examine what it is possible to do in terms of expert evaluation at early design stages. They report a methodology for collecting observed real road behavioural data in a controlled fashion. They go on to explain the need for clearly identified criteria to enable such observed behaviour to be compared to desired norms. They show that their method is capable of identifying certain driving styles, and relate the potential of RTI systems to fulfil a tutoring function.

The contributions show that there is no simple answer to the question of how to establish whether the introduction of a new information system in a vehicle will adversely affect driver behaviour.

References

Parkes, A. M., 1991, Data capture techniques for RTI usability evaluation, *Advanced Telematics in Road Transport*, Vol II, Proceedings of DRIVE Conference, Brussels, pp. 1440–56, Amsterdam: Elsevier.
Zaidel, D., 1991, *Specification of a methodology for investigating the human factors of advanced driver information systems*, Transport Canada Publication No. TP 11199(E), Ottowa, Canada.

33

The use of physiological measures to validate driver monitoring

K. A. Brookhuis

Introduction

A growing number of traffic safety studies shows that human error is the major contributing factor in the majority of traffic accidents. In a large study in the US, Treat *et al.* (1977) found that the leading causes of accidents are improper lookout, excessive speed and inattention. Smiley (1989) concludes that some 90 per cent of all traffic accidents can be attributed to human failure. Brookhuis and Brown (1992) argue that an ergonomic approach to behavioural change via engineering measures, in the form of electronic driving aids, should be adopted in order to improve road safety, transport efficiency and environmental quality.

The main purpose of implementing electronic driving aids in vehicles, however, should be to improve road safety. This can be attained, for instance, by detecting decrements in driver performance. It is repeatedly shown that performance decrements in general may be due to common factors such as alcohol, medication, illness, temperature, sleep deprivation, fatigue, and with respect to driving performance specifically, driver underload or overload (Yabuta *et al.*, 1985; for an overview see Thomas *et al.*, 1989). These factors all affect driver status, which in fact is causatively related to the performance decrement, in the sense that changes in driver status will inevitably lead to changes in driving performance. Changes in driver status are reflected in changes in relevant physiological parameters such as electroencephalogram (EEG), electrocardiogram (ECG), galvanic skin response (GSR), electromyogram (EMG), etc. (see Brookhuis and de Waard, 1991). Therefore, to predict rather than to detect (serious) decrements in driver performance, the electronic driving aid should be able to detect detrimental changes in driver status. A device intended to monitor driver status, however, must be able to detect detrimental changes, irrespective of source (internal or external influences), from non-obtrusive in-vehicle parameters alone. After all, it is highly unlikely that measuring physiological parameters while driving a motor-vehicle will ever be standard. The consequence of this state of affairs is that physiological measures must be considered from the point of view of validating non-obtrusive in-vehicle measures that will be used to monitor driver status continuously through vehicle parameters.

In order to determine performance with respect to driving a car from driving, i.e. vehicle parameters alone, it is necessary to establish the relationship between vehicle parameters and generally occurring changes in driver status from different sources. Driver status is not some unitary, fixed phenomenon, even within each individual. It varies with time-of-day, age, subjective feelings (mood), and also with time-on-task and all kinds of external influences, such as traffic environment and situational task-load, alcohol and (medicinal) drugs. The optimum with respect to driving performance should be located (cf. Wiener *et al.*, 1984), and signs of deviations from the optimum should be detected, first from the individual's physiology, as a reflection of driver status foreboding behaviour, and then from the relevant behavioural parameters themselves. To assess an individual's optimum and deviations from that optimum, with respect to the relationship between physiology and performance, careful research should be carried out under controlled circumstances. Having established this relationship, it should be possible to predict deviations from optimal behaviour in the actual driving environment.

Alcohol and drugs

Alcohol and drug consumption may serve as first examples of sources of changes in driver status. It is well known that alcohol and, probably to a lesser degree, most drugs have a detrimental effect on driving performance (Seppala *et al.*, 1979; Smiley and Brookhuis, 1987). Epidemiological studies on alcohol and accident involvement have made a strong case for the alcohol part of this assertion. In an experimental study, a relationship between the amount of alcohol, blood alcohol concentration (BAC), and the amount of weaving, as measured by the standard deviation (SD) of lateral position, has been established in the actual driving environment by Louwerens *et al.* (1987), who reported that performance with respect to control over the vehicle's lateral position started to deviate from 'baseline' significantly at 0.06 per cent BAC. Laboratory studies showed various impairments of driving related skills after alcohol or drug intake, especially reaction time and detection performance deteriorate as a function of alcohol dose (Peeke *et al.*, 1980; Gustafson, 1986a, 1986b; Rohrbaugh *et al.*, 1987), or degree of sedation by (mostly medicinal) drugs (Hink *et al.*, 1978; Gaillard and Verduin, 1983). It was generally found that relatively low levels of alcohol impaired sensory-perceptual tasks more than psychomotor tasks (Levine *et al.*, 1973). Weaving, or lateral position control, belongs largely to the latter category of tasks, implying that a field test of sensitivity for alcohol effects must not be copied indiscriminately from this weaving test, but should also incorporate sensory-perceptual variables (Smiley and Brookhuis, 1987; Brookhuis *et al.*, 1987). The impression might otherwise arise that the effects of alcohol on driving performance (and on traffic safety in general) are not so negative below 0.06 per cent BAC. In accordance with this assertion, De Waard and Brookhuis (1991) found that at a (during the test decreasing) BAC between 0.046 per cent and 0.035 per cent, subjects responded more slowly to speed changes of a leading car. Consumption of alcohol, leading to a BAC below the legal limit in many countries of 0.05 per cent, resulted in an impairment of 19 per cent compared to baseline-delay. Moreover, in the same experiment De Waard and Brookhuis found an increase in the SD of the lateral position of more than 5 cm, i.e. 30 per cent impairment, whereas Louwerens *et al.*, reported such an impairment after a BAC over 0.10 per cent.

A number of physiological changes have been reported as a result of alcohol and drug intake. Gaillard and Trumbo (1976) reported increased heart rate and heart rate variability after amphetamine; barbiturates were found to decrease heart rate but increase variability. Gaillard and Verduin (1983) found a similar effect after azatadine, an antihistamine. Heart rate increased after alcohol (Peeke *et al.*, 1980). Brookhuis *et al.* (1986) found increased energy in the alpha (4–8 Hz) and theta range (8–13 Hz) of the EEG power density spectrum after several antidepressants. Alcohol dose-related changes in event-related potential components (Peeke *et al.*, 1980; Rohrbaugh *et al.*, 1987) are an indication of changes in central information processing capacity.

ECG
Heart Rate

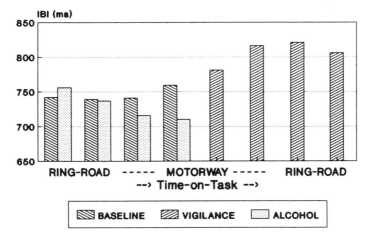

Figure 33.1 Inter-beat-intervals of the ECG while driving on a busy ring-road and quiet motorway, separated in to and from a turning point, with and without alcohol.

Riemersma *et al.*, (1977) found that heart rate decreased and heart rate variability increased in eight hours of night-time driving. The data from De Waard and Brookhuis (1991) confirmed this finding, however, after taking alcohol the effect was reversed (see Figure 33.1). Initially heart rate was slightly decreased, but gradually increased as the BAC decreased from 0.046 per cent to 0.035 per cent. Increases in energy in the alpha- and theta-band of the EEG, indicative for decreases in subject's activation (cf. Akerstedt and Thorsvall, 1984), were found in an on-the-road experiment, coinciding with increases in amount of weaving (Brookhuis *et al.*, 1985; Brookhuis *et al.*, 1986). In Figure 33.2 deviation effects of four different antidepressant drugs on weaving (in cm SD lateral position) are depicted relative to placebo as a function of energy in the alpha- and theta-band. The combined effects on alpha and theta energy, particularly, show that decreased activation towards falling asleep caused by the antidepressant drugs is related to a decrement in driving performance (see also

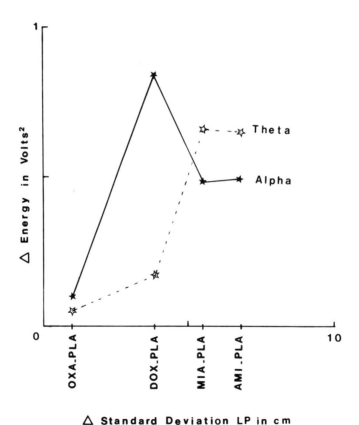

Figure 33.2   Mean changes (relative to placebo) of energy in the alpha- and theta-frequency bands of the EEG, corresponding to mean changes in SD lateral position (in cm) in each of four antidepressant drug conditions.

Akerstedt *et al.*, 1991). In this particular experiment, six out of twenty subjects were not able to finish their test ride in the condition that showed the highest effect on performance (AMI = amitryptiline in Figure 33.2). de Waard and Brookhuis (1991) found further highly significant effects of alcohol on alpha and theta energy in the EEG and on the amount of weaving (SD lateral position increase of about 30 per cent). However, in the alcohol condition steering wheel movements (both standard deviation and number of steering wheel reversals) were not affected. Petit *et al.* (1990), in a driving simulator study, reported considerable effects of alcohol on both alpha energy in the EEG and steering wheel movements, however, their subjects were administered a whole bottle of wine.

Vigilance

Monotony is an important aspect of the driving task. The reason for this is that driving a car under relatively quiet conditions is, at least for experienced drivers, relatively simple and monotonous. Therefore, driving a car on a quiet highway is often considered to be a vigilance task (Wertheim, 1978), resulting from underload. The time-on-task related deactivation makes the driver accident-prone (Lisper *et al.*, 1986). Underload will result in a deviation from the top of the inverted-U curve towards low activation and poor performance (Wiener *et al.*, 1984). This is nicely illustrated in two experiments that combine performance measures and physiology (Brookhuis *et al.*, 1986; Thorsvall and Akerstedt, 1987). After prolonged driving of a train or car, driver's activation tends to diminish rapidly, as may be found by means of spectral analysis of the EEG. An increase in power in the theta- and alpha-band of the EEG power density spectrum has been found to coincide with performance decrement while driving a car (Brookhuis *et al.*, 1986). Figure 33.3 illustrates this coincidence in high correlations of power in the alpha- and theta-band with the amount of weaving, measured as SD of lateral position.

Petit *et al.* (1990) demonstrated a relationship between the handling of the steering wheel and the occurrence of alpha waves in the EEG. The state of vigilance, as indicated by the power in the alpha range, was found to correlate highly with specially developed steering wheel functions in most cases. De Waard and Brookhuis (1991) also found effects on steering wheel behaviour with time-on-task (150 minutes of continuous driving). The standard deviation of the steering wheel movements increased and the number of steering wheel reversals per minute decreased, both highly significant. The amount of weaving also increased after prolonged driving, by 30 per cent. Subjects' activation, measured by a relative energy parameter ((alpha + theta)/beta), gradually diminished from the start of the experiment. In Figure 33.4 both the relative EEG energy parameter and the amount of weaving are depicted, showing that physiological signs of changes in driver status are readily followed by changes in driver behaviour. Alcohol and medicinal drugs mostly

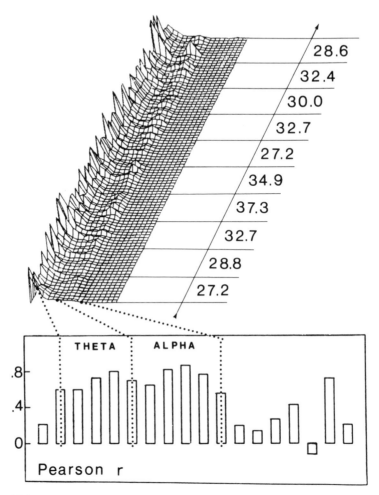

Figure 33.3   Top: EEG power density spectra (x-axis) for each 1 kilometer segment in a 100 kilometer driving test (y-axis) next to 10 SDs of lateral position in cm over 10 kilometer each. Bottom: bargraph indicating product-moment correlations of the SD lateral position and each 1 Hz band averaged across 10 kilometer.

aggravate these effects (Peeke *et al.*, 1980; Brookhuis *et al.*, 1985; Gawron and Ranney, 1988), although this is not necessarily the case, depending on dose and type of drug (Fitzpatrick *et al.*, 1988).

Heart rate and heart rate variability can also serve as an index of mental load (Mulder, 1985). Time-on-task effects were found on performance, heart rate, heart rate variability and the 0.10 Hz component of the power spectrum of inter-beat-interval sequences (Coles, 1983; Brouwer and van Wolffelaar, 1985). Riemersma *et al.* (1977) found decreases in heart rate and increases in heart rate variability in eight hours of night-time driving. They concluded that

EEG and SDLP
Vigilance Condition

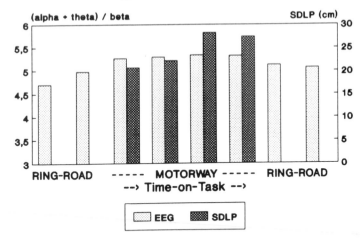

Figure 33.4 Average standard deviation lateral position (amount of weaving) and EEG ((alpha + theta)/beta) energy during 150 minutes of continuous driving.

the observed decreases of heart rate reflected adaptation, including decreasing stress and habituation to sensory stimulation. The increases in heart rate variability are in line with this explanation.

Task-load

Although the primary concern in driving performance research has been with mental underload and alcohol, high workload situations resulting in mental overload might well be a major cause of, for instance, intersection related accidents (Smiley and Smith, 1985; Smiley and Brookhuis, 1987). Overload emerges when two or more aspects of the driving task compete for attention (Michon, 1985), and this is often the case at intersections. Secondary, i.e. subsidiary, tasks have been used to measure spare mental capacity while driving a motor vehicle (Brown *et al.*, 1969; Wetherell, 1981; Lisper *et al.*, 1986). The results from this type of measurement do not look very promising. In so far as the experiments involved relevant aspects of the driving task situation, they indicated that various levels of workload, e.g. induced by levels of steering difficulty, can be demonstrated weakly by performance on the secondary task.

Heart rate and heart rate variability might be more promising. In particular heart rate variability and the earlier mentioned 0.10 Hz component are reported to be reliable indices of increased workload (Egelund, 1985; Pruyn

Driving Test
heart rate

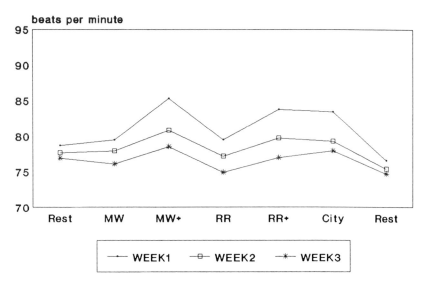

Figure 33.5   Heart rate (beats per minute), averaged across three weeks of practice, with (+) telephoning while driving and normal driving, separately for quiet motorway (MW), busier ring-road (RR) and city.

et al., 1985; Aasman *et al.*, 1987). Brookhuis *et al.*, (1991) found that heart rate increased when subjects in an on-the-road driving test carried out a subsidiary telephone task. In Figure 33.5 this effect is illustrated, showing effects of handling a telephone while driving, whereas training shows a considerable habituating effect in 3 weeks of daily practice. Heart rate variability and its 0.10 Hz component showed a similar, systematic decrease in amplitude, indicating increasing workload. Effects of telephoning while driving were also visible in the standard deviation of the steering wheel movements, but only on the busier parts of the test track, not on the quiet motorway.

Discussion

Demonstrating the feasibility of monitoring driver status by monitoring vehicle parameters depends on demonstrating co-occurrence of changes in physiology and changes in behaviour. Signs of changing driver status may be manyfold from many sources arising either fast or slow. A discrimination in this sense can be found in phasic versus tonic changes in the driver's physiology, EEG versus ECG. In the study of De Waard and Brookhuis (1991) it was found that the tonic changes (in the more global ECG parameters) co-occurred with gradual changes in behavioural parameters, such as the SD of the lateral position, which is in line with the findings of Riemersma *et al.* (1977) and Petit *et al.* (1990). In this way the effects of a (potential) driver underload are also replicated/simulated.

The effect of external factors (alcohol and drugs) are shown to be deleterious in many respects by many researchers. These factors not only cause gradual changes, but prompt a different offset as well. A monitoring device should be able to cope with this (see De Waard and Brookhuis, 1991), and it is necessary that each driver, together with his or her capabilities and driving style, should be identified by a monitoring device at the moment he or she enters the vehicle. This could be accomplished by, for instance, a driver's license on a smart-card, that has to be read by the monitoring device before starting a ride. This card should contain a record of vehicle parameters of a drivers' normal performance. In this way, individual differences between drivers are taken care of at the same time.

From several studies it appears that phasic changes (as reflected in EEG energy parameters) precede changes in behaviour, indicating that (even small) changes in driver status are followed by changes in relatively easily measured vehicle parameters such as steering wheel movements or, less easily measured, lateral position control.

Another important factor is driver overload. Phasic measurements are necessary to detect a driver's overload, since in traffic overload usually arises fast. Only few studies have shown potential measurement of relevant driver behaviour in this respect. Additional information from the world outside the moving vehicle will be a necessary prerequisite for the monitoring device to predict potential overload. Interactions between the driver and the information given to the driver will become important then.

In conclusion, combined measurements of a driver's physiology and behaviour have demonstrated that it is feasible to develop a monitoring device on the basis of a limited number of unobtrusive vehicle parameters alone. However, since no valid data exist on which estimates of the risks of particular traffic behaviours can be based, it is very difficult, if not impossible to forecast the savings of death and disability that might result from the introduction of risk-reducing driving aid systems (Brookhuis and Brown, 1992). There also remains the problem of acceptance of a device that monitors driver behaviour. Since voluntary adoption of such a device by the majority of the car drivers should not be expected to happen in the very near future, it seems reasonable to consider the possibility of imposing a driver monitoring system through legislation. Initially, in view of the safety potential of a driver monitor, installation should be enforced upon commercial vehicles and recidivist offenders of, for instance, the legal BAC limit.

References

Aasman, J., Mulder, G. and Mulder, L. J. M., 1987, Operator effort and the measurement of heart-rate variability, *Human Factors*, **29**, 161–70.

Akerstedt, T. and Thorsvall, L., 1984, Continuous electrophysiological recordings, in Cullen, J. J. Siegriest, J. (Eds) *Breakdown in Human Adaptation to Stress. Towards a multidisciplinary approach. Vol I*, pp. 567–84. The Hague: Martinus Nijhoff.

Akerstedt, T., Kecklund, G., Sigurdsson, K., Anderzén, I. and Gillberg, M., 1991 Methodological aspects on ambulatory monitoring of sleepiness, in Proceedings of the workshop *Psychophysiological Measures in Transport Operations*, pp. 1–21, Köln: DLR.

Brookhuis, K. A. and Brown, I. D., 1992, Ergonomics and road safety. *Impact of Science on Society*, **165**, 35–40.

Brookhuis, K. A. and Waard, D. de, 1991, The use of psychophysiology to assess driver status, in Proceedings of the workshop *Psychophysiological Measures in Transport Operations*, pp. 283–307, Köln: DLR.

Brookhuis, K. A., Louwerens, J. W., and O'Hanlon, J. F., 1985, The effect of several antidepressants on EEG and performance in a prolonged car driving task, in Koella, W. P., Rüther, E. and Schulz, H. (Eds) *Sleep '84*, pp. 129–31, Stuttgart: Gustav Fischer Verlag.

Brookhuis, K. A., Louwerens, J. W. and O'Hanlon, J. F., 1986, EEG energy-density spectra and driving performance under the influence of some antidepressant drugs, in O'Hanlon, J. F. and de Gier, J. J. (Eds) *Drugs and Driving*, pp. 213–21, London, Philadelphia: Taylor & Francis.

Brookhuis, K. A., Volkerts, E. R. and O'Hanlon, J. F., 1987, The effects of some anxiolytics on car-following in real traffic, in Noordzij, P. C. and Roszbach, R. (Eds) *Alcohol, Drugs and Traffic Safety—T86*, Amsterdam: Excerpta Medica.

Brookhuis, K. A., Vries, G. de and Waard, D. de, 1991, The effects of mobile telephoning on driving performance, *Accident Analysis & Prevention*, **23**, 4, 309–316.

Brouwer, W.H. and Wolffelaar, P.C. van, 1985, Sustained attention and sustained effort after closed head injury: detection and 0.10 Hz heart rate variability in a low event rate vigilance task, *Cortex*, **21**, 111–9.

Brown, I. D., Tickner, A. H. and Simmonds, D. C. V., 1969, Interference between concurrent tasks of driving and telephoning, *Journal of Applied Psychology*, **53**, 419–24.

Coles, M. G. H., 1983, Situational determinants and psychological significance of heart rate change, in Gale, A. and Edwards, J. A. (Eds) *Physiological correlates of human behaviour vol. II: Attention and performance*, pp. 171–85, London: Academic Press.

Egelund, N., 1985, Heart rate and heart rate variability as indicators of driver work load in traffic situations, in Orlebeke, J. F., Mulder, G. and Vandoornen, L. J. P. (Eds) *Psychophysiology of cardiovascular control*, New York, London: Plenum Press.

Fitzpatrick, P., Klorman, R., Brumaghim, J. T. and Keefover, R. W., 1988, Effects of methylphenidate on stimulus evaluation and response processes: evidence from performance and event-related potentials, *Psychophysiology*, **25**, 292–304.

Gaillard, A. W. K. and Trumbo, D. A., 1976, Drug effects on heart rate and heart rate variability during a prolonged reaction task, *Ergonomics*, **19**, 611–22.

Gaillard, A. W. K. and Verduin, J., 1983, The combined effects of an antihistamine and pseudoephedrine on human performance, *Tijdschrift voor Geneesmiddelenonderzoek/ Journal of Drug Research*, **8**, 1929–36.

Gawron, V. J. and Ranney, T. A., 1988, The effects of alcohol dosing on driving performance on a closed course and in a driving simulator, *Ergonomics*, **31**, 1213–44.

Levine, J. M., Greenbaum, G. D. and Notkin, E. R., 1973, *The effects of Alcohol on Human Performance: A Classification and Integration of Research Findings*, American Institutes for Research.

Lisper, H. O., Laurell, H. and Vanloon, J., 1986, Relation between time to falling asleep behind the wheel on a closed track and changes in subsidiary reaction time during prolonged driving on a motorway, *Ergonomics*, **29**, 445–53.

Louwerens, J. W., Gloerich, A. B. M., De Vries, G., Brookhuis, K. A. and O'Hanlon, J. F., 1987, The relationship between drivers' blood alcohol concentration (BAC) and actual driving performance during high speed travel, in Noordzij, P. C. and Roszbach, R. (Eds) *Alcohol, Drugs and Traffic Safety—T86*, Amsterdam: Excerpta Medica.

Michon, J. A., 1985, A critical view of driver behavior models: what do we know, what should we do? in Evans, L. and Schwing, R. C. (Eds) *Human behavior and traffic safety*, New York: Plenum Press.

Peeke, S. C. Callaway, E., Jones, R. T., Stone, G. C. and Doyle, J., 1980, Combined effects of alcohol and sleep deprivation in normal young adults, *Psychopharmacology*, **67**, 279–87.

Petit, C., Chaput, D., Tarrière, C., LeCoz, J. Y. and Planque, S., 1990, Research to prevent the driver from falling asleep behind the wheel, in *Proceedings of the 34th AAAM conference*, pp. 505–23.

Pruyn, A., Aasman, J. and Wyers, B., 1985, Social influences on mental processes and cardiovascular activity, in Orlebeke, J. F., Mulder, G. and Vandoornen, L. J. P. (Eds) *Psychophysiology of Cardiovascular Control*, pp. 865–77, New York, London: Plenum Press.

Riemersma, J. B. J., Sanders, A. F., Wildervanck, C. and Gaillard, A. W. K., 1977, Performance decrement during prolonged night driving, in Mackie, R. R. (Ed.) *Vigilance: Theory, operational performance, and physiological correlates* pp. 41–58. New York: Plenum Press.

Rohrbaugh, J. W., Stapleton, J. M., Parasuraman, R., Zubovic, E. A., Frowein, H. W., Varner, J. L., Adinoff, B., Lane, E. A., Eckardt, M. J. and Linnoila, M., 1987, Dose-related effects of ethanol on visual sustained attention and event-related potentials, *Alcohol*, **4**, 293–300.

Seppala, T., Linnoila, M. and Mattila, M. J., 1979, Drugs, alcohol and driving, *Drugs*, **17**, 389–408.

Smiley, A. and Smith, R. L., 1985, *Driving task monitoring to reduce inattention-related accidents*, Technical Report, National Highway Traffic Safety Administration.

Smiley, A. and Brookhuis, K. A., 1987, Alcohol, drugs and traffic safety, in Rothengatter, J. A. and de Bruin, R. A. (Eds) *Road users and traffic safety*, pp. 83–105, Assen: Van Gorcum.

Smiley, A., 1989, Cognitieve vaardigheden van autobestuurders (cognitive skills of car drivers) in van Knippenberg, C. W. F., Rothengatter, J. A. and Michon, J. A. (Eds) *Handboek Sociale Verkeerskunde*, pp. 351–71, Assen: Van Gorcum.

Thomas, D. B., Brookhuis, K. A., Muzet, A. G., Estève, D., Tarrière, C., Poilvert, C., Schrievers, G., Norin, F. and Wyon, D. P., 1989, *Monitoring driver status: the state of the art*. Technical report to the Commission of the European Communities, DRIVE Project V1004 (DREAM), Köln: TÜV-Rheinland.

Treat, J. R., Tumbas, N. S., McDonald, S. T., Shinar, D., Hume, R. D., Mayer, R. E., Stansifer, R. L. and Castellan, N. J., 1977, *Tri-level study of the causes of traffic accidents*, Report DOT-HS-034-3-535-77 (TAC), Indiana University, USA.

Waard, D. de and Brookhuis, K. A., 1991, Assessing driver status: a demonstration experiment on the road, *Accident Analysis & Prevention*, **23**, 297–307.

Wertheim, A. H., 1978, Explaining highway hypnosis; experimental evidence for the role of eye movements, *Accident Analysis & Prevention*, **10**, 111–29.

Wetherell, A., 1981, The efficacy of some auditory-vocal subsidiary tasks as measures of the mental load on male and female drivers, *Ergonomics*, **24**, 197–214.

Wiener, E. L., Curry, R. E. and Faustina, M. L., 1984, Vigilance and task load: in search of the inverted-U, *Human Factors*, **26**, 215–22.

Yabuta, K., Iizuka, H., Yanagishima, T., Kataika, Y. and Seno, T., 1985, The development of drowsiness warning devices, Paper presented at the Tenth International Technical Conference on Experimental Safety Vehicles, Oxford, UK.

34

Psychophysiological measures of workload and stress

S. Fairclough

Introduction

Psychophysiological techniques have been applied to the study of driving behaviour for over twenty-five years. The rationale behind this particular technique is rooted in the Yerkes-Dodson model which describes the relationship between performance and arousal (Yerkes and Dodson, 1908). According to the Yerkes-Dodson model, optimal task performance is achieved when the human evokes an optimal 'quantity' of psychophysiological arousal. Any deviation above or below this psychophysiological 'ideal' results in a decline of human performance. If arousal rises above the optimal level, performance deteriorates due to the operator being overloaded or under stress. Conversely if arousal falls below the optimal, the operator becomes bored and/or fatigued resulting in a similar decline in task performance. It does not take a great deal of imagination to see the implications of this model within the context of driving behaviour. The decline of driving performance due to the influence of psychophysiological arousal was characterized by Wildervanck *et al.* (1978) in terms of different traffic environments. These authors assumed that monotonous highway driving deteriorated due to boredom or 'underload', whereas busy town driving declined due to stress or 'overload'.

The scope of the Yerkes-Dodson model may be broadened to include the deleterious effects of alcohol, medication and sleep deprivation on driving performance. This category of performance breakdown where the driver is impaired by excessive stress, fatigue or the influence of drugs undoubtedly contributes to the total number of accidents that occur on the road. Consequently, a substantial research effort has been directed towards the investigation of the feasibility of a monitoring device designed to detect the influence of either fatigue or stress on driving behaviour. This research includes the DREAM project under the first EC-funded DRIVE initiative and DETER in the current programme. The PROMETHEUS programme also contains a driver status monitoring component in the form of the Proper Vehicle Operation CED (Common European Demonstrator). Whilst these research programmes continue, a number of driver monitoring systems are already available in the marketplace. Thomas *et al.* (1989) report that more than 200 such devices have been patented in the USA such as the 'Stay Awake Alarm' that predicts driver arousal on the basis of steering wheel movement. Psychophysiological techniques are obviously valuable as a correlate of physiological arousal to be used during the design, testing and implementation of driver state monitoring systems. Other devices directly tap the psychophysiology of the driver using changes in the electrodermal activity of the skin (see later for full explanation) in the case of the 'Dormalert' or the ambitious 'Magneto Encephalo Grammetry' that aims to capture the electrical activity of the brain (see Thomas *et al.*, 1989 for full review).

The role of psychophysiological variables (PPV) as a means of quantifying the level of arousal of a driver is self-explanatory, however other applications of PPV capture may be less apparent. In the history of driving research, PPV have been used to assess the impact of road type (Brown and Huffman, 1972), road curve radii (Babkov, 1973), driving experience (Taylor, 1964), driving manoeuvres such as braking (Helander, 1978) and overtaking (Rutley and Mace, 1970), and other factors that contribute to the perceived difficulty of the driving task (see Robertson, 1987 and Fairclough, 1990 for reviews). Recent development in microelectronics and communications technology has rekindled research interest in the quantification of driver mental workload. The proposed addition of computer technology into the vehicle environment (e.g. route guidance systems) must be evaluated in terms of driver safety. The development of measurement techniques (such as PPV) to assess driver mental workload are important to predict how the implementation of in-vehicle technologies might affect driving ability (Parkes, 1991). Naturally opinion over the precise role and value of psychophysiological research within the area of driver behaviour is divided. Armour (1978) assessed a number of techniques for the assessment of driver mental workload and concluded that:

'. . . problems with analysis, interpretation and interference from factors outside the driving task would exclude physiological measures from being useful (to assess driver task demand).'

It is difficult to deny the intuitive logic of this statement, however (in theory) a reliable PPV correlate of driver mental workload has a number of advantages over more traditional techniques of workload assessment (such as primary and secondary task methodologies). Using unobtrusive psychophysiological techniques it should be possible to capture data without disrupting the driving task (unlike secondary task methodologies). Additionally, the 'real time' nature of psychophysiological data (i.e. data capture and task performance are carried out simultaneously) contains none of the post-hoc distortion that may occur during the administration of subjective questionnaires. The theoretical 'pros' of PPV are offset by the pragmatic 'cons' involved in data collection and interpretation. Movement artifacts originating from the physical activities of steering and changing gears plus environmental variables, such as ambient temperatures, culminate as 'biological noise' which may confound psychophysiological data.

The following sections represent an attempt to review and evaluate major studies from the existing literature that have utilized psychophysiological techniques in the vehicle environment. The emphasis is on research conducted within the vehicle during either real-road or closed-circuit studies.

Electrodermal (EDR) measures

The category of electrodermal measures includes all PPV that describe changes in the electrical conductance of the skin. These are captured by placing two electrodes on the skin (usually the palms of the hands, the inside or outside of the wrist, the arch of the foot or the forehead) and passing a small current between them. The skin behaves as a resistor and as arousal increases, the resistance of the skin decreases due to changes in skin secretions such as perspiration.

Electrodermal (EDR) activity can be measured in terms of long-term ('tonic') units or short-term ('phasic') units. The basal level of skin conductance (tonic EDR or skin conductance level, SCL) is usually captured on a minute-to-minute basis to quantify long-term arousal. Most EDR research has concentrated on the phasic EDR measures known as the Galvanic Skin Response (GSR). The GSR takes the form of a characteristic waveform occurring approximately two seconds after the presentation of a stimulus and reaching a peak amplitude after three seconds (Lockhart, 1973). The GSR data may be represented as either, (i) a frequency measure, i.e. the number of GSRs occurring during an experimental trial, or (ii) a measure of degree of responsivity, i.e. which stimuli induce the largest amplitude GSRs. Edelberg (1970) makes the cautious point that these units of GSR data should not be used interchangeably as they may function independently of one another. For instance (he claims), at consistently high levels of arousal GSR frequency is consistent whereas the amplitude of the GSR exhibits a characteristic decrease.

As with the other PPV discussed in this paper, EDR studies of driving behaviour may be categorized as one of two types of study. The most common are 'stimulus/situational variable' studies (Stern *et al.*, 1980) in which differential psychophysiological activity is examined (usually as a correlation) as a function of different types of stimuli in the traffic environment. The second type of investigation uses psychophysiological techniques to discriminate between different subject groups or 'subject variables' (Stern *et al.*, 1980), i.e. male versus female drivers or experienced versus inexperienced drivers. The 'hypersensitivity' problem is particularly relevant to 'stimulus/situational variables' EDR studies within the vehicle. EDR activity is governed by the excitatory influence of the sympathetic branch of the autonomic nervous system, therefore GSRs may be evoked by any number of indiscriminate stimuli in the roadway environment. For example, Hulbert (1957) noted that male drivers GSRs coincided with the appearance of acquaintances, pretty girls and policemen who might be encountered en-route.

Psychophysiological investigations into the relationship between traffic volume and driver mental workload are amongst the earliest attempts to apply PPV within the vehicle. In 1962, Michaels measured the amplitude of the GSR as a function of traffic intensity. He found a linear relationship between GSR magnitude and traffic intensity until a certain point (approximately 2500 vehicles per hour), when the pattern exhibited a sharp, exponential increase. Michaels explained the latter effect as indicative of an increase of 'instream conflict' situations as higher densities of traffic change and merge lanes (this finding was replicated by Torres in 1971). In a similar fashion, Brown and Huffman (1972) measured SCL as drivers travelled through four types of traffic environment (residential, rural 2-lane highway, 4-lane expressway and 4-lane business district). They reported that SCL was lower for the rural condition where there was less traffic and fewer lanes.

We have already noted the distinct forms and measurement units that may be used to represent EDR activity. In 1974, Helander carried out a large study, involving over sixty subjects, that applied three measures of EDR in conjunction with vehicle parameters to investigate the relationship between driving performance and discrete traffic events. Helander employed a methodology in which trained observers accompanied the subjects in the vehicle in order to note the occurrence of traffic events, such as overtaking, traffic merging and the appearance of pedestrians and cyclists. Helander discovered that GSR frequency significantly correlated with the appearance of a cyclist or when the vehicle was either overtaking or being overtaken. By contrast, the tonic SCL measure correlated with steering wheel activity and lateral acceleration. This latter finding should be treated with some caution as it could be attributed to the physical component of steering the vehicle (Helander used the dorsal side of the hand as his EDR electrode site). The important feature of this study was that the SCL and the GSR magnitude measures were not significantly correlated (see Table 34.1). This data provides evidence for Edelberg's (1970) claim that EDR measures can be used to differentiate between long-term

Table 34.1   Correlation matrix of three measures of EDR activity captured during driving (adapted from Helander, 1974).
* = statistical significance (p < 0.05).

	SCL	GSR amplitude	GSR frequency
SCL (tonic GSR)	1.00	0.32	0.47*
GSR amplitude		1.00	0.63*
GSR frequency (phasic GSR)			1.00

physiological arousal (as represented by SCL) and specific responsivity to discrete events (as represented by the amplitude of the GSR).

The problem with Helander's (1974) study (and the others cited in the preceding paragraphs) is exemplified by the significant correlation between tonic SRL and steering wheel movement; in other words, the scope for potential confounding of PPVs by the physical activity of controlling the vehicle. Dependent variables (i.e. traffic density, overtaking another vehicle) that are often used in roadway studies simultaneously increase aspects of physical (i.e. steering, changing gears) and mental (i.e. judging distance, making decisions) driver workload. In brief, the researcher is faced with two choices. He or she may expand their conceptual framework of workload to include the physical as well as mental aspects of the driving task or attempt to 'filter out' the effects of physical workload. Helander (1978) returned to this problem during a field study investigating aspects of braking behaviour. He concluded that since the GSR peaked three seconds after introduction of the stimulus, then one could perform a time sequence analysis to differentiate between the onset of the GSR, brake pressure and the muscular activity of the right leg used to activate the brake pedal. Helander discovered that GSR activity anticipated the muscular activity of the leg before braking behaviour. He concludes that '. . . the recorded muscular activity had little if any influence on the electrodermal response.' (i.e. GSR) '. . . We propose that an EDR is evoked when there is a relative increase in task demand.'

Other researchers were interested in the relationship between GSR responsivity and fixed features of the roadway that involved little physical activity. Taylor (1964) had twenty subjects drive over a number of route sections during daytime and night time. As a dependent variable, he used an integrated measure of GSR which combined both frequency and amplitude of phasic EDR activity. Taylor's findings failed to reveal any significant difference of GSR responsivity during daytime and night time driving however he discovered a number of significant correlations between GSRs and other roadway variables. Taylor (1964) explains his finding via the following hypothesis:

'A major restriction on speed is due to other vehicles, the drivers of whom may be expected to behave in much the same way, therefore driving could be called a "collectively self-paced task". This pacing factor, or in servo terms, the error signal may be the GSR rate: if this is raised by larger and more frequent GSRs, a slowing of pace is called for; if there are few hazards, the pace is quickened (in order to achieve the object of travelling) until they reappear . . .'

This particular study has a number of important implications. In the first instance, it is often cited as experimental evidence of the phenomenon of 'risk compensation' (Wilde, 1988). Within this context the drivers' subjective appraisal of risk is linked to perceived level of mental workload or stress and both are measured by EDR activity. The problem with these findings is illustrative of the limitations which typify psychophysiological investigations of

Table 34.2   Correlation matrix of integrated GSR and traffic variables (from Taylor, 1964).

GSR per mile			
0.61	Accident rate per mile		
0.67	0.68	Side turnings per mile	
−0.75	−0.67	−0.8	Average speed

driving behaviour in the real road environment. Taylor placed tremendous emphasis on the positive correlation between EDR activity and accident rate and the negative correlation between EDR and average speed. Unfortunately less attention is given to those inter-correlations that exist between the independent variables of accident rate, side turning and speed irrespective of EDR activity. The positive correlation between accident rate and side turnings is not a new finding (Manning, 1949), furthermore a negative correlation between speed and number of side turnings may be interpreted as a decrease of speed in anticipation of other vehicles joining the road. Therefore, we may also conclude that EDR activity is due to either subjective appraisal of risk (as represented by accident rate per mile) or by those sources of 'uncertainty' inherent in the static features of the roadway environment (as represented by number of side turnings per mile). Under the circumstances of the study, both interpretations are equally valid and the end result, totally ambiguous.

Researchers have had more success using EDR to differentiate between demographic groups of the driving population. Helander (1977) compared GSR frequency and amplitude scores between experienced (driven more than 175 000 kms during the previous five years) and inexperienced (less than 30 000 kms during the previous five years) groups of drivers. Helander reported that GSR frequency to external traffic events and roadway features was significantly higher for the inexperienced group. This finding is fairly unexceptional, however Helander also reported that his two subject groups responded differentially to features of the roadway environment. For instance, the experienced group of drivers exhibited a higher amplitude of GSR (i.e. intensity of response) to a complex intersection than the inexperienced group. Conversely, the inexperienced drivers produced higher GSR amplitudes when approaching a narrow bridge compared to the experienced group. This study is important as it illustrates not only the importance of demographic characteristics to behavioural studies of driving as a whole, but also that EDR activity may be used to differentiate between covert sources of mental workload that exist in the roadway environment.

Electrocardiogram (ECG) measures

The electrical activity of the heart may be recorded at the surface of the skin via electrocardiography (ECG or occasionally known as EKG from the German spelling). The monitoring equipment used to capture ECG measures is less obtrusive than EDR apparatus and therefore more applicable in the vehicle environment. Electrocardiogram records may be expressed as, (i) heart rate, which is defined as the number of beats per time interval, usually beats per minute (bpm), and (ii) interbeat interval measured as the time between each heart beat (expressed either as mean unit time or the variance of the interbeat interval). There has been extensive research investigating changes in driver heart rate as a function of traffic events and the demands of the driving task. As with EDR measures, heart rate (HR) may be measured as a tonic (long-term) or a phasic (short-term) measure. The heart rate measure is less sensitive to independent variables in the environment than EDR measures as it is affected by the combined influence of both sympathetic (excitatory) and parasympathetic (inhibitory) branches of the human autonomic nervous system.

However, there is the usual problem of filtering the influence of physical workload from HR correlates of driver mental workload. Wyss (1971) examined this question via a comparison of the 'O$_2$ pulse' (ratio of oxygen consumption and simultaneous heart rate) values obtained from subjects driving a vehicle and working on a bicycle ergometer. The amount of O$_2$ consumption corresponds to the amount of oxygen intake by the heart and hence, the physical workload of the task. Wyss reported that for equal O$_2$ consumption, the O$_2$ pulse values during driving were always lower despite a higher value of heart rate. Wyss interpreted that the higher heart rate was due to the mental activity evoked by the driving task.

On a theoretical level, Lacey and Lacey (1967) posited that HR exhibits a characteristic decrease during periods of 'environmental intake', and shows a corresponding increase during the filtering out of 'irrelevant stimuli that will have a distraction value for the organism'. These characteristic changes in HR are phasic, i.e. they respond to transient stimuli in the external environment. Rutley and Mace (1970) attempted to correlate the phasic HR changes of four drivers with aspects of the driving task and discrete events occurring in the traffic environment. They found that phasic increases in HR coincided with so-called 'action points' from the drivers' perspective (i.e. pulling out of lane to overtake the vehicle in front), whereas phasic HR decreases were correlated with preparation for the 'action points' (i.e. slowing the vehicle when approaching a roundabout) and the completion of the action (i.e. stopping the vehicle). Rutley and Mace (1970) attribute the phasic increases of HR to the cumulative influence of both mental and physical workload. By contrast Hunt *et al.* (1968) noted a phasic increase of HR by (relatively) inexperienced police drivers whilst being overtaken by another vehicle, a passive action involving no overt, physical activity.

This latter finding resulted from a large study into the differences in HR patterns between two groups of police drivers (Hunt *et al.*, 1968). These researchers divided their subjects into two demographic groups, (i) experienced (class 1 police drivers), and (ii) inexperienced (class 4). Hunt *et al.* (1968) report that the experienced group had a lower mean HR when driving the vehicle compared to the inexperienced group. They also found that the experienced group showed less phasic HR response to 'unexpected hazards' and 'conflict situations' than the inexperienced group. As seen previously (in Helander's 1977 study), the psychophysiological index of workload and stress is influenced by subject variables such as driving experience.

HR measures have also been used to investigate the effects of driver-alerting devices installed on the roadway to decrease approach speed to an intersection. Watts (1974) used mean HR to compare two driver-alerting devices, a rumble strip (providing visual, auditory and proprioceptive cues) and a grid of yellow bars (providing visual cues only). He reported that significant increases of HR were obtained for the rumble strip, whereas this effect was not so pronounced for the painted grid. In a similar fashion HR measures have been used to provide a physiological index of workload and stress when evaluating the effects of in-vehicle technology on driver behaviour. Fairclough and Parkes (1990) reported how HR increased when drivers negotiated an unfamiliar route using a complex map compared to simple, electronic text display. Fairclough *et al.* (1990) also used HR to distinguish between an auditory secondary task which was conducted either, (i) within the vehicle to an experimenter in the passenger seat, or (ii) via a handsfree carphone to an experimenter in an office. These researchers reported that HR was significantly higher when the secondary task was conducted via the carphone.

Heart rate variability (HRV) is defined as the variability of the inter-beat intervals of the heart rate over a specified period of time. HRV may be measured either as a standard deviation or coefficient (Zeier, 1979), or subjected to a power spectrum analysis (Hyndman and Gregory, 1975) to yield the 0.1 Hz component (also known as sinus arrhythmia) which varies systematically as a function of mental workload (also see Brookhuis in this volume).

The 0.1 Hz sinus arrhythmia measure is reported to be free from the confounding influence of physical workload and therefore appears to be most appropriate for the investigation of driver behaviour. In 1982, Egelund reanalysed the data of Nygaard and Schiotz (1975) using the 0.1 Hz HRV, the standard deviation of the heart rate and the mean heart rate as indicators of driver fatigue (reflecting an increase of mental workload due to boredom or 'underload'). He found that the 0.1 Hz component steadily increased (indicative of a decrease of mental activity) in line with the distance driven, whilst HR and the standard deviation of the heart rate showed no direct relationship with fatigue.

Van Winsum (1987) used the 0.1 Hz HRV to distinguish between two types of route guidance systems. One system included a component of decision support, whereas the other did not. Unfortunately, details are scarce as the report is only available in abstract form at the time of writing. However, it is reported that the 0.1 Hz HRV increased when the subject used the system which contained the decision support component and when the subjects were familiar with the route.

Electromyography (EMG)

Electromyography (EMG) techniques are used to measure the electrical activity of specific groups of muscles. EMG may be used to provide a measure of muscular tension indicative of stress (when measured at the frontalis muscle situated on the forehead), or to filter out the effects of muscular movements by quantifying the physical workload of a given task. This latter technique was employed by Helander (1977, 1978) to differentiate the onset of mental activity from muscular activation. Zeier (1979) measured EMG activity in the frontalis muscle, but failed to differentiate between drivers using either manual or automatic gear transmission.

Electrooculography (EOG)

The visual activity of the driver may be captured in terms of eye movements or eyeblinks by placing electrodes around the eyes. Miura (1986) demonstrated that gaze duration decreases as the situational demands of the driving task increase. In the area of driver state monitoring, Neculau *et al.* (1986) discovered a good correlation between eye movements and steering behaviour, the latter being an indicator of driver arousal. Similarly Beideman and Stern (1977) monitored eyeblinks during a simulated driving task as affected by alcohol consumption and time-on task. They reported that blink frequency and duration increased as a function of time on task.

The electroencephalogram (EEG)

The electroencephalogram (EEG) record comprises a history of the electrical activity of the brain. Within this signal, several distinctive waveform may be identified and correlated to specific psychological states. For example, an increase of power in the *alpha* bandwidth of the EEG has been correlated with a performance decrement induced by medicinal drugs by Brookhuis *et al.* (1986). Similarly de Waard and Brookhuis (1991) demonstrated that EEG activation decreased under the influence of alcohol and correlated with increases in the lateral position of the vehicle.

Summary

The application of psychophysiological measurement techniques to driving behaviour on the real road creates a dilemma for the researcher. Ethical scientific investigation seeks to investigate the primary task of driving without significantly disrupting the driving task, and placing subjects and other road users in any explicit danger. The psychophysiological paradigm represents a means of quantifying covert changes in driver mental workload as a response to dependent variables inherent in the task without forcing the driver to perform any secondary task or significantly alter their normal driving 'style'.

Psychophysiological techniques may contribute to the study of driver mental workload and stress by providing a covert index of the motivation or 'effort' facet of the performance equation. For instance, the field study carried out at HUSAT (Fairclough *et al.*, 1990) employed vehicle parameters (primary measures), a modified NASA – Task Load Index questionnaire (subjective measure) and heart rate (PPV) to discriminate changes in driver behaviour when, (a) carrying out a verbal task on a handsfree carphone, (b) carrying out the same verbal task with a passenger in the vehicle, and (c) normal driving. All three categories of measures successfully discriminated between the two secondary task conditions and the control, but only the heart rate data discriminated between the two secondary task conditions. This example illustrates how PPVs provide additional information that increases the sensitivity of the experimental study. As Wastell (1990) points out:

> '. . . it is the potential for *disagreement* (rather than regularities) between the physiological, phenomenological and behavioural domains that gives psychophysiology its unique importance.'

This 'unique importance' is called into question by the problems of experimental artifacts (i.e. physical workload), confounding variables ('hypersensitivity' of PPVs), data interpretation (i.e. the non-causal nature of the psychophysiological correlates) and human individual differences (i.e. the influence of driving experience). The present chapter has provided a number of examples of how *not* to apply psychophysiological techniques in the vehicle environment. On the basis of this work, it is recommended that future field investigations:

- use PPVs which may be captured at sites that minimize any physical restrictions on the driver in order to minimize the invasiveness and physical artifacts, i.e. ECG measures captured via two chest electrodes.
- employ long-term (tonic) psychophysiological measures as phasic measures are usually confounded by the influence of many roadway variables.

The future use and applicability of PPV in the vehicle will depend, to a large degree, on the homogeneity of future research findings. If a sub-set of the PPV category could be formalized and demonstrated as a reliable index of driver arousal or mental workload, then a range of possible applications would emerge. For example, changes in drivers' psychophysiology may be tapped in order to detect the onset of fatigue or overload and provide appropriate warning messages. The problem with this approach is that the psychophysiological monitoring process must be unobtrusive. It is unlikely that this category of in-vehicle system (often viewed as coercive and autocratic from the point-of-view of many drivers) would be acceptable if the driver had to 'plug' electrodes from the body into an interface on the dashboard. Therefore, any PPV being used to directly tap the psychological state of the driver must be captured surreptitiously as in the case of electrodes on the steering wheel to monitor EDR activity, e.g. the 'Dormalert' system. If such a system could be realized (and successfully marketed), then the data could be used to quantify both the presence and *degree* of overload or fatigue and react accordingly. The output from a driver monitoring system might vary depending on the degree of impairment registered by the psychophysiological sensors. Naturally a monitoring system might provide feedback to the driver according to a 'schedule' or hierarchy of messages or responses that differ according to their severity. At one end of the continuum, the system may detect a mildly-fatigued driver and provide a warning message that advises him or her to take a break. If the system detects a driver who is obviously heavily intoxicated then it may take more stringent action, such as providing feedback to the driver to pull over, to switch on the vehicle's emergency lights (to warn other drivers) and at the most autocratic extreme, communicate with an IT network to inform the authorities.

This development of hierarchies or 'adaptive' interfaces (i.e. where the system 'adapts' to the perceived needs of the user) could benefit from the input of psychophysiological data. The PPV data may be used to supplement other data sources originating either in-vehicle or on-site at the side of the roadway to provide adaptive driver information support. The pragmatic problems of developing sensitive and sufficiently discreet monitoring devices should be solved as the technology develops over the next decade or so. The uncertainty over their application lies with the variability of the PPV themselves and the acceptability of these devices to the driving public.

If PPV could not be applied directly within the vehicle environment then they may be used as behavioural correlates of driver arousal. Within this context PPV are essential to the basic research that aims to identify reliable changes in driver behaviour due to variations in the drivers' psychological state. This approach relies on PPV as behavioural correlates of driver state, and captures the relationship between PPV and driving parameters such as steering wheel use, lateral position etc. (see DREAM report by Thomas *et al.*, 1989). PPV may also be used during the pre-market evaluation of in-vehicle systems by interpreting changes in driver arousal in conjunction with vehicle measures and subjective questionnaires as suggested by Parkes (1991).

To summarize, it is difficult to dismiss the usefulness of psychophysiological techniques on the basis of previous experimental studies of driving behaviour. The area is in a constant flux as new variables, monitoring apparatus and analysis techniques are developed and tested as part of the continuing search for a reliable physiological index of mental workload. In theory, the psychophysiological approach provides 'a method of triangulation that brings depth to our description and understanding of psychological phenomena' (Wastell, 1990). In pragmatic terms, the psychophysiological paradigm has a long way to go before it completely fulfils the hopes and expectations of behavioural researchers from this area of human investigation.

References

Armour, M., 1978, Practical measures of driver task demand, *ARRB Proceedings*, Vol 9, Part 5.

Babkov, V. F., 1973, La perception de la route par le conducteur—nouveau facteur dans la conception des projets de routes. *Proceedings of the First International Conference on Driver Behaviour*, Zurich, (cited in Helander, 1976).

Beideman, L. R. and Stern, J., 1977, Aspects of eyeblink during simulated driving as a function of alcohol, *Human Factors*, **19**, 73–7.

Brookhuis, K. A., Lowerens, J. W. and O'Hanlon, J. F., 1986, EEG energy-spectra and driving performance under the influence of some antidepressant drugs, in O'Hanlon, J. F. and de Gier, J. J. (Eds) *Drugs and Driving*, pp. 212–21, London: Taylor & Francis.

Brown, J. D. and Huffman, W. S., 1972, Psychophysiological measures of drivers under actual driving conditions, *Journal of Safety Research*, **4**, 172–8.

Edelberg, R., 1972, Electrical activity of the skin—its measurement and uses in psychophysiology, in Greenfield, N. S. and Sternbach, R. A. (Eds) *Handbook of Psychophysiology*, New York: Holt, Rinehart and Winston Inc.

Egelund, N., 1982, Spectral Analysis of heart rate variability as an indicator of driver fatigue, *Ergonomics*, **25**, 7, 663–72.

Fairclough, S., 1990, A review of psychophysiological measurement of driver mental workload, Report no. 22. DRIVE Project V1017 (BERTIE), HUSAT Research Institute, UK.

Fairclough, S. and Parkes, A., 1990, Drivers visual behaviour and in-vehicle information, Report no. 29. DRIVE Project V1017 (BERTIE), HUSAT Research Institute, U.K.

Fairclough, S., Ashby, M., Ross, T. and Parkes, A., 1991, Effects of handsfree telephone use on driving behaviour, *Proceedings of ISATA*, Florence, pp. 403–8, Automotive Automation Ltd.

Helander, M., 1974, Drivers physiological reactions and control operations as influenced by traffic events, *Zeitschrift für Verkehrssicherheit*, **20**, 3.

Helander, M., 1976, Driver's reactions to road conditions: a pychophysiological approach, Chalmers University of Technology, Göteberg Report.

Helander, M., 1977, Vehicle control and driving experience: a psychophysiological approach, *Zeitschrift für Verkehrssicherheit*, **23**, 1, 6–11.

Helander, M. G., 1978, Applicability of drivers electrodermal response to the design of the traffic environment, *Journal of Applied Psychology*, **63**, 4, 481–8.

Hulbert, S. F., 1957, Drivers GSRs in traffic, *Perceptual and Motor Skills*, **7**, 304–15.

Hunt, T., Dix, B. and May, P. I., 1968, A preliminary investigation into a physiological assessment of driving stress, Accident research unit, Metropolitan police, Tintagel House, Albert Embankment, London, (cited in Robertson, 1980).

Hyndman, B. W. and Gregory, J. R., 1975, Spectral Analysis of sinus arrhythmia during mental loading, *Ergonomics*, **18**, 3, 255.

Jex, H. R., 1988, Measuring mental workload: problems, progress and promises, in Hancock, P. A. and Meshkati, N. (Eds) *Human Mental Workload*, pp. 5–36, Amsterdam: North Holland.

Lacy, J. I. and Lacey, B. C., 1967, Somatic response patterning and stress: some revisions of activation theory, in Appley and Turnbull (Eds) *Psychological Stress*, New York: Appleton-Century-Crafts (cited in Perry *et al.*, 1980).

Lockhart, R. A., 1972, Interrelations between amplitude, latency rise time and the Edelberg recovery measure of the galvanic skin response, *Psychophysiology*, **9**, 437–42, (cited by Helander, 1978).

Manning, J. R., 1949, Accident rates on classified roads in Buckinghamshire, Department of Scientific and Industrial Research, Road Research Laboratory, Research Note No. RN/1094/JRM (Harmondsworth: unpublished) (cited by Taylor, 1964).

Michaels, R. M., 1962, Effect of expressway design on driver tension responses, *Highway Research Board Bulletin*, **330**, 16–25 (cited by Helander 1974).

Miura, T., 1986, Coping with situational demands: a study of eye movements and peripheral vision performance, in Gale, A. G. *et al.* (Eds) *Vision In Vehicles*, pp. 205–16, Amsterdam: North-Holland/Elsevier Science Publishers B.V. (cited by Thomas *et al.*, 1989).

Neculau, M., Sepher, M. and Kramer, U., 1986, Sacaddic eye movements and the workload of the car driver, in Gale, A. G. *et al.* (Eds) *Vision In Vehicles*, pp. 195–204, Amsterdam: North-Holland/Elsevier Science Publishers B.V. (cited by Thomas *et al.*, 1989).

Nygaard, B. and Schiotz, I., 1975, Monotoni på motorveje, *Rådet for Trafiksikkerheds Forsknung*, Notat 135, København (cited in Egelund, 1982).

Parkes, A. M., 1991, Data capture techniques for RTI usability evaluation, *Advanced Telematics In Road Transport*, Proceedings of the DRIVE Technical Conference, Brussels, pp. 1440–56, Amsterdam: Elsevier.

Perry, D. R., Armour, M. and Jackson, G. D., 1980, Driver task demand: theory and measurement, in *Ergonomics in Practice*, Proceedings of the 17th Annual Conference of the Ergonomics Society of Australia and New Zealand, pp. 75–89.

Robertson, S. A., 1987, *Literature Review on Driver Stress*, TSU, Oxford.

Rutley, K. S. and Mace, D. G. W., 1970, Heart rate as a measure in road layout design, Road Research Laboratory Report LR347.

Stern, R., Ray, W. and Davis, C., 1980, *Psychophysiological recording*, Oxford: Oxford University Press.

Taylor, D. H., 1964, Drivers galvanic skin response and the risk of accident, *Ergonomics*, **7**, 4, 438–51.

Thomas, D., Brookhuis, K., Muzet, A., Esteve, D., Tarriere., C., Norin, F., Wyon, D. and Schrievers, G., 1989, *Monitoring driver status: the state of the art*, DRIVE project V1004 (DREAM).

Torres, J. F., 1971, Objective measurements of driver-vehicle effort under field conditions and some relationships, *Proceedings of Symposium on Psychological Aspects of Driver Behaviour*, SWOV Voorburg, The Netherlands.

Waard, D. de and Brookhuis, K. A., 1991, Assessing driver status: a demonstration experiment on the road, *Accident Analysis & Prevention*, **23**, 4, 297–307.

Wastell, D., 1990, Mental effort and task performance: towards a psychophysiology of human-computer interaction, in Diaper, D. *et al.* (Eds) *Human-Computer Interaction—INTERACT'90*, pp. 107–12, Amsterdam: Elsevier Science Publishers.

Watts, G. R., 1974, The assessment and comparison of the covert responses to two driver-alerting devices, MSc thesis, Faculty of Engineering, University of London, London (cited in Helander, 1974).

Wilde, G., 1988, Risk homeostasis theory and traffic accidents: propositions, deductions and discussion of dissension in recent reactions, *Ergonomics*, **31**, 4, 441–68.

Wildervrank, C., Mulder, J. A. and Michon, J. A., 1978, Mapping mental load in car driving, *Ergonomics*, **21**, 3, 225–9.

Winsum, W. van, 1987, Mental workload of car driving, *Vetkeerskundig studiecentrum*, abstract only.

Wyss, V., 1971, Investigation on comparative O_2 pulse values measured in car driving and in equivalent work on a bicycle ergonometer, *Proceedings on Psychological Aspects of Driver Behaviour*, Institute for road safety research SWOV Voorburg, The Netherlands.

Yerkes, R. M. and Dodson, J. D., 1908, The relation of strength of stimulus to rapidity of habit formation, *Journal of Comparative Neurology of Psychology*, **18**, 459–82.

Zeier, H., 1979, Concurrent physiological activity of driver and passenger when driving with and without automatic transmission in heavy traffic, *Ergonomics*, **22**, 7, 799–810.

35

The use of subjective rating in deciding RTI success

C. Lindh and P. Gårder

Introduction and aim

The following work has been carried out within the Swedish research of PROMETHEUS programme 76000 Evaluation. Introducing RTI aims at ensuring a higher level of traffic safety than at present, while simultaneously reducing environmental impacts, travel times and vehicle operating costs. One of our goals, in society in general, is to create a healthy environment. A large percentage of the population spends a great portion of their time on the road, both during working hours and at leisure. It is therefore reasonable to demand a 'healthy' traffic system. Health, according to the World Health Organization (WHO), is a state of complete physical, psychological, and social well-being, and not only an absence of sickness and weakness. In other words: traffic safety, measured in the number of accidents (or conflicts), is not a sufficient indication of how healthy the system is. Every day stress may also take a toll on health. Therefore, another prerequisite for introducing RTI is that this can be done without reducing the subjective well-being of the road-users.

Stress is said to be one of the most serious problems in society, causing both traffic risks and traffic accidents. In a medical context stress is sometimes defined to have positive implications as well as negative ones. In this context, we define stress as negative stress, whether it is 'background stress', 'time stress', 'traffic stress', or some other stress. Many leading researchers in the field, e.g. Cox (1988) and Johansson (1989), define negative stress as 'a Cognitive Experiential Process'. It is generally agreed that a continuous subjective perception of the situation, and of the perceived ability to cope with its demands, will influence whether a temporary or more permanent state of stress is developed.

Negative stress on a motorist may occur from a combination of the following or from each one of the following examples:

- more or less permanent 'background stress' caused by difficult working conditions, family conditions, etc.,
- temporary background stress from, e.g. time-pressure caused by uncertainty of whether one will arrive in time for something important,
- 'traffic stress' or acute stress, due to difficult or non-influenceable driving tasks, such as fear of running into an accident, or getting stuck in a traffic congestion and not knowing for how long.

The three types of stress outlined above are thought to interact to a combined level of stress. Furthermore, they are difficult to distinguish between. One of our hypotheses is that temporary background stress, as defined above, hardly occurs in a person that has no permanent background stress; and situations leading to acute stress are more likely to occur among drivers with high levels of background stress.

Another way of looking at stress in traffic is that of Tavris (1989) in her book *Anger, the Misunderstood Emotion*. In discussing 'traffic anger' as seen in the USA she mentions the following contributing factors to the problem:

- the influx into a community of people whose driving rules differ from the natives. Every city has its unspoken pace and rhythms, and visitors violate them at their peril . . .
- increasingly crowded driving conditions and clogged traffic cause physiological arousal—increased blood pressure and heart rate, signs of tension. Drivers feel confined, unable to control their pace and movement, and unable to get where they are going, which adds to their impatience. If, in addition, they *used* to be able to get where they were going more quickly—if they remember the 'good old days' before traffic jams—they have the psychological burden of diminished expectations, which are annoying in their own right . . .
- drivers rehearse a 'mental script' in which they interpret the behaviour of other drivers as intentionally rude or insulting . . .
- the invisibility of other drivers and one's own anonymity. If you accidentally step on someone's foot in an elevator, you are likely to say 'Excuse me.' . . . the more distant we are from others, and the more anonymous they are to us or we to them, the less restrained, courteous, and empathic we become.

The problem is not unique to traffic situations, but is rather a special case of the 'crowding-and-anger' question, a category of interference in getting where YOU are going.

Tavris (1989) also says: 'Driving a car is stimulating, especially in thick traffic, because it demands constant attention and split-second decisions . . . The same high arousal that . . . fast drivers find exciting, however, can turn rapidly to anger (an intense anger, at that) when their movements are blocked by someone who can be blamed. Yelling . . . or honking the horn loudly is an effort to restore control, reduce the sense of danger, and retaliate against the offending person in your way'.

She continues 'Of course, these reactions . . . do not relieve your anger. They make it worse, both by pumping up your blood pressure and by causing the target of your wrath to yell back at you, which in turn makes you feel angrier yet.'

In summary: it is obvious that anger may create traffic safety risks, and that negative stress and anger are closely linked. Whether we look at traffic from the stress research point or from the anger point we see that the last few decades have meant increased traffic problems; and that these problems are likely to grow to intolerable levels in more and more areas around the world. Certain RTI-functions may be able to counteract this development.

It is obvious that a high level of stress is detrimental to the health, but stress might have other negative outputs. Stress leads to a modified behaviour, for example, less cautious or faster driving. A natural hypothesis is, therefore, that there is a relationship between the level of stress and the risk of meeting with an accident. It is not clear if such a relationship would be of a functional type—for example linear, exponential or U-shaped—or if there is a threshold value, so that a stress level below this value is 'safe', while a higher stress level gives a high risk. But as long as we only deal with negative stress, it ought to be safe to assume that a higher level of stress gives a higher risk. (This does not mean that we deny that driving a car can be soothing on the nerves for a stressed person.) Therefore, one of our prerequisites for an RTI is that it is at least not allowed to increase the level of stress. A question that remains to be answered is how the level of stress is to be measured in a valid and reliable way when evaluating alternative RTIs.

Alternative RTIs

Different types of RTIs have to be outlined in order to solve different problems, e.g. the following ones which would:

(a) decrease the uncertainty in deciding arrival time,
(b) inform the drivers of road conditions,
(c) keep a sleepy driver alert,
(d) soothe a 'stressed' driver,
(e) help the driver to overtake other vehicles safely, and thereby reduce travel time,
(f) enable communication outside the vehicle.

Thus RTI functions potentially have many positive implications. Among these are the supposed effect of reducing stress and/or anger. Whether stress and/or anger is caused by 'background stress', 'time stress' or 'traffic stress' may not be necessary to analyse. In other words, different RTI functions solve different problems. Many of these RTIs are thought to have positive safety implications, though some of them might also have adverse effects in certain situations. It is especially difficult to estimate the effects of functions that still do not exist even as prototypes. Sometimes deterministic models can be used to estimate the expected (average) effects in a theoretical way. At other times computer simulation can be of help. But often 'field simulation' is needed. This means that RTI functions are simulated, and their effects measured in real life situations.

The method of subjective rating of stress

Cox (1985) comes to the conclusion that it is not possible to measure stress directly with physiological techniques. What one can get is physiological correlates. The questions of reliability, validity or 'fairness' should therefore be questioned in all types of measurements, not only when it comes to subjective ratings. The first question that has to be answered is whether we can use some kind of expert judgement. If not, can we ask the drivers how they feel, how they are influenced by the surrounding traffic, and how they react to this traffic and the traffic environment in general? Cox (1985) says that stress refers to a complex psychological state coming from the person's cognitive appraisal of the demands from the environment, e.g. the traffic situation. Many people think that stress exists through the persons inability to meet the demands of a situation. But the important point is the discrepancy between demands and ability. The first step is to ask the person about their feelings in relation to traffic and traffic environment. If subjective rating methods are to be accepted, they have to be reliable, valid and 'fair'. One dimension of the subjective state of emotion relates to the feeling of uneasiness/well-being, or 'hedonic tone'. Another dimension relates to alertness/drowsiness. The first stage is often called stress and the second arousal. The specific subjective rating scale that is used in this pilot-study is described below.

Adjectives describing negative stress and subjective rating scales

The measure we have used here combines ratings into a '*stress-index*'. The rating scale was developed by Kjellberg and Iwanowski (1989), and the purpose is to measure subjective stress using a mood-adjective checklist. They describe how a short list of 6 + 6 adjectives was developed from a starting list of 47 adjectives. Factor analyses indicated that these ratings could be reduced to two fundamental dimensions: 'Energy' and 'Stress', or 'eustress' and 'distress'. The analysis was carried out in a work condition context, but can be applied to other fields, such as traffic situations. The resulting adjectives from the study mentioned above, translated into English, are presented in

Figure 35.1. Here we can see that each of the chosen adjectives measures only one of the two aspects (eustress or distress), i.e. these two aspects can be clearly separated using these adjectives.

In our pilot study, we have used a sample of the 12 adjectives above plus the word 'concentrated', giving a model according to Figure 35.2.

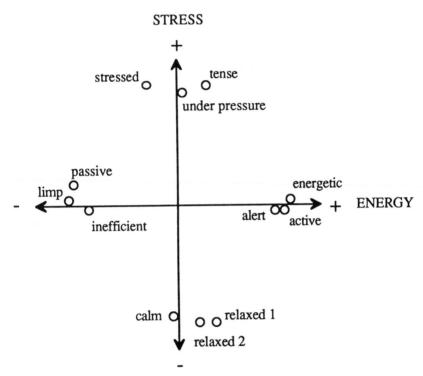

Figure 35.1 *Test of adjectives measuring the degree of eustress and distress.*

	Demands	
	low	high
low	passive inefficient	stressed tense under pressure
Self Control		
high	calm relaxed	energetic concentrated active

Figure 35.2 *Model of Demand/Self Control measuring adjectives*

Examples:

In a traffic situation this means that 'caught in a traffic jam' leads to low self control. If the driver at this time is under a lot of 'time stress' it will mean 'high demand' simultaneously, which is the most negative type of combination.

A high degree of self control and high demands may correspond to every day driving with high speeds as long as the traffic density is not too high; while high self control and low demands may correspond to a typical picnic -trip in low traffic density.

The level of stress has, in our pilot study, been estimated by having the road-user rate his degree of feeling 'stressed', 'under pressure', 'tense', 'calm', 'relaxed', and 'relaxed'[1].

Each adjective was rated in a six degree scale, as:

$$0 = \text{not at all} \qquad 1 = \text{hardly at all}$$
$$2 = \text{somewhat} \qquad 3 = \text{rather}$$
$$4 = \text{very} \qquad 5 = \text{very, very}$$

A *stress index* was then calculated as follows:

$$\text{Stress index} = 1/6 \{v_{stressed} + v_{pressured} + v_{tense} + (5 - v_{calm}) + (5 - v_{relaxed\ 1}) + (5 - v_{relaxed\ 2})\}$$

where $v_{stressed}$ = value on the degree of being 'stressed'

$v_{pressured}$ = value on the degree of being 'under pressure'

v_{tense} = value on the degree of being 'tense'

v_{calm} = value on the degree of being 'calm'

$v_{relaxed\ 1}$ = value on the degree of being 'relaxed' (in Swedish 'avspänd')

$v_{relaxed\ 2}$ = value on the degree of being 'relaxed' (in Swedish 'avslappnad').

By taking 5 minus the value for the 'positive' adjectives, they all become 'one-dimensional' with increasing values meaning more 'stressed'.

This way of calculating a stress index is fully in line with the way Kjellberg at the Swedish National Institute of Occupational Health (Arbetsmiljöinstitutet) has used the measure in his research on stress in work conditions. Therefore, we have extensive data sets for comparison purposes if we want to judge whether a particular stress level is high or not.

The pilot-study—a simulated RTI-function

We have simulated an 'RTI function' by letting the road itself 'imitate' a more sophisticated tool in guiding overtaking of other vehicles (function (e) p. 393: an RTI that helps the driver to overtake other vehicles safely, and thereby reduces travel time).

[1] The Swedish synonymns were 'stressad', 'pressad', 'spönd', 'lugn', 'auspönd' and 'auslappnad.'

In Sweden, where we have a large network of two-lane roads that are 13 meter wide and few kilometers of multi-lane expressways, it might be of interest to repaint 13 meter two-lane roads (with wide shoulders) into 13 meter three-lane ones, in order to get a higher level of information with respect to overtaking possibilities. Does this give reduced levels of stress? We have addressed this question in a pilot study comparing the levels of stress among randomly chosen drivers along two such segments.

The situation without RTI is, in other words, simulated by driving on a traditional Swedish (13m) wide two-lane road with wide shoulders, which are allowed to be used (but do not have to be used) for driving on when being overtaken. Here overtaking is often done in spite of short sight distances, and with oncoming traffic present. The overtaking vehicle has to use part of the opposite lane. Oncoming drivers might possibly not see an overtaking vehicle. This is thought to create stress among the road users (risks occur to those oncoming as well as those being overtaken and those overtaking).

The situation with RTI is simulated by driving on a road with the same width and traffic volumes, but with three lanes instead of two. Two lanes in one direction enables safe overtaking, simulating an RTI saying 'overtaking allowed'; while only one lane in the opposite direction hinders overtaking, simulating the RTI saying 'no overtaking'. Combined with the knowledge that overtaking will be allowed in less than two kilometers (when the one lane direction gets two lanes), one simulates this qualitative information as well. The layout of the two types of sections are illustrated in Figure 35.3.

The study was carried out in the autumn of 1990 on Highway 40 between Borås and Göteborg, in the west of Sweden. The road has a ten kilometer

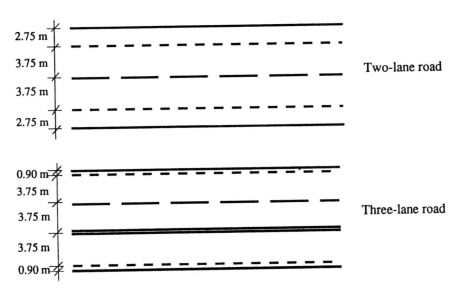

Figure 35.3 Layout of the studied sections

stretch with the two-lane section and another part of the road, somewhat longer, with three lanes, one in one direction and two in the other. In the one-lane direction overtaking is impossible (without violating the rule on using the opposite direction lane). After approximately two kilometers the overtaking possibilities are changed to the opposite direction. This can mean that a fast driver will have to wait for about two kilometers before he can overtake. On the other hand, it means that he is always assured of being able to overtake once the three-lane section is reached.

In total, 92 persons were stopped. Out of these, one refused to answer any questions, and two were foreigners, unfamiliar with the Swedish language. Forty-five drivers were interviewed on the two-lane road, and 44 on the three-lane one. They were asked to answer three questions:

Task 1:   Rate your level of stress and relaxation on a six level scales. (Six adjectives were added into one 'stress-index' as described above.)

Task 2:   Distribute your stress according to source.
          1. Road and traffic
          2. Intervention from police
          3. Other sources, like time-pressure etc.

Task 3:   Do you consider the three-lane road less stressful than the two-lane one?

All subjects interviewed excluding the two foreigners, could answer the questions without any great effort, and understood the chosen adjectives so they could give them values without obvious hesitation.

Results

Our pilot study showed that the stress index was reduced from 1.28 on the two-lane road to 1.14 with the simulated RTI. On the two-lane road 16 persons had a stress-index of 1.5 or more, on the three-lane road only 11 interviewed reached this level. The tendency seems clear, but the numbers are not significantly different, so there is a risk that the difference could be due to random variation. (So far we do not know which levels of stress are acceptable. Future research will aim at establishing relationships between a given stress-index value and behaviour related to safety.)

With the direct question: 'Do you consider the three-lane road less stressful than the two-lane one?' the following results were obtained. On both roads the answer shows that around 87 per cent were in favour of the three-lane road. Of the remaining 13 per cent (11 persons) some thought they gave equal stress, and others that the three-lane one gave more stress. The latter case (more stress) was motivated by not being able to overtake immediately, but this was not a very commonly expressed opinion.

If we take Task 2 into consideration 'Distribute your stress according to source' we get the following results:

	Two-lane	Three-lane
1. Road & traffic	42%	32%
2. Intervention from police	25%	38%
3. Other sources, like time-pressure etc.	32%	29%

If we multiply these percentages with the total average stress-index in the respective section, we find that the 'road and traffic contribution' is 0.53 for the two-lane road and 0.36 for the three-lane one.

Conclusions and further studies

It is hard to say whether 1.5 is a level that starts causing a safety problem or not. The expert earlier referred to (Kjellberg) stated in a telephone interview that a level of 1.4 or 1.5 is normal in regular work conditions and that a 'high' stress index would rather be above 2 or even 2.5. This shows that none of the road segments are experienced as 'stressful'. However, Kjellberg also is of the opinion that background stress and traffic stress are additives, i.e. that any contribution from the traffic environment can cause a hazard among drivers with high background levels.

One strength of the experimental design used was that 'normal' road-users were randomly stopped along the road. This means that normal variations should exist in temporary background stress level. This is never the case when paid subjects are used. (It is impossible to study how time-pressure influences stress realistically when the subject is paid for the time he is driving, and knows how long the test will go on for.)

A negative aspect with the subjective rating method is that it can not be used as a tool for automatically triggering an aid (co-driver assistance) when a driver reaches unacceptable levels of stress. An alternative method of measuring the level of stress will be needed if this is requested.

Another drawback with the method of using subjective rating scales (in general) is the scepticism towards this method among road and traffic researchers. This scepticism does not exist among pure stress researchers, who consider the method one of the most valid. In order to meet the scepticism of the former group, we plan to combine the subjective rating with physiological methods, tentatively with Respiratory Sinus Arrythmia (RSA) measurements. The RSA is based on the fact that the pulse-rate normally increases during inhaling, and decreases during exhaling. This is true for all humans (as well as other mammals), as long as the person is not subject to heavy stress. If the person is subjected to heavy stress, then the pulse-rate does not decrease while he exhales. In other words, if the pulse-rate is measured continuously one can see if the person is 'stressed' or not. The RSA seems to have an advantage in that it reacts almost instantly, and that one can get an estimate of the stress level for specific situations, and not only end (or mean) values as the subjective rating method gives (Grossman, Stemmler and Meinhardt, 1990). A disadvantage with the RSA method is that the measurement in practice can be obtained only when the person is inhaling/exhaling deeply, i.e has been told: 'inhale deeply now', so called induced RA measurement. We now plan to continue with further testing of the RSA method, in spite of the shortcoming that it has to involve 'paid' subjects.

References

Cox, T., 1988, in Fisher, S. and Reason, J. (Eds) *Handbook of Life Stress, Cognition and Health*, New York: John Wiley.

Grossman, P., Stemmler, G. and Meinhardt, E., 1990, Paced respiratory sinus arrhythmia as an index of cardiac parasympathetic tone during varying behavioral tasks, *Psychophysiology*, **27**, 4.

Kjellberg, A. and Iwanowski, S., 1989, Stress/Energi-formuläret: Utveckling av en metod för skattning av sinnesstämning i arbetet, *Enheten för Psykofysiologi, Arbetsmiljöinstitutet*, Stockholm,

Tavris, C., 1989, *Anger—The Misunderstood Emotion*, New York: Simon & Schuster.

36

Contributions and limitations of simulator studies to driver behaviour research

L. Nilsson

Introduction

If we observe real life traffic of today, we are faced with a system containing an almost infinite number of situations. These traffic situations are neither static nor similar, but vary in different respects. For example, the driving task can be everything from driving alone on a six-lane motorway, to critical collision avoidance manoeuvres at intersections. The various driving tasks may be complicated by: adverse driving conditions such as darkness, fog and low friction; by difficult road lay-out and infrastructure; and by too much information. On top of that, the traffic environment also contains chance events or various degrees of uncertainty. Individual drivers with different abilities and limitations are exposed to, and expected to cope with, the described traffic system. If they do not succeed in their task the result may, in the worst case, be an accident.

Thus, accident statistics are the ultimate criterion in traffic safety research. But statistics reflect only an aggregated view. When we want to study specific traffic situations there are too few (!) accidents to permit safe conclusions about causative factors. Nevertheless, knowledge about the causes of accidents is very important in striving for improved traffic safety. Therefore, we have to study the behaviour of individual drivers, and also to point out the variations (variability range) in behaviour between drivers. Then we have to interpret the results in terms of traffic safety and risk. Field studies in real traffic, and simulator studies are two different ways for studying behavioural aspects of individual drivers.

Simulator requirements

When simulator studies are discussed, as an alternative to field studies in real traffic, it must be remembered that there is a broad scale of simulators with various characteristics and capabilities. The scale goes from very simple static mock-ups, over video game solutions, and somewhat more advanced equipment with some control possibilities, to very sophisticated driving simulators like the VTI simulator (Nilsson, 1989; Nordmark, 1990), and the Daimler Benz simulator (Drosdol and Panik, 1986). To study the complete driving task, and when the driver's reactions and manoeuvring play a central role, an advanced simulator is preferable. For such simulators it is obvious that certain demanding design criteria must be fulfilled:

- The simulation must run in real time.
- The model describing the vehicle must be complete in the sense that all the subsystems such as engine, brakes, transmission, steering system, suspensions etc. must be modelled.
- The model describing the vehicle characteristics must be able to represent a broad spectrum of handling qualities, for example front wheel drive, rear wheel drive, and different levels of understeer and oversteer.
- The vehicle model must also be able to reproduce the effects of the interface between vehicle and road, for example slippery surfaces and gravel roads.
- The simulator must have a wide angle visual system, preferably in colour. The picture should contain enough detail to give a realistic driving impression.
- The simulator must have a moving base system for the simulation of inertia forces.
- It is crucial that the time delay introduced by the simulator is short compared to the lags of real vehicles (100–250 ms). That is, the driver must not experience a delay between a certain manoeuvre (for example a steering wheel turn) and the corresponding change of the visual scene.

Advantages

It has been said that 'measuring influences on driving performance in a real driving situation is the best representation of what happens on the road' (de Gier and Nelemans, 1981). Then, why do we need simulators and when to use them?

One advantage of simulators, which is important in behavioural research, is that they allow driving conditions and environmental conditions to be kept constant. All subjects participating in a study can be exposed to exactly the same conditions. In field experiments, it is obvious that the factors that influence driver behaviour (except those studied) are usually too many and too unpredictable (in occurrence and in effect) to be appropriately controlled.

Besides, interactions between the factors can make the interpretation of experimental results from field studies very difficult. Therefore, driving simulators are well suited for comparative effect studies where one or a couple of factors are systematically varied in steps, while all other factors are kept constant.

Example. The effects of handsfree mobile telephones on safety related driving behaviour have been investigated in a DRIVE project, using an advanced driving simulator (Alm and Nilsson, 1990; Nilsson and Alm, 1991a). The driving behaviour of subjects performing a specific telephone task while driving was compared to the driving behaviour of subjects who only drove. Data were collected from exactly the same road sections for all subjects (driving with, as well as without, the mobile telephone). The mobile telephone also rang in exactly the same positions along the road for all the subjects in the 'telephone group'. Effects were obtained on speed choice, steering ability, brake reaction time and workload. Due to the methodology used, these effects could be concluded to be caused by the mobile telephone use.

Because the environmental and driving conditions can be kept constant in a simulator, the required number of subjects can be much smaller, compared to field studies with preservation of the same statistical 'strength' in the obtained data. Another significant advantage of driving simulators is that no matter how seriously degraded the driver performance is (by mental overload, fatigue, drugs, alcohol etc.), there is no danger whatsoever to the driver or other road users. Driver behaviour in specific situations and under specific conditions can thus be tested to the point of catastrophic failure.

Example 1. The driving performance of subjects with habitual sleep spells while driving, that is drivers suffering from the sleep apnéa syndrome, has been compared to the driving performance of control subjects (Haraldsson *et al.*, 1990). Brake reaction time and tracking deviations were significantly increased. The driving simulator was thus found to be a sensitive and 'safe' tool to ascertain vigilance impairment in quantitative terms. It was also considered to be a valuable tool to evaluate the efficiency of surgical treatment on driving behaviour of selected sleep apnéa syndrome patients.

Example 2. A system for detecting drops in driver wakefulness, based on steering wheel movements, has been evaluated in an advanced simulator (Petit *et al.*, 1990). In 28 out of 30 experiments the system was capable of discriminating between two levels of wakefulness. Furthermore, in 12 of the experiments the subjects fell asleep and drove off the road. The dozing off could have been prevented by the system if the alarm had been actuated. If so, the resulting 'accident' would probably not have happened.

In both the DRIVE and PROMETHEUS programmes large research and development efforts are made to use modern information technology to support drivers in their task. In the light of this fact another advantage with simulators is obvious, namely that it is possible to test and evaluate proposed in-vehicle systems during the conceptual phase of the development.

Example. Behavioural effects and driver attitudes of a simulated vision enhancement system, intended to be used in poor visibility (fog), have recently been studied within a PROMETHEUS project (Nilsson and Alm, 1991b). Driving performances in clear conditions, in fog, and in fog with the vision enhancement system available were compared. The 'system' consisted of a monitor positioned on the hood near the windscreen. On the monitor a 'clear' picture of the road and its environment was presented to the driver. Speed, lateral position on the road, and brake reaction time were all found to be influenced by the presence of the support system.

Limitations

There are situations where physical laws limit the use of a driving simulator, even if it is advanced. The forces needed to perform certain manoeuvres may be impossible to create by a moving base system through rotations (roll and pitch) and linear motions.

The possible motion and velocity ranges in a driving simulator are limited, as are the simulated accelerations. Therefore, some scaling is necessary. Because of the scaling, or long time lags, a moving base simulator may fail to coordinate the visual and moving parts to give a realistic impression of the accelerations (forces) generated while driving. Even if the time lag in the moving base is negligible, it is possible that the effect of scaling might be interpreted by the driver as a vehicle that is less responsive or slower in reaction, compared to a real vehicle. The driver experiences a lack of correspondence between visual and motion impressions, which can lead to uncertainty and abnormal behaviour. Considering 'normal' driving, the described limitations rarely introduce problems. They are more likely to appear during experiments that contain many sharp bends, and in experiments that require fast and hard manoeuvres.

Another simulator problem is, of course, that it is impossible to fully represent the real traffic environment, with all its components and huge variability. In most applications a complete representation is not necessary, but in some cases a low level of realism, in the presented driving task, can cause abnormal driver behaviour, which in turn may lead to interpretation problems. Route finding (navigation) is an example of a driving task which is unsuitable to study in a simulator for this reason. When drivers try to find an unknown destination, in an unfamiliar environment, they make consecutive route choice decisions at a series of choice points, mainly intersections of different but often high, complexity. If driver behaviour during such route finding is to be studied, and correct interpretations made, it is crucial that the presented road configurations are extremely realistic. Otherwise, the decision making task may not appear sensible. The presentation of realistic landmarks is also important. So far, the requirements mentioned for the navigation task have not been satisfactorily met in simulators. The problems identified imply that the effects (the potential) of different designs of route guidance systems ought not to be studied and compared in driving simulators.

Nausea

Nausea is not considered to be a great problem in simulator studies, even if it occurs now and then. Because of its rare occurrence it is difficult to know the causes of simulator nausea. It is not unusual for the same individual to experience severe problems on one occasion, but little or no problems at all on another occasion. One reason for nausea may be the discussed lack of correspondence between visual and motion impressions during, for instance, hard manoeuvres and driving through frequently occurring sharp bends.

Validity

Because of the enormous variety of conditions in real traffic, and the limitations in reproducing them in a simulator, it is impossible to generalize from simulator driving to real car driving and vice versa. Every new situation needs to be validated, and it is always necessary to evaluate the measures used so they really reflect driver behaviour relevant to traffic safety. The question of validity is not satisfactorily answered yet, but is continuously considered.

Example. Eye movements (number and position of fixations, scan path) for curve negotiation tasks have been compared between field tests and simulator tests (Laya, 1990). How well the manoeuvres were performed (speed choice, steering pattern) as well as driving experience were taken into account in the analysis. The results show some correlation, although several discrepancies were also obtained. The reasons for these are not clear and demand further research for clarification.

Measures

Recalling the description of the traffic system, the statement made by Crawford (1961) that 'it has proved extremely difficult to define what is meant by driving performance and to develop adequate techniques of measuring it', seems to be relevant and to hold true even today. The fact that measurements from the real driving situation are the best representation of what happens on the road is probably also true. The problem is, what measurements to take from a complex real situation and how to interpret them. Something that has implication for traffic safety must be 'measured', and this something is not valid just because the measurements are carried out whilst driving in real traffic. If a meaningful interpretation of results is to be obtained, the choice of measurements, to reflect the traffic safety aspects of driving behaviour, is critical, independent of the method or tool used. This choice is probably easier for the more standardized simulator world.

One type of measurement, which is frequently used is driving performance measures, e.g. steering wheel movements. Such movements have been reported to become less frequent as the time on the task increases and the level of wakefulness decreases. On the other hand, it has also been demonstrated that experienced drivers make less steering wheel movements compared to inexperienced drivers. Thus, in the first case few steering wheel movements are taken as an indication of degraded performance, while in the second case the same behaviour is interpreted as improved performance.

In studies of driver behaviour it can be valuable to relate the driving performance to the condition of the driver. Thus, physiological measurements may be considered. The prerequisites to take physiological measurements are probably better in a simulator than in a real road setting. The advantages are mainly of a technical nature. For instance power supply and noise suppression are easier to provide in the simulator environment.

Discussion and conclusions

The interest in using driving simulators in behavioural research has increased markedly, especially during the last three years. The reason may be that driving simulators have been found to provide a safe and repeatable research method for studying drivers' reactions and vehicle handling, even under unstable driving conditions when the driver loses control of the car. Also, the focusing on development of various technical support systems for in-car use have increased the demand for methods to test system concepts, in a realistic way. From the increasing number of application studies, it is obvious that a driving simulator offers such a great variety of possibilities that the problem is not to find suitable issues for its use. The problem is rather to select which studies to perform, and which measurements to choose from the innumerable number of available variables, in order to optimally utilize the possibilities. Therefore, it is important that the prospective simulator user realizes the limitations, and understands that certain types of studies are less suitable, because the limitations are too obvious to the subjects, making the interpretation of results difficult.

The importance of a moving base has been a constant source for discussion. It has been argued that a lot of research, for instance concerning MMI design, can be done just as well in static simulators. This may be true, at least for certain purposes. Certain aspects of display and control design (colour, size etc.) can, of course be studied in the laboratory or in a static simulator; but when the topic to investigate is the 'total' driving task and behavioural effects, e.g., speed choice and steering patterns under different conditions, the ('true') motion feedback from a moving base simulator, or a real vehicle, is probably of paramount importance for the relevance of the results.

In view of the fact that driving to a large extent consists of visual scanning, the significance of the bad side view and the lack of rear view in most simulators must be clarified. Our knowledge in this area is too limited, but a reasonable hypothesis is that the use of rear view mirrors depends heavily on the complexity of the driving task. For example, when the level of workload is high due to the traffic situation ahead, the driver probably looks in the rear mirror less frequently.

When driving simulators are used to study driver behaviour, in order to gain knowledge that can be used to increase traffic safety, the main advantages are that a lot of subjects can drive under equal conditions, and that studies can be performed which are not ethically defensible in a real traffic situation. The main drawback with driving simulators is that the question of validity is not satisfactorily answered.

References

Alm, H. and Nilsson, L., 1990, *Changes in driver behaviour as a function of handsfree mobile telephones—a simulator study*, DRIVE Project V1017 (BERTIE), Report No. 47.

Crawford, A., 1961, Fatigue and driving, *Ergonomics*, **4**, 143–54.

Drosdol, J. and Panik, F., 1986, The Daimler-Benz driving simulator—a tool for vehicle development, Society for Automotive Engineers.

Gier, J. J. de and Nelemans, F.A., 1981, Driving performance of patients receiving Diazepam medication, in Goldberg, L. (Ed.) *Alcohol Drugs and Traffic Safety*, Vol. 3, pp. 1009–23, Gothenburg: Graphic Systems.

Haraldsson, P.-O., Carenfelt, C., Laurell, H. and Törnros, J., 1990, Driving vigilance simulator test, *Acta Otolaryngological.* (Stockholm), **110**, 136–40.

Laya, O., 1990, Eye movements in actual and simulated curve negotiation tasks, Symposium *Vision in Vehicles*, 22nd International Congress of Applied Psychology, Kyoto, Japan, July 22–27.

Nilsson, L., 1989, *The VTI driving simulator—description of a research tool*, DRIVE Project V1017 (BERTIE), Report No. 24. VTI, Linkoping Sweden.

Nilsson, L. and Alm, H., 1991a, *Effects of mobile telephone use on elderly drivers' behaviour*, DRIVE Project V1017 (BERTIE), Report No. 53. VTI, Linkoping Sweden.

Nilsson, L. and Alm, H., 1991b, *Effects of a simulated vision enhancement system on driver behaviour and driver attitudes*, PROMETHEUS Report VTI, Linkoping Sweden.

Nordmark, S., 1990, The VTI driving simulator—trends and experiences, *Proceedings from Road Safety and Traffic Environment in Europe*, Gothenburg, Sweden, September 26–28.

Petit, C., Chaput, D., Tarrière, C., Le Coz, J.-Y. and Planque, S., 1990, Research to prevent the driver from falling asleep behind the wheel, *34th Annual Proceedings Association for the Advancement of Automotive Medicine*, Scotsdale, Arizona, October 1–3.

37

Human and technical reliability

G. Reichart

Introduction

Some readers might wonder why a contribution reflecting aspects of human and technical reliability is included in a book on advanced RTI-systems. Although the link to RTI-systems is not obvious it is more important than one might think.

Many years of successful research have been spent on the improvement of passive safety, which means the mitigation of consequences of an accident. With the current research programmes, like PROMETHEUS and DRIVE, strong emphasis has been placed on active safety, which means the prevention of accidents. A number of RTI-systems with varying degrees of automation are recommended as remedies for the existing road traffic safety problems. This seems to be justified by the traffic accident statistics which, in all countries, point to the human errors as the dominant factor of accident causation. Up to 90 per cent of all road traffic accidents, with injuries, are attributed to human error (VDA, 1989), and the widespread view now considers the human as the weak link of the complex multi-man, multi-machine system 'road traffic'. Many of the RTI-systems under development in the above mentioned research projects aim at an improvement or, in the long-term, even a partial replacement of this weak link by autonomous systems.

However, a closer look at current practices of accident recording and analysis, as well as experience in other technical fields, question this traditional paradigm. Moreover, results of traffic conflict research reveal a surprisingly high human reliability in road traffic. It will be shown in this contribution that even technical systems require a considerable design effort to meet or surpass the level of reliability of a human driver in a given driving task. Thus appropriate forms of driver support are required, which will not take away driver responsibility, and which allow her/him to learn from errors made in a system context, which provides some degree of fault-tolerance. This change in the view of the human role in road traffic has therefore immediate implications for the kind and design of driver assistance systems. What we need are RTI-systems based on a concept of a human centred automation. In this paper an attempt is made to discuss some aspects of the underlying assumptions of this position and to outline some of its implications.

Reliability—some definitions

To aid an understanding of the arguments outlined in this contribution a few basic definitions and concepts of reliability engineering have to be introduced. The interested reader may refer to one of the many textbooks available (SAE, 1987; Visvanadham, 1987; MBB, 1986) to gain a deeper insight into this field. Reliability engineering has gained growing importance with the increasing complexity of systems. What does the term reliability mean? According to (MIL-STD 721) and (IEC-Publication 271) reliability can be used as a synonym for the 'success probability', this means the probability that an item (system, component or part) successfully meets the requirements placed on the behaviour of its characteristics during a specified (mission) duration. The complement of the success probability to one is the failure probability.

The failure probability p_F of an item is defined by the probability that the item fails at least once during the time t, which is equivalent to the probability that the duration T of successful performance ('life time') is less than t. The term 'life time' means the time from the beginning of the component life up to its failing. Collecting life time data on a sample of identical components will yield different life times for most of the components. Its expected value is the mean time to failure (MTTF). The distribution of the life time can take various forms, for electronic systems the assumption of an exponentially distributed life time is quite often justified. The reciprocal of MTTF is in case of an exponential distribution the so-called failure rate λ. Thus the failure probability p_F can be expressed as:

$$p_F = 1 - e^{-\lambda t}$$

The success probability p_s corresponds to

$$p_s = 1 - p_F = e^{-\lambda t}$$

The unavailability u of an item is defined as the probability, that the item cannot fulfil its function at a given point within the time-interval T_1 to T_2. If there is no repair time after a failure and the demands on the function are uniformly distributed over the time interval the average unavailability $\bar{u}$ corresponds to the failure probability at the time $(T_2 - T_1)/2$.

$$\bar{u} = \frac{1}{T_2 - T_1} \times \int_{T_1}^{T_2} (1 - e^{-\lambda t}) \, dt \approx \lambda \times \frac{T_2 - T_1}{2}$$

There is another way to estimate the probability of a human error in a given task, and this is based on the definition, that a good estimate of the probability

p_e of an event can be derived from the relation of the number of the positive respective negative outcomes n of an experiment after N trials:

$$p_e = n/N$$

The greater N the better will p_e approach the 'true' probability.

Applied to human errors one can estimate the probability of an error p_H of a given type in a specified task by:

$$p_H = \frac{\text{number of errors of a given type}}{\text{number of opportunities for an error of this type}}$$

For instance, if in 500 observations of dialling a 6-digit phone number 15 wrong inputs have occurred, this would yield an estimate for the probability of the incorrect dialling of a 6-digit number by

$$p_H = \frac{15}{500} = 3 \times 10^{-2}$$

The same approach can be used to infer, from the number of observed 'traffic-conflicts' and the number of resulting accidents, the probability of the transition p from a given 'traffic-conflict' to an accident:

$$p_T = \frac{\text{number of accidents}}{\text{number of traffic conflicts}}$$

Human reliability in road traffic

The current practice of accident recording and analysis has influenced a segment of the public to consider the human as the unreliable element of the road traffic system. However, this prevailing opinion has to be questioned for several reasons. The accident recording by the police is oriented towards liability of persons involved, rather than towards accident causation (Mitschke, 1983). Moreover, the research on accident causation has to face a number of methodological problems. Accidents are statistically rather rare events, with a more or less unique pattern of contributing factors in each individual case. Also the recording of required data is still a rather unresolved issue despite a few exceptions (Wanderer, 1976). Little is usually known about the pre-crash-phase, which would be the most interesting source of information for an analysis of the causal mechanisms of an accident. Even if one considers the human role in the road traffic from a risk perspective, risk understood as the product between frequency of an accident times the level of damage, there is no reason to assume a low human reliability. It is important to note that there is a rather low individual risk, but a considerable risk within the whole population due to the many vehicles in operation. At least the low individual risk of being killed due to participation in road traffic, which is about $R = 2 \times 10^{-4}$ per year, yields little evidence for the assumption of a low human reliability in the road traffic, taking into account the skills, rules and knowledge needed by a driver for a safe traffic participation. Table 37.1 gives a rough idea of the number of observations and decisions a driver has to make, and shows how low the related accident probability is. It has, therefore, become obvious that simple monocausal cause-effect relations are of no help, especially for the achievement of progress in the field of active safety, which means the prevention of accidents (BASt, 1989).

Various models and procedures have been proposed for the analysis and assessment of human reliability, which means the probability that a person performs some system-required activity (in a required time period—if time is a limiting factor) within specified limits of acceptability. These approaches, developed mainly for manufacturing and process plants, can be applied as long as well defined proceduralized tasks have to be assessed for which a normative sequence of acts is available as a reference (Rasmussen, 1990). But for the behaviour of road traffic participants, for which a concise theory is still missing, such approaches are hardly feasible. Nevertheless, it seems to be very useful to consider some of the basic assumptions of this human reliability research and to transfer it to the road traffic.

Human error is basically understood as a human output which is outside the tolerances established by the system requirements in which the person operates (Swain, 1983; Rigby, 1970). The tolerance limits range from material constraints, e.g. guard rails in road traffic over legal constraints, e.g. traffic signs; to 'after-the-fact' established limits, e.g. 'non-adapted speed'. Only to the degree of consciousness and insight into these limits can a responsibility be attributed to a person in cases where these limits have not been observed. In these cases one should speak of violations rather than human errors. The system 'road traffic' is generally characterized by rather weakly defined tolerance limits. This fact is reflected by the still missing concept of a normative model for safe driving behaviour. Moreover, even if some rules exist like 'do not drive over 80 km/h, if the road surface is wet', the question 'what means wet and how can it be correctly assessed by the driver' remains unanswered. Another problem is due to the fact that some limits are not at all

Table 37.1   Driver-related events in traffic and their frequency (Platt, F. N., 1986).

Events	Frequency
Observations	200 per mile
Decisions	20 per mile
Errors	1 per 2 miles
Near-collisions	1 per 500 miles
Collisions	1 per 61 000 miles
Personal injury accidents	1 per 430 000 miles
Fatal accidents	1 per 16 000 000 miles

in accordance with a driver's experience, and not observing them rarely results in any negative consequences. Sometimes even examples can be found in which the limits are conflicting, e.g. at night the reflective markers along a rural road indicate clearly the road boundaries and the course, but they give a misleading cue for the driver's visual ability to see objects ahead. Behavioural tendencies often conflict with established limits, e.g. speed limits along wide roads will be less observed than along small roads, in which the difficulty of the situation is clearly experienced by the driver. This problem of only weakly defined limits is likely to be one of the basic problems in traffic accident causation. Consequently, it seems to be advantageous to make the existing limits more and earlier 'visible' with the help of RTI systems.

Unlike tasks in other industrial settings, participation in road traffic, e.g. by driving a vehicle, does not consist of a well-structured proceduralized sequence of activities. It is characterized by a partly self-paced, partly system-paced sequence of situation-dependent activities in a complex dynamic environment, which leads to severe problems in the development of suitable analytic tools. A lot of psychological theories and concepts have been applied to the study and explanation of human error in road traffic accidents—see *Ergonomics* Vol. 34 (1991) for an overview—which provide various potential explanations of the causal mechanisms involved. But after many years of accident statistics and research a convincing and comprehensive concept is not at hand yet. However, it has been demonstrated that much can be gained from the observation and analysis of traffic conflicts (Brown, 1990; Gstalter, 1983). The traffic conflict research has yielded a lot of valuable insights even into the relation between conflicts and accidents. As a rough estimate, transition probabilities from conflicts to accidents in the range of 10^{-4} till 10^{-5} can be assumed. Traffic conflicts are defined as 'situations where interaction between several road users (or between a vehicle and the environment) would result in a collision, unless at least one of these involved takes evasive action; it is the success of the action that determines the final result—conflict or collision' (Malaterre, 1980). Thus the transition probabilities can be regarded as the success probability for taking an evasive action.

Human and technical reliability—some comparisons

The transition probabilities mentioned above allow us to compare technical and human reliability in road traffic. If a technical system is to be introduced, which automatically shall initiate and perform an evasive action, it has at least to meet or better to surpass the human success probability. For such a technical system two reliability characteristics are of main interest:

- the probability of initiation of the correct action, when demanded (availability), and
- the probability of correct performance.

Since the duration of the required performance is usually very short, the probability of failure during this phase can often be neglected. Thus the basic characteristic is the average unavailability of the system, when demanded. This average unavailability has to be less than $\bar{u} = 10^{-5}$ to yield any significant benefit in terms of risk reduction compared to the human performance according to the results of the traffic conflict research. This is a rather strong requirement, which cannot easily be met unless specific design principles of reliability engineering are applied. This can be shown by a simple example. Given a system with a failure rate of $\lambda = 10^{-5}$/h the mean unavailability for a time T of 300 h (roughly the operation hours of a private car per year) can be assessed to $\bar{u} = 1.5 \times 10^{-3}$, a value two orders of magnitude greater than required. Redundancy techniques have to be applied to reach the required low failure probability.

With the assumptions that the system can be fully tested after each start of the engine, and has to run in the worst case for 10 hours between two tests; and with the further assumption that potential demands of the system functions are uniformly distributed within this interval, one derives a average unavailability of $\bar{u} = 5 \times 10^{-6}$, which is just somewhat better than the human. However, full-scale functional tests are hardly feasible, thus a very careful reliability and safety analysis and an appropriate system design are of utmost importance.

Some consequences for the automation of driver tasks

The above argument strongly suggests that we do not replace the driver, but assist him with intelligent technical means in the driving task. This support is not restricted to measures inside of the car, but is also related to improvements in driver training as well as in the infrastructure, either by technical systems in the infrastructure or by improvements of the road lay-out. This rather comprehensive approach is necessary, since a variety of factors related to all elements of the system, driver-vehicle-environment, are contributing to the occurrence of accidents, and is in accordance with results from other areas of the safety research; e.g. nuclear safety where various researchers have demonstrated that factors of the work situation, like design of display and controls, the type of tasks and organizational aspects, have a dominant impact on the probability of human error. Swain (1983) has called these influences 'performance shaping factors' (PSFs) and he has chosen the terms 'error prone' for situations in which the PSFs are such as to increase strongly the likelihood of human error, and 'accident-likely' if the errors will result in damage or injury. The same author estimates that in process-plants about 85 per cent of all human errors are situation-caused rather than human-caused.

Table 37.2   Requirements for a human oriented automation.

- Maintain or enlarge the control over the environment
- Tolerate human errors
- Maintain the validity of mental models
- Resolve the problems resulting from nonlinear system behaviour
- Reduce duration for learning by 'teach-back'-functions
- Make the system adaptable to various driver populations
- Assure a consistent and transparent system behaviour
- Design for reliability and safety
- Guarantee controllability in case of malfunction

Any attempt to simply replace the drivers by automated systems is likely to miss the intended goals of safety improvements, as long as such implications are not considered. This should be interpreted as an argumentation against automation in general, but it is a call for a human centred automation concept of vehicle driving, which has been outlined, to some extent, in another paper (Reichart, 1989). The main characteristics of this concept are shown in Table 37.2.

The decisions, what and when to automate tasks, which have been performed by drivers up to now cannot be successfully based on a purely technological imperative. It has to be seen that reliability is only one, but an important aspect in deciding on an appropriate automation concept. The debate on function allocation between man and machine has already a rather long tradition. Many proposals and concepts have been developed in the various technical fields.

For future RTI-systems it is very interesting to look at the experience in aerospace-technology. After many critical comments about traditional cockpit-automation concepts, NASA has developed a concept for a human centred automation based on a scientific investigation of appropriate automation philosophy (Norman *et al.*, 1988; Norman and Orlady, 1989). The basic view of this concept is to provide the pilot and the crew respectively with a number of functions for information-management and decision-making support. The automation is basically seen as a tool for the crew, which can be used situation-dependent on demand by the crew. This is well in accordance with results of the engineering psychology, that human control over technical systems and over the environment plays an important role for operator motivation and acceptance as well as for the safe use of these systems (Seligman, 1975; Wiener and Curry, 1980; Frese *et al.*, 1987).

The implications of automation are manifold. We still lack a concise theoretical approach towards a human centred automation. A lot of publications (Bainbridge, 1982; Wickens, 1983; Reason, 1990) have pointed towards the problems (see Table 37.3 for a summary of major problem areas which can arise, if these implications are not considered very carefully).

Table 37.3   Potential problems of automation

- Responsibility/Authority double bind
- Increased monitoring load
- False sense of safety
- Out-of-the-loop familiarity
- Higher level of operator error
- Sensitivity to design errors
- Greater influence of maintenance errors
- Loss of skills
- Reduced learning
- Reduced system transparency
- Reduced coherence of remaining operator activities

Nevertheless automation concepts in vehicles which have been implemented rather successfully include:

- automatic transmission,
- ABS (anti-lock brakes),
- ASC (drive slip control),
- automatic climate control,
- automatic search of radio stations.

These automations have addressed either secondary functions, or are active in very limited circumstances in which the driver still has the prime control. The driver has been unburdened from unnecessary and distracting tasks, and the systems optimize the driver's control actions but leave him or her as the ultimate authority for the initiation and, to a large degree, the performance of the control function. This automation philosophy is consistent with the requirements mentioned before, and if the underlying principles are further pursued benefits for the active safety can be expected.

Conclusion

The list of automated functions given above does not intend to say that this kind of automation is the only way to proceed. Systems, which assist the driver in his lane keeping performance, which help her/him to maintain a safe distance between the vehicle in front, or which assist her/him in the information collection from the environment, will be well in accordance with the presented concept, as long as the driver has the ultimate authority. It might also be possible to develop systems, which can intervene, if the driver does not react properly, and a collision would otherwise be impending. But there is utmost care needed to achieve high levels of safety and reliability of such systems as well as to avoid negative impacts on the driver behaviour due to overreliance or responsibility shift. It is the conviction of the author that a human centred automation cooperating with the human element will lead to an optimized overall system performance. For the ongoing development of driver support systems some consequences should be drawn:

- High levels of systems reliability and safety are a prerequisite for any improvements of the active safety.
- The kind of support delivered by these systems has to be tailored towards the needs of the drivers.
- The behaviour of the RTI-systems has to be consistent with the drivers expectations.
- The RTI-systems should not replace the driver but should cooperate with him.

This means that we need to develop a much better understanding of what errors, under what circumstances and with what frequency are likely to occur. This is very similar to what (Reason, 1990) has stated: 'Modern technology has now reached a point where improved safety can only be achieved through a better understanding of human error mechanisms.'

References

Bast, 1989, Bundesanstalt für das Straßenwesen (Hrsg.), Verkehrsmobilität und Unfallrisiko in der Bundesrepublik Deutschland, *Unfall- und Sicherheitsforschung*, Heft 72.
Bainbridge, L., 1982, *Ironies of automation*, IFAC/IFIP/IFORS/IES Conference on Analysis, Design and Evaluation of Man-Machine-Systems, Baden-Baden.
Brown, I., 1990, Drivers margins of safety considered as a focus for research on error, *Ergonomics*, **33**, 10 & 11.
Frese, M. *et al.*, 1987, Industrielle Psychopathologie, *Schriften zur Arbeitspsychologie* Bd. 23, Hrsg. Ulich E., Bern: Hans Huber Verlag.
Gstalter, H., 1983, Der Verkehrskonflikt als Kenngröße zur Beurteilung von Verkehrsabläufen und Verkehrsanlagen, Dissertation TU Braunschweig.
Mitschke, M., 1983, Definition kritischer Situationen im Kraftfahrzeugverkehr—Eine Pilotstudie, *Automobilindustrie*, Heft 3.
Malaterre, G. and Mühlrad, N., 1980, Conflicts and accidents as tools for safety diagnosis, in, Older, S. J. and Shippey, J. (Eds) *Proceedings of the 2nd international traffic conflict workshop*, TRRL Supplementary Report 557.
Norman, S., Billings, C. E., Nagel, D., Palme, E., Wiener, E. L., and Woods, D. D., 1988, *Philosophy of Flight Deck Automation*, Carmel Valley, CA.
Norman, S. and Orlady, H. W., 1989, *Flight Deck Automation: Promises and Realities*, Final report of a NASA/FAA/Industry Workshop held at Carmel Valley, CA, NASA Ames Research Center, Moffett Field, CA 94035.
Platt, F. N., 1986, cited from Braess, H. H. and Frank, D., Mensch-Auto-Zukunft, *ATZ Automobiltechnische Zeitschrift*, **88**.
Rasmussen, J., 1990, The role of error in organizing behaviour, *Ergonomics*, **33**, 10 & 11.
Reason, J., 1990, *Human error*, Cambridge: Cambridge University Press.
Reichart, G., 1989, Autofahren und Überwachungstätigkeit—Möglichkeiten und Grenzen, in *Kolloqium Leitwarten*, TÜV Rheinland, Köln.

Rigby, L., 1970, The nature of human error, *Annual Technical Conference Transactions*, ASQC, Milwaukee, WI.

Society of Automotive Engineers, 1987, *Automotive Electronics Reliability Handbook*, Warrendale, USA.

Seligman, M. E. P., 1975, *Helplessness. On depression, development and death*, San Francisco: Freeman.

Swain, A. D. and Guttman H. E., 1983, *Handbook of Human Reliability*, Analysis with Emphasis on Nuclear Power Plant Applications NUREG/CR-1278, Nuclear Regulatory Commission, Washington DC.

Visvanadham N., Sarma V. V. S. and Singh M. G., 1987, *Reliability of Computer and Control Systems*, Amsterdam: North-Holland.

Wanderer U. and Weber H., 1976, Verkehrsunfälle unter der Lupe-Unfallforschung am Unfallort, *VDI-Nachrichten* Nr. 6.

Wiener E. and Curry R., 1980, *Fight Deck Automation: Promises and Problems*, NASA Technical Memorandum 81206, Ames Research Center, Moffet Field, CA.

Wickens, C., 1991, *Engineering Psychology*, Columbus, OH: Charles E. Merrill.

38

Can artificial intelligence help traffic policing?

J. G. Harper

Introduction

This paper seeks to establish a level of plausibility for increasing the automation of traffic violation detection and deterrence. It should not be taken for
granted that artificial intelligence (AI) can make an important contribution to
improving enforcement effectiveness. Indeed, why bother to introduce AI into
the solution at all? Given the current state of AI development, there are few
convincing *a priori* arguments for favouring its approaches over traditional
structured analysis, excepting a philosophical disposition to solve problems
using mechanisms modelled upon human reasoning abilities. Moreover, the
utility of AI techniques in this particular problem domain is rendered somewhat ambiguous by the faint separation between conventional technology used
in a novel way and the application of a novel methodology to a conventional
problem. If the solution to a problem requires manipulating knowledge of
facts and rules then it is the second course which predominates. In this
scenario artificial intelligence is a very useful analytical tool. Once a problem
has been decomposed correctly into its AI segments, various computational
interpretations are available. The computation of the AI analysis allows us to
test various hypotheses without committing ourselves to a final more practical
implementation. It is important that this point be grasped clearly. All too
often conjuncts of the form *artificial intelligence and* . . . suggest a
problematic synergy. This paper does not claim that a knowledge-based
analysis has a direct on-site implementation, but that it contains important
heuristic guides for prototype development.

The problem: compliance in perspective

Traffic law enforcement would be unnecessary if traffic law compliance was observed. Of course, the fact is that this principle is more honoured in the breach than the observance. At least one component of the puzzle is the failure by drivers to treat traffic violations as manifestations of criminal behaviour. This presumed dissonance significantly contributes to undermining enforcement programmes (Whitlock, 1971; Grime, 1987). However, from the traffic authorities vantage, deterring 'bad' driving must proceed on the back of a detection programme (more acutely, the *threat* of detection must be subjectively appreciated by drivers). Consequently, a punitive enforcement methodology ensures that while drivers are punished for violatory driving, they remain unrewarded for 'good' driving. Whether this approach fosters ambivalence towards traffic laws is an issue beyond the scope of this paper. What is not in doubt is the public tolerance of ambivalent attitudes. To subvert such ambivalence many enforcement programmes are accompanied by relevant publicity campaigns (Shinar and McKnight, 1985; Roojiers, 1988; Reidel *et al.*, 1988; Gundy, 1988). Both British studies on seat-belt compliance and the Irish study on the effects of the breathalyser demonstrate that a joint programme-publicity strategy can be very effective (Grime, 1987; Walsh, 1987). Other studies acknowledge the utility of a joint strategy, but equally conclude that without a constant tangible level of enforcement it is insufficient (Shinar and McKnight, 1985; Makinen, 1988; Roojiers, 1988; Rothengatter *et al.*, 1989).

Enforcement strategy

Effective enforcement operates on a *deterrence through detection* principle. Violation detection in itself is not sufficient to guarantee compliance. This is the responsibility of the post-detection legal process. If offences earned no punitive judicial scrutiny, it is difficult to imagine a useful future for enforcement programmes. The 'halo' or group effects of a large number of programmes have been measured in various studies and commented upon (Shinar and McKnight, 1985; Aberg, 1988; Makinen, 1988). The studies by Makinen (1988) and Aberg (1988) reveal that the rate of compliance was extremely low despite intense enforcement. Aberg points out that 60 per cent of drivers persisted in violating a 30 km/h limit. All the studies draw attention to disappointing short-term halo effects. Such observations prompt Rothengatter (1988) to ask whether violation behaviour can be effectively managed by current enforcement strategies for any significant period of time? Tilting towards a negative answer is the banal fact that in a survey by Ostvik *et al.* (1989) of police forces in Ireland, The Netherlands, Norway and Spain, understaffing was cited as one of the critical debilitating factors in enforcement.

Increasing drivers' subjective assessments of detection requires persistent visible enforcement. Taking the constraints above plus environmental variables into account, the argument pursued here is that current police resourced programmes are not adequate to the enforcement task. The development of semi-automatic monitoring systems over the last twenty-five years has laid a foundation for innovative fully automatic systems. One contention for investigation is that an automatic system with some borrowings from AI can make an effective and sensitive intervention tool.

Crime and punishment

The commonest means for punishing offenders are fines and license revocation. Superficially, the utility of such means appears unchallengeable, but the period of elapsed time between detection and conviction can obscure the serious criminality of an offence. Some researchers argue that the time lag has diluted the reaction to judicial response to an unacceptable level (Whitlock, 1971; Rothengatter, 1988). Furthermore, they point out that the punishments (fines) are frequently paltry. Traffic offences are perceived as minimal transgressions of the social contract, and therefore minimum police intervention is expected by the driving public. Will a significant reduction in elapsed time improve compliance rates? There is not enough evidence to support an incontrovertible answer. Extrapolations from available studies, however, such as Lamm and Kloeckner (1984), support the tentative conclusion that immediate punishment reduces noncompliance. In an automatic policing system introducing immediacy does not pose a problem.

Monitoring technology

Current traffic monitoring technology is focused almost exclusively around the use of induction loops, radars and cameras. The loop and radar can be used as sensors triggering the action of a camera. Radar, however, is more commonly used to both detect and measure speeding. The reports of Shinar and McKnight (1985) and Chin (1989) demonstrate with a high degree of confidence the effectiveness of camera-sensor circuits. Unfortunately, the analysis of a photograph still requires a human eye. Monitoring technology of a global nature is contained in AUTOGUIDE and CACS (Jeffrey *et al.*, 1987; Shibata, 1989). These systems provide navigational information to a driver. Since each requires a vehicle's identity and displacement as input, logically each fulfils the main functions of an automatic monitoring system.

Non-navigational monitoring is also practical as the following examples show. The PREMID automatic toll collection system in Norway relies on a microwave beam to reflect a vehicle's identity from a window mounted card. As a vehicle crosses into the toll section its toll transaction is automatically recorded, with appropriate deductions from the driver's account with the toll company. If a vehicle's identity is not reflected, a video camera automatically records the rear-end of the vehicle as it leaves the detection area[1]. Using UHF technology, the Lojack Corporation of Massachusetts produces a system

---

[1] I am extremely grateful to Mr Richard Muskau of the Institute of Transport Economics (TOI), Oslo for bringing this system to my attention and generously offering to translate the technical literature.

which broadcasts a 'find me' signal in the event of vehicle theft. Upon purchase of a system, the purchaser is given a unique code number. Should the vehicle be stolen, the code number is given to the traffic police who broadcast it and activate the vehicle's transmitter. Using a special receiver, a police vehicle can focus on the source of the signal and (hopefully) recover the vehicle.

Automatic policing?

Two salient inferences may be drawn from the above sections. Firstly, the transiency and uneven coverage of current enforcement programmes. It renders them largely ineffectual on a meaningful scale. Secondly, both semi-automatic and automatic systems have been shown to effect significant reductions in violation rates. Given the state of current technology the question is not whether automatic policing is feasible, but whether it can be rendered acceptable. Harper (1991) outlines the technological scenario, with hypothesized behavioural effects, which could at present house an automatic policing system[2].

The main tasks assigned to an automatic system would encompass the detection of speeding offences and hazardous driving. Speed detection is almost trivial, and provided each vehicle is equipped with some means for broadcasting its identity, attributing an offence to a vehicle is feasible (we assume single-lane systems; multi-lane systems introduce a measure of complexity which we have not fully explored)[3]. Hazardous driving can arise due to alcohol consumption, carelessness and inexperience. No system could predict the cause of the phenomenon, but merely recognize it. In fact the principle underlying the detection of this type of behaviour is simply that the system must record violations, not reason as to how they come about. If it were the case that the recording of a violation became contingent upon the intentional state of the driver, the judicial process would be set aside. Moreover, it is doubtful that any envisioned technology could consistently both identify intentional states and associate them with behaviours. A division of responsibility between proposed monitoring and policing functions of an automatic policing system must be accepted.

A present role for artificial intelligence?

From an untutored perspective it seems most natural to introduce an AI

---

[2]I am no longer as convinced as I was in 1989 that the arguments advanced for AI in automatic policing hold water (Harper, 1991). My current view is that AI has much to contribute once the linkage between knowledge representation and driver modelling is solved. I do not foresee a satisfactory solution for quite some time to come, however.
[3]I am not assuming that the 'some means' is technically trivial, but experiences with beacon-transponder systems and microwave systems evidence that the technology has moved well beyond working prototypes.

component by representing the traffic laws as a set of production rules. Each rule is structured as: **if <condition1, ...> then <action1, ...>**. The resulting *expert system* theoretically harbours all the knowledge necessary to judge whether a violation has occurred. This characterization is a misconception. More correctly, it is a misapplication of a useful software methodology. Expert systems are very effective where a problem requires reasoning among alternative conclusions (explanations). Traffic violation detection does not require such sensitive processing of input. Consequently the traffic laws, particularly those pertaining to speed, are more readily represented as nested conditional statements used by any procedural language, e.g. as found in PASCAL and C. For instance, if the detected speed is greater than the legal threshold, then an offence has occurred. There is no mystery about how the conclusion was reached, no detailed chain of reasoning to be followed and certainly no requirement for an expert system.

Separating the technology from the knowledge required for it to be useful raises two questions: (a) what knowledge is at issue; and (b) how should it be structured. The knowledge is defined by the traffic laws and the 'rules of the road'. The combined rule set is composed from a series of *<condition>* ... *<action>* pairs. In terms of detection requirements the rules can be expressed as conditional statements in almost any standard computer language. Hence, the AI influence on the structuring of the practical knowledge is non-significant. This should not be taken to imply that the rich AI formalisms for knowledge representation have nothing to offer the task at hand, but that the level of representation required does not involve sophisticated schemes for cognitive processing. When all has been said and done, we are talking about the satisfaction of straightforward boolean conditions.

A future role for artificial intelligence?

Intelligent behaviour is adaptive to changes in the environment. A vehicle travelling at 10 kmh on a motorway is generally violating a rule which says something like 'you must drive at a speed which does not hinder other road users'. But these are under ordinary circumstances. The traffic rules by and large provide a *standard interpretation* of road user behaviour. Where AI may have a future role is in determining the application of traffic rules under non-standard conditions. But even in this scenario it is doubtful if anything more than changing an assignment statement for a speed threshold is involved.

Based on our own research we are willing to conjecture that, even in current applications where vehicle monitoring is undertaken by human controllers, the amount of tacit or presumed knowledge is actually quite small. If a controller 'can do very little' about a problem situation, they tend not to use sophisticated reasoning to repair the problem (Harper and Fuller, 1993). In the context of traffic problems the number of recovery strategies available are quite small. Hence, though shutting down a lane may cause congestion, the action is the result of quite uncomplicated reasoning. Obviously, it is assumed that some *worst case* analysis has been performed prior to the action, but practical traffic flow problems limit the utility of an optimum response.

In the case of automatic policing, the points above are equally valid. The functions of the system must be to detect and deter traffic violations. The ordering of the functions is essential. If a system is weak on detection then it is highly unlikely to be a good deterrent. The facts remain, however, that the response strategies available are small in number. Arguably the best that can be expected is immediate detection of a violation, with perhaps a voice warning on an in-vehicle radio stating that legal processing has or will begin. By and large, the objectives of the AUTOPOLIS project were to establish that this strategy was feasible and likely to produce the desired behavioural effect (Harper, 1991).

Another possible avenue for AI involves user-modelling. In essence, it is an attempt to create a computationally tractable psychological profile of a user in terms of their knowledge, skills and attitudes relative to a particular task. Kass and Finin (1988) upon completing a review of other work in the area offer a comprehensive General User Modelling Schema (GUMS). The generality of such systems, integrating belief models, reasoning and 'detected' user skills, obscures the lack of any sustainable computational representation of the very idea itself. In fact Kass and Finin readily admit that in precisely those areas where user modelling would be very useful, the critical nature of the work prevents their application. For instance, nuclear reactor management. In certain environments the penalty for error must be so low as to be negligible. Unhappily for proponents of user-models the highway is a far too dynamic environment for their application. In the first place personality traits, derived from age-experience profiles, may be so general as to be vacuous. Secondly, there is no evidence to suggest that detecting pathological driving behaviour will be beneficially influenced by driver-models. Finally, convictions are secured on the basis of observed behaviour, not conjectured intentional states.

Conclusion

The AUTOPOLIS project established that conventional technology could successfully perform the functions of an automatic policing system. More speculatively it tried to evaluate the usefulness of AI techniques in any such system. Certainly, at the level of software simulation they have something to offer (Harper *et al.*, 1990), but in a practical implementation their utility is diminished considerably. The answer to the title question of this paper is therefore given in the negative.

This is not to imply that other areas of traffic science will not benefit from AI practices, but that automatic policing at least for the foreseeable future will not. Answering the title question in the negative is important for other strategic reasons. By ruling out one speculative avenue for further development, we enlarge our understanding of the practical dimensions of a problem. Whether automatic policing will become a reality is less a practical than a political question.

The issue of risk homeostasis is addressed in Harper (1991). In the absence of experimental data it is impossible to resolve anxieties that drivers may engage in risky compensatory behaviours. A related notion for discussion is whether drivers' judgements about safe driving will be shifted onto the automatic system. In other words, would drivers engage in *safe* risky behaviour? The position supported here is that as there is a tangible punitive dimension to an automatic traffic policing system, driver inattentiveness is unlikely to decrease. In the absence of field trials, we leave aside other conjectures for another day.

References

Aberg, L., 1988, Driver detection and probability of detection on roads with temporary 30 kmh speed limit, in Rothengatter, T. and de Bruin, R. (Eds) *Road user behaviour: theory and research*, 572–7, Assen: Van Gorcum.

Chin, H. C., 1989, Effect of automatic red-light cameras on red-running, *Traffic Engineering and Control*, **30**, 175–9.

Grime, G., 1987, *Handbook of road safety research*, Guildford: Butterworth.

Gundy, C. M., 1988, The effectiveness of a combination of police enforcement and public information for improving seat-belt use, in Rothengatter, T. and de Bruin, R. (Eds) *Road user behaviour: theory and research*, 595–600. Assen: Van Gorcum.

Harper, J., 1991, Traffic violation detection and deterrence: implications for automatic policing, *Applied Ergonomics*, **22**, 189–97.

Harper, J. and Fuller, R. G. C., (1993), Knowledge acquisition strategies in a design of a bus schedule monitoring system, (unpublished).

Harper, J., McLoughlin, H. B. and Webster, E. M., 1990, *Intelligent road traffic informatics: SIMWAY a prototype simulated system*, SCS Eastern Multi-conference, AI & Simulation, April, Nashville, TN.

Jeffrey, D. J., Russam, K. and Robertson, D. I., 1987, Electronic route guidance by AUTOGUIDE: the research background, *Traffic Engineering and Control*, **28**, 525–9.

Kass, R. and Finin, T, 1989, The role of user-models in cooperative interactive systems, *International Journal of Intelligent Systems*, **4**, 81–112.

Lamm, R. and Kloeckner, J. H., 1984, *Increase of traffic safety by surveillance of speed limits with automatic radar devices on a dangerous section of a German autobahn*. Transportation Research record 974, Transportation Research Board, Washington, DC.

Makinen, T, 1988, Enforcement studies in Finland, in Rothengatter, T. and de Bruin, R. (Eds) *Road user behaviour: theory and research*, 584–8, Assen: Van Gorcum.

Ostvik, E., Harper, J. and Vaa, T., 1989, *Police surveillance techniques and strategies*, Report 1033/D2, Traffic Research Centre, Haren, The Netherlands.

Reidel, W., Rothengatter, T. and de Bruin, R., 1988, Selective enforcement of speeding behaviour, in Rothengatter, T. & de Bruin, R. (Eds) *Road user behaviour: theory and research*, 578–83. Assen: Van Gorcum.

Roojiers, T., 1988, Effects of different public information techniques in reducing driving speed, in Rothengatter, T. & de Bruin, R. (Eds) *Road user behaviour: theory and research*, 589–94. Assen: Van Gorcum.

Rothengatter, T., 1988, The effects of police surveillance and law enforcement on driver behaviour *Current Psychological Review*, **2**, 349–58.

Rothengatter, T., de Bruin, R. and Roojiers, T., 1989, *The effects of publicity campaigns and police surveillance on the attitude-behaviour relationship in different groups of road users*, Paper presented at the Second European Workshop on Recent Developments in Road Safety Research, Paris.

Shibata, M., 1989, Development of a road/automobile communication system, *Transportation Research*, **23**, 63–71.

Shinar, D. and McKnight, A. J. 1985, The effects of enforcement and public information on compliance, in Evans, L. & Schwing, R. C. (Eds), *Human behaviour and traffic safety*. New York: Plenum Press.

Walsh, B. M. 1987, Do excise taxes save lives? The Irish experience with alcohol taxation, *Accident Analysis and Prevention*, **19**, 433–448.

Whitlock, F. A., 1971, *Death on the road: a study in social violence*, Tavistock Publications.

39

Behavioural studies of accident causation

R. Risser and C. Hydén

Introduction

When writing about 'behavioural studies of accident causation' in the frame of a book about 'driving future vehicles', the authors have DRIVE and PROMETHEUS and new RTI-systems in mind, which are going to be developed as a consequence of work with these programmes. This means that the aim of this contribution is to point out aspects in behaviour which should be supported, influenced, or modified by new systems and functions in order to achieve those traffic characteristics these two big European projects are aiming at with respect to traffic safety.

A look at what was done in the past concerning safety

Traditional safety strategy in road traffic can, a little ironically, be summarized as follows:

(a) we develop some equipment (start with the first car, or with the first road constructed for cars).
(b) at the same time we develop some laws which 'guarantee' that 'no accidents will happen if all traffic movements are performed according to law'.
´c) we 'forget' to ask ourselves what is the probability that road users will keep to the law, and which aspects of the traffic system will actually prevent them from complying.
d) we find that many things do not work out, nevertheless: technical failures (in the beginning) and human 'failures' (ever since the beginning) caused and cause problems.
e) we do not adequately reflect, and/or control, the fact that wishes for higher speeds (and/or shorter travelling times) and improved mobility develop in parallel with technical development.

(f) we refuse to recognize the relationship between 'unlimited' mobility and speed—and their attraction for car users—on one hand, and lack of safety on the other hand; however, the industry has mainly been interested in this kind of attraction and its relation to product selling.

(g) we tend to think that we have to accept conditions, such as increasing speed, the industry's sales strategies, the industry's interests, car drivers' interests, and others when we pursue our safety work.

(h) we realize that accident numbers per capita in our countries have grown, at first, and then more or less remained stable, without any relevant reduction of numbers of people injured or killed in road traffic during the last 15 years.

(i) we collect accident data hoping that this will give us the information on what went wrong, and why so many people are injured or killed in road traffic.

(j) we have lived for a long time with the fact—knowingly or not—that accident data does not give us satisfactory information on behaviour and interaction, and their relation to traffic as a social system. At the same time, methods for collecting and interpreting accident data became more and more sophisticated (in-depth studies, etc.).

(k) we still try to correct the system and to 'correct' the road users based on the knowledge we gathered from accident data, in spite of all short-comings connected to accident analysis.

(l) we have learned that, in practice, data describing road users' behaviour and interaction (from observations, interviews, etc.) are not accepted as indicators for the necessary countermeasures by authorities, road constructors, industry, and others, because their validity (= correlation with accident data as criteria!) cannot be proved.

(m) we know that in other areas of traffic (railway, aviation, public transportation in towns and cities) a system as described above—to collect accident data and then to try to prevent accidents—would never be accepted.

(n) we explain this difference by saying that in road traffic it is people themselves who have to decide in a democracy—forgetting how many people are injured or killed by other people, without getting a chance to decide themselves.

(o) we also explain this difference by relying on the assumption that in road traffic accident consequences are borne by the responsible car drivers themselves, which is an erroneous assumption, of course.

(p) we have no adequate philosophy and models for tackling the 'freedom' aspect which has infiltrated road traffic: that 'in road traffic people should decide and choose themselves and authorities should not interfere' (see (n) as well).

(q) we do not have sufficient knowledge to tackle this aspect of freedom with respect to how people behave under certain environment circumstances, how they use certain equipment, how they react to other road users, etc;

to what extent should new systems take away decisions from car drivers, thus reducing what we call their 'freedom'.

New technologies could be a starting point for new strategies

The description given here is of course not the only possible way to look at things. But that is not the aim of this paper. What is important is the question: how can we improve? In connection with the two European programmes PROMETHEUS and DRIVE we will get the chance to show that we can do better.

  Why is that? The new equipment that will be developed within the named programmes has not yet been 'involved' in accidents. Further, it is very doubtful that the industry will wait for accidents. Customers are more aware today of their rights than in former times, and the probability that new equipment will be questioned in the courts in case of accidents is higher today. So producers must make sure that their equipment does not negatively affect safety so they ask traffic-safety specialists for help. (History has taught us, though, that such 'calls for help' might be an alibi.)

What can traffic-safety specialists do in such a case?

One could say that we have to redefine safety: we must not accept a prolongation of the philosophy that safety somehow means a lack of accidents. Safety is, according to one way of redefinition, a characteristic of the behaviour and interaction of road users. Behaviour and interaction, on the other hand, can be observed and judged with regard to the safety or unsafety they reflect. Of course one can say that that is not the ordinary scientific way. But that is only valid if one wants natural-scientific accuracy in behavioural science. We know today that it is quite a useless effort to apply natural-scientific accuracy standards when we try to understand human behaviour. How do you quantify 'to be offended', or 'to feel unsafe'? Moreover, analysing accident data and behaving as if they were giving the information we need is not the scientific way either, as many decades of accident based safety analysis have shown us.

  From a psychological point of view, two perspectives for judging (un)safety of behaviour and interaction are very valuable:

- *The passenger perspective.* As passengers we are much more critical towards the behaviour of people in road traffic than when we drive ourselves (any theory to be developed around this phenomenon will have to deal with the 'locus of control'—questions we know from attribution psychology);
- *The results of behavioural research in road traffic.* Over a period of many years we have found plausible indicators that certain types of behaviour and interaction are dangerous, and that certain types of interactional events reflect the existence of danger (see Chaloupka, 1990).

No proof and no certainty, but some understanding

The problem is that many institutions and many safety experts are still looking for the final 'proof' of the hypothesis that the use of the two perspectives named above helps us identify behaviours and interactions that 'really' are dangerous. What is even worse is that this request for a final objective proof is connected to the attitude that 'until we have found this final proof we cannot proceed to implementing countermeasures'. We know very well that for many principles by which we live and which unanimously are judged useful, there is no final proof in a natural-scientific sense, e.g. if we say, 'follow certain rules when educating your children and they will not become physically aggressive'—we have no final proof. Still, certain principles are included in the laws of almost all European countries, e.g. you are not allowed to beat your children. The same is certainly valid for road traffic. The assumption is that it is possible to recognize behaviour that is safe, both from the individual's and from the system's point of view, and—as a necessary complement—to recognize unsafe behaviour. The problem lies within the description, operationalization, and quantification.

We will relate the following thoughts to the problem of speed. One type of behaviour that happens so often that it has to be looked upon as 'normal' is speeding; another one is travelling with inadequate speed. These are two very different aspects. Speeding could be looked upon as breaking of an abstract rule, where the immediate necessity for road safety might not be transparent; whereas 'inadequate' speed should be recognizable as risky behaviour in the moment it happens—otherwise it would not be 'inadequate'. However, the concept of 'inadequate' causes some definition problems. We tend to support the opinion that adequate speed is a speed where an observer (see 'passenger perspective' above) gets the impression that the driver can react safely to any 'normal' activity of other road users (but not, e.g., to somebody who wants to commit suicide), and that other road users, especially the unprotected ones, do not feel unsafe because of that. The ability to anticipate the possible behaviour of other road users, for instance, should include the knowledge that children do not behave 'normal'. Any speed limit, if it is well adapted, can theoretically be interpreted as an effort from the authorities to provide for adequate speed in the above defined sense. So we state that inadequate speed is recognizable and thus describable. However, we know how bad an instrument language is to describe, e.g., body-language aspects. To say, that roughly 'the car driver has travelled so fast that he forced the pedestrian to stop', is much easier than to explain the process of 'forcing a pedestrian to stop' in detail. So we have to work on methods to operationalize interaction of road users better than in the past, in order to understand safety problems better.

So how could one proceed in practice?

New equipment is mostly developed to support car drivers' skills. However, we must not forget that car drivers might use the equipment in such a way that this support of skills ends up with negative effects: delegation of responsibility, reduction of communication (and thus increase in risk where the individual has to improvise), imitation by non-equipped or unskilled drivers, transfer of habits to areas where an equipment might not help, etc.

What we have to accept is that such notions of behaviour modification are based on experts' opinion only and that we do not have any accident data to support this assessment. But expert assessment is the only method we have to evaluate new equipment. Thus:

1. One has to choose criteria for deciding if behaviour is desired or not. Two criteria have been used in the past in this respect (see Risser, 1988):
   (a) What is behaviour like according to the laws? A new equipment must not turn out to encourage law breaking.
   (b) Do collision potentials result as a consequence of stubbornness, lack of skills, or behaviour that is misunderstood by other road users? One can understand that changes of speed are essential signals in this connection.
2. One has to choose methods to analyse (a) and (b). Many studies have been done already in order to develop and practically use such methods (see Risser, 1985). Without going into more detail, the following methods can be named:

- behaviour observation from inside the car (by trained observers),
- video registrations,
- automatic registration of car movements (datalog),
- interviews after test rides.

If types of cars or equipment are sufficiently common and easily identified one can carry out stationary observations on the road as well.

Within DRIVE and PROMETHEUS, and even before these two programmes started, a lot of theoretical work dealing with the question, what 'desired behaviour' actually is, had been done. So the aim in future evaluation work without accident data is not to discover 'the model' for safe traffic. Know-how rather has to be collected and summarized in a way that it can function as an adequate base for evaluation work. Moreover, interaction aspects and social or socialization principles have to be included in one's assessments. The above mentioned discussion is the background to a series of behavioural observations, done in Austria, which will be described on the following pages.

A behavioural study in Vienna

During the years 1989 and 1990 a large study was carried out in Vienna. The aim was to identify risks in road traffic by comparing observed road users' behaviour (out of observed subjects' cars) to traffic-conflict data (stationary observation) and accident data along a standardized route. This study was financed by the Scientific Research Fund in Vienna.

One of the first results of the work, after the pretests, was that the concept of 'behaviour' had to be extended. The behaviour to be observed was, to a large extent, coordinated with the behaviour of other road users. Problems seemed to arise when coordination was 'bad' in the widest sense. Our observation instruments were adapted to this notion after an ample theoretical discussion.

Interaction = communication as central aspects in traffic safety

To make later argument easier, we proceed by calling all types of behaviour coinvolving more than one individual 'communication' from now on. The usually indispensable options for individuals to solve conflicts, in all kinds of social areas, by interpersonal communication are reduced in road traffic. Communication is utterly ambiguous there, because of reduced means for communication, which makes it almost impossible to give clear negative or positive feedback, which is otherwise the basis for human beings' social learning. No wonder it is so difficult to make vehicle drivers behave in a socially acceptable way. The character and the 'efficiency' of communication between traffic participants reflect the social climate in road traffic as a most important context variable of traffic safety. Considering the fact that more than 70 per cent of all accidents are accidents between two or more traffic participants (KfV, 1988; PRI, 1984), and that this portion is still considerably higher in densely inhabited areas, the relationship of communication between road users to traffic-safety aspects seems clear.

Starting from a totally different perspective, Hydén (1987) describes severe traffic conflicts—which, in theory, are closely related to accidents—as 'breakdown of interaction'.

To our team it seemed to be more of a 'dead end' to look upon behaviour in road traffic as the behaviour of isolated individuals, only 'interacting' with the physical road conditions and/or with their 'machine', independent of the social environment. Why did this attitude prevail in the past?

Because of speed and physical obstacles (physical separation, speed differences) the preconditions for communication between vehicle drivers and other road user groups are not at all satisfactory. We found out that in many respects communication is so distorted that one has difficulties in recognizing even its existence and its impact on the character of road traffic The presence and the function of interpersonal communication have not been considered adequately. The necessity to be able to communicate with other road users, to correctly interpret the communicative meaning of one's own behaviour and the behaviour of others, and above all the willingness to communicate, have been badly neglected as far as their importance for the social climate in traffic and for traffic safety is concerned.

The main hypotheses for the described study

Summarizing the theoretical arguments above and including pretests results (15 observation rides) the following hypotheses headed the project work:

Hypothesis 1
Traffic participants' behaviour is strongly influenced by the existence and the behaviour of other road users—not forgetting modelling effects—and not just by road characteristics, traffic education, driving school, traffic laws and law enforcement. (One can even look at single accidents as consequences of a behaviour shaped by socialization, where communication, of course, plays an important role.)

Hypothesis 2
A lot of problems in road traffic emerge, because motivation for correct behaviour based on social motives, like fairness, politeness, consideration for others, etc. is lacking. Lack of social motives is both supported and reflected by communication in traffic.

Hypothesis 3
The theoretical starting-point for the project described above was the hypothesis that special types of behaviour or interaction, respectively labelled as critical by the observers, should also be identified as relevant factors for the traffic conflicts one can observe on site and for the accident genesis, as it is reconstructed in police records.

Procedure

The first and initial step in the frame of the described project was to discuss and redefine all possible types of interaction which, up to now, had been described in terms of individual, stimulus-controlled behaviour, more or less independent of social aspects. This was done within the group working on the project. The collection of empirical data followed.

1. Registration of driving behaviour variables (main criteria)
Using a partly new and partly modified Austrian observation method ('Vienna driving test' or 'Wiener Fahrprobe', Risser *et al.*, 1982, 1983, 1988), the behaviour of 150 volunteers driving along a standardized route was recorded with special regard to communication with other traffic participants. Driving tests were carried out in a way that different types of traffic throughout the week were met: rush-hour on working days, light traffic on working days, heavy traffic at weekends, and light traffic at weekends.

Typical observation variables were as follows:

- insisting on one's own right of way, even if it is very dangerous;
- badgering pedestrians by driving past them within a small distance, e.g. in left or right turn situations;

- hampering pedestrians by obstructing their way when trying to drive over the zebra-crossing before them;
- overtaking other traffic participants on the right side (in right hand traffic) on city highways;
- and many others.

2. Registration of traffic conflicts on site

On ten selected sections of the test route traffic conflict registrations on site were carried out. These ten sections were selected with the help of a cluster analysis in such a way that they would represent typical aspects of the test route. Traffic-conflict registration was done according to the German technique, i.e. a technique based on psychological methods of behaviour observation, where behaviour aspects are not quantified (like time to collision, etc.), but only labelled according to the observer's impression (e.g., 'very dangerous', 'slightly dangerous'). Former studies have shown that observers can be trained in a way that very reliable results are obtained when using this semantical method (Erke and Gstalter, 1985; Risser, 1985; ICTCT, 1990).

3. Registration of accident data

Accident data were collected for all sections (e.g. crossroads, road sections between crossroads, motorway entrances and motorway exits). We were allowed to use the accident data base of the city of Vienna which consists of a compilation of all injury accidents registered by police within the area of Vienna (Ö St Z, 1989, 1990).

4. Comparison of the collected data

The results of our driving observations were then compared with on-site traffic-conflict registrations and with the accident records. As statistical methods correlation-analysis, regression-analysis, discriminance-analysis and variance-analysis were used, according to the distribution of the data (see Chaloupka *et al.*, 1991).

Main results

One of the main results was that car drivers, as a rule, behave under normal circumstances in a way that would be described as 'extremely impolite'. The outcome is often a genuine physical threat for other road users, especially the unprotected ones.

An 'impolite' communication style and consequences for safety

The general 'impolite' attitude—which has to be operationalized more precisely—in our project was interpreted as one of the main reasons for risks in traffic. Interpretation of one's own behaviour by others and consequences for others, are obviously either not well understood or neglected by car drivers. Other road users, their interests, their interpretation of events, and the consequences they more or less have to accept, are not sufficiently taken into consideration in practice. In the frame of a traffic system an awkward precondition arises where individuals not only have to decide on their behaviour, but also have to share the 'territory' with others.

The characteristics of communication influence the immediate reactions of both interacting partners, and will probably influence their future behaviour as well. This is indicated by spontaneous inconsiderate, stubborn or even aggressive actions and reactions that cannot be explained differently. This vicious circle will present traffic problems for the future.

Last but not least, it became quite clear that avoidance of communication (e.g. avoidance of eye-contact) and lack of efforts to gain sufficient understanding for a traffic situation are, for a great part, responsible for problems in traffic. This is reflected by an overall tendency to swerve and drive past critical situations without changing speed, instead of decelerating or braking.

Bad habits of car drivers?

To our observers it seemed clear that a large part of this type of behaviour was due to bad habits; or as it was put above—very 'impolite' style of communication prevailed that allowed one to reach 'egoistic goals' in spite of the presence of other road users. This lack of social control allows such a communication style to become a habit when people drive a car. The mentioned lack of control is certainly due to physical isolation (coachwork), and speed differences between traffic participants.

However, there is no doubt that part of the problems are also due to overload and/or lack of ability. According to our study this is mainly true for lane changing in city traffic, gap choice on crossroads, and interaction with pedestrians on complicated crossroads (many traffic streams, many road users of different type, etc.). In many cases drivers are not conscious of the fact that they could easily solve their problem by reducing their speed, but there seems to be a strong resistance towards reducing speed in many situations.

Four phenomena summarizing the results of the study

When discussing safety of new equipment, we have to deal with four phenomena:

1. Lack of ability (like low stress tolerance reflected by nervous behaviour, etc.) that tends to become a problem in complex situations. This problem can probably be solved by typical DRIVE- and PROMETHEUS-functions.
2. Relaxation (little traffic, few pedestrians, wide roads) and higher speeds—speeds seem to go up automatically as soon as the situation is 'safe'.
3. Lack of ability or preparedness on the car drivers' side to communicate with other road users in a cooperative way. Either they do not know of the

consequences of their behaviour—both as far as social climate and safety are concerned—or they do not care. This is probably mainly due to the social-psychological characteristics connected with car driving (e.g., no social feedback). In many cases intelligent electronic feedback could solve problems (headway control, warnings for the presence of pedestrians and cyclists before right and left turns in city traffic, etc.). However, there is some probability that warnings will be neglected. In those cases the system will have to take over. This is especially valid for all cases where the car drivers should reduce speed (see 4, as well).

4. A strong resistance to adapt speed to a lower level, and especially to brake. It seems quite clear that speeds—both in relation to speed limits and situational aspects—will have to be controlled automatically in a future traffic system, and that drivers need support in those situations where they should reduce speed in order to gather all information they need for tackling any situation adequately.

These four phenomena are logically connected to the principles modifying behaviour and communication that are discussed together with the PRO-GEN traffic-safety checklist (see Broughton *et al.*, 1991; Risser and Hydén, 1990; Chaloupka *et al.*, 1990).

Two strategies for RTI

In connection with the introduction of RTI-systems there are two main types of interaction between the drivers and the system to be discussed.

Tutoring

Feed-back to the driver after he has performed poorly could be one solution, but then it is very difficult, under any condition, to systematically tell people how they *should* have behaved, without annoying them or making them aggressive. And there is no guarantee that tutoring may change drivers' behaviour, in view of the four phenomena discussed above. The conclusion seems to be that the system must somehow intervene with the drivers' own decisions.

A system take over

Developing more and more systems that take away decisions from drivers, letting 'the system take over' (via consequent law enforcement by police, or via automatic equipment) certainly interferes with the characteristics of an individual traffic system. In aviation and railway traffic, mechanisms to control behaviour of the operating persons are the rule and not the exception. In connection with road traffic a control of behaviour is difficult to achieve, in many cases there is even opposition to control mechanisms from the car drivers or their representatives.

There are, however, obvious ways to approach the problem. Ways that combine the tutoring principle with system take-overs. Just look at the general speeds on our roads: we know that speeding is extremely common. We also know that a reduction of speeds so that every driver always complies with the speed limits would in itself produce very considerable benefits. Estimations in Sweden say that such a speed reduction would reduce injuries by at least 15 per cent. Taking this as a starting point we could, of course, develop a system that combines the principles of tutoring and intervention: a system that interferes with the driving only in the sense that it prevents the driver from speeding. The tutoring part of such a system would very quickly be learned by the driver, i.e. the driver will easily recognize when he tries speeding! We realize that such a system would not be 'automatically accepted' by all drivers. The 'freedom' aspect of today's traffic that we have mentioned before tells us that, but on the other hand, what are the real arguments against such a system?

In principle, most of the severe traffic problems have something to do with the speed adaptation problem. Inadequate speed always turns up as one of the major problem-related factors in accident analyses. Besides, problems in connection with, e.g. overtaking and too short headways have a strong link to speed adaptation problems a priori. One obvious way forward is therefore to start with a simple 'speed limit tutor' as mentioned before, and then develop it in the direction of stronger links to the speed adaptation problem. This would demand more and more intelligent input regarding the dynamic inter-action between drivers and other road users, as well as the environment. In principle, such a system could also be conceived as a tutoring system. If the driver does not comply with speed adaptation criteria he is tutored in such a way that finally he is complying with the rules by his own initiative. Ideally the system should work in such a way that the driver learns to act properly without any need of the system to intervene. The system take-over is functioning as an 'aversive stimulus' in such a scheme.

If a 'speed limit and speed adaptation tutor' can be intelligently developed, we are quite convinced that the safety benefits will be very great, especially compared to traditional measures that in most cases fail to demonstrate any considerable effect. Such a system would also produce strong synergy effects in the sense that communication/interaction problems in general will be easened by a system that assures a proper speed adaptation.

A typical contradiction with which to finish

The discussion about how to induce good behaviour via new equipment will have to go on. For the time being, nobody can tell if systems will be developed that give the car drivers some direct or indirect information about how they should behave, and which, at the same time, really succeed in motivating car drivers to behave according to information. We have only indicated one possible 'basic system'. The point is, however, that as long as we can agree on the criteria, i.e. the kind of behavioural changes that are desired, 'our system' and others can be tested and compared with regard to these criteria.

There is one problem however. Psychological rules are not well respected. 'Tell the driver how he should use equipment; if he does not comply, it is his fault.' Because of this philosophy we have many types of cars which can travel at more than 200 km/h, and many streets which allow such speeds. At the same time, traffic safety experts discuss how the problems of inadequate speed should be handled. We really should try to eliminate such contradictions in the future.

References

Bauer, T., Risser, R., Soche, P., Teske, W. and Vaughan, C., 1981, Kommunikation im Straßenverkehr, *Zeitschrift für Verkehrsrecht*, **26**.

Broughton, J., Colk, H., Dryselius, B., Fontaine, H., Gastaldi, G., Hydén, C., Klöckner, J. H., Neumann, L., Risser, R. and Veling, I., 1991, The PRO-GEN traffic safety checklist, PROMETHEUS Stuttgart.

Chaloupka, C., 1990, How to identify risks by observing human behaviour and interaction, ICTCT, Wien.

Chaloupka, C., Hydén, C. and Risser, R., 1990, Die PRO-GEN Verkehrssicherheits-Checkliste, ZVS 1/I.

Chaloupka, C., Risser, R. and Roest, F., 1991, *Identification and analysis of risky behaviour in road traffic*, Report about a project sponsored by the Fund for Scientific Research in Vienna.

Dorsch, F., 1982, *Psychologisches Wörterbuch*, Vienna: Huber-Verlag.

Erke, H. and Gstalter, H., 1985, *Verkehrskonflikttechnik*, Handbuch für die Durchführung und Auswertung von Erhebungen, Unfall- und Sicherheitsforschung Straßenverkehr 52, Bundesanstalt für Straßenwesen, Bergisch Gladbach.

Hartfiel, G. and Hillmann, K. H., 1972, *Wörterbuch der Soziologie*, Stuttgart: Alfred Kröner Verlag.

Hydén, C., 1987, The Development of a Method for Traffic Safety Evaluation: The Swedish Traffic-Conflicts-Technique, Dissertation. The Technical University of Lund.

ICTCT, 1990, Theoretical aspects and examples for practical use of traffic conflicts and other interactional safety criteria in several industrial and developing countries, *Lund Institute of Technology, Bulletin* **86**.

KfV, 1988, Unfallstatistik 1987, *Verkehr in Österreich*, Heft 1, Austrian Road Safety Board, Vienna.

Merten, K., 1977, Kommunikationsprozesse im Straßenverkehr, *Unfall- und Sicherheitsforschung im Straßenverkehr* **14**.

Michel, C. and Novak, F., 1975, *Kleines psychologisches Wörterbuch*, Freiburg, Basel, Wien: Herder Verlag.

Onions, C. T., 1974, *The Oxford Universal Dictionary Illustrated*, London: Caxton Publishing Company.

Ö St Z (Österreichisches Statistisches Zentralamt), 1989, *Straßenverkehrssicherheit im Jahre 1988*, Beiträge zur österreichischen Statistik.

Ö St Z (Österreichisches statistisches Zentralamt), 1990, *Straßenverkehrssicherheit im Jahre 1989*, Beiträge zur österreichischen Statistik.

Persson, H., 1988, Kommunikation mellan fotgängare och bilförare, Litteraturstudie, Technical University of Lund.

PRI (Prevention Routiere International), 1984, Road Safety in Future—Social and Economic Impact, World Congress, Vienna.

Risser, R., 1985, Behaviour in traffic-conflict situations, *Accident Analysis and Prevention*, **17**, 2.

Risser, R., 1988, Kommunikation und Kultur des Straßenverkehrs, Literas Universitätsverlag, Vienna.

Risser, R. and Bukasa, B., 1985, Versicherungsdaten über das Unfallgeschehen und ihre Beziehungen zu Fahrproben-, Explorations-, Leistungs- und Persönlich-keitsdaten aus der verkehrspsychologischen Begutachtung, *Zeitschrift für Verkehrsrecht*, **10**.

Risser, R. and Hydén, C., 1990, The Traffic Safety Checklist, *Proceedings of the PRO-GEN Workshop in London*, November 1990.

Risser, R., Schmidt, L., Brandstätter, C., Bukasa, B. and Wenninger, U., 1983, *Verkehrspsychologische Testverfahren und Kriterien des Fahrverhaltens*, Austrian Road Safety Board, Vienna.

Risser, R., Steinbauer, J., Amann, A., Roest, F., Anderle, F. G., Schmidt, G. A., Lipovitz, G. and Teske, W., 1988, Probleme älterer Personen bei der Teilnahme am Straßenverkehr, Literas Universitätsverlag, Vienna.

Risser, R., Teske, W., Vaughan, C. and Brandstätter, C., 1982, *Verkehrsverhalten in Konfliktsituationen*, Austrian Road Safety Board, Vienna.

Savigny, E., 1980, *Die Signalsprache der Autofahrer*, München: dtv junior.

Scherer, K. R. and Wallbott, H. G. (Eds), 1979, *Nonverbale Kommunikation: Forschungsbericht zum Interaktionsverhalten*, Weinheim: Beltz Verlag.

Watzlawik, P., Beavin, J. H. and Jackson, D. D., 1974, *Pragmatics of human communication*, New York: Norton.

Appendix: DRIVE PROJECTS

DRIVE 1 Projects

1004 DREAM:
A Feasibility Study for Monitoring Driver Status.
TUV RHEINLAND (D)
Contact   Herberg, K. W.
Telephone   49 221 83932053
Fax   49 221 8393114

1006 DRIVEAGE:
Factors in Elderly People's Driving Abilities Stage I and Stage II.
KINGS COLLEGE LONDON (UK)
Contact   Warnes, A.
Telephone   44 71 8365454
Fax   44 71 8361799

1016 INFOSAFE:
An Information System for Improved Road User Safety and Traffic.
TFK/VTI TRANSPORT FORSCHUNG (D)
Contact   Kallstrom, L.
Telephone   49 40 331070
Fax   49 40 335587

1017 BERTIE:
Changes in Driver Behaviour due to the Introduction of RTI Systems.
HUSAT RESEARCH INSTITUTE (UK)
Contact   Parkes, A.
Telephone   44 509 611088
Fax   44 509 234651

1031:   An Intelligent Traffic System for Vulnerable
                                        Road Users.
                                        ITS, UNIVERSITY OF LEEDS (UK)
                                        Contact   Carsten, O. M. J.
                                        Telephone   44 532 335348
                                        Fax   44 532 335334

1033 AUTOPOLIS: Automatic Policing Information Systems.
                                        RIJKSUNIVERSITEIT GRONINGEN (NL)
                                        Contact   Rothengatter, J. A.
                                        Telephone   31 50 636778
                                        Fax   31 50 636784

1037 STAMMI: Definition of Standards for In-Vehicle Man
                                        Machine Interface.
                                        HUSAT RESEARCH INSTITUTE (UK)
                                        Contact   Parkes, A.
                                        Telephone   44 509 611088
                                        Fax   44 509 234651

1040:   Safety Scenario: Identification of Hazards
                                        UNIVERSITY OF NOTTINGHAM (UK)
                                        Contact   Howarth, I.
                                        Telephone   44 602 484848
                                        Fax   44 602 590339

1041 GIDS: Generic Intelligent Driver Support Systems7
                                        RIJKSUNIVERSITEIT GRONINGEN (NL)
                                        Contact   Michon, J. A.
                                        Telephone   31 50 636778
                                        Fax   31 50 636784

1042:   Accident Data Collection and Analysis
                                        TECHNISCHE UNIV. MUNCHEN (D)
                                        Contact   Keller, H.
                                        Telephone   49 89 21052456-38
                                        Fax   49 89 21052000

1050 DRACO: Driving and Accident Coordination Observer
                                        QUEEN MARY COLLEGE (UK)
                                        Contact   Fincham, W.
                                        Telephone   44 71 975344
                                        Fax   44 71 9810259

1051:	Procedures for Safety Submissions for RTI Systems
	TUV RHEINLAND (D)
	Contact Herberg, G.
	Telephone 49 221 83932053
	Fax 49 221 8393114

1061 PUSSYCATS:	Improvement of Pedestrian Safety at Traffic Lights
	CETE (F)
	Contact Chretiennot, J.
	Telephone 33 7841 8125
	Fax 33 7826 4039

1062:	Multi-Layered Safety Objectives
	LUND UNIVERSITY(S)
	Contact Hydén, C.
	Telephone 46 46 109125
	Fax 46 46 123272

DRIVE 2 Projects

2002 HOPES:	Horizontal Project for the Evaluation of Safety
	SWEDISH NATIONAL ROAD ADMIN. (S)
	Contact Dryselius, B.
	Telephone 46 8 7576620
	Fax 46 8 983030

2004 ARIADNE:	Application of a Real-time Intelligent Aid for Driving and Navigation Enhancement
	ROVER GROUP (UK)
	Contact MacCaulay, B.
	Telephone 44 203 675511
	Fax 44 203 69418

2006 EMMIS:	Evaluation of Man Machine Interface by Simulation Techniques
	CENTRO RICERCHE FIAT (I)
	Contact Gay, P.
	Telephone 39 11 9023291
	Fax 39 11 9023673

2007 SAMOVAR:	Safety Assessment Monitoring On-Vehicle with Automatic Recording
	QUEEN MARY & WESTFIELD COLLEGE (UK)
	Contact Fincham, W.
	Telephone 44 71 9755344
	Fax 44 81 9810259

V2008 HARDIE:	Harmonisation of ATT Roadside and Driver Information in Europe
	TRAFFIC RESEARCH LABORATORY (UK)
	Contact Stevens, A.
	Telephone 44 344 770945
	Fax 44 344 770356

V2009 DETER:	Detection, Enforcement & Tutoring for Error Reduction
	TRC, UNIVERSITY OF GRONINGEN (NL)
	Contact Brookhuis, K.
	Telephone 31 50 636758
	Fax 31 50 636786

2010 TESCO:	Test on Co-operative Driving
	CSST (I)
	Contact Morrello, E.
	Telephone 39 11 878033/8397385
	Fax 39 11 8122832

2012 PROMISE:	PROMETHEUS CED 10 Mobile and Portable Information Systems in Europe
	VOLVO (S)
	Contact Hellaker, J.
	Telephone 46 31 7724075
	Fax 46 31 7724086

V2013 SOCRATES:	System of Cellular Radio for Traffic Efficiency and Safety
	IAN CATLING CONSULTANCY (UK)
	Contact Catling, I.
	Telephone 44 81 6434451
	Fax 44 81 6434452

2031 EDDIT:	Elderly and Disabled Drivers and Information Telematics
	CRANFIELD CENTRE FOR LOGISTICS (UK)
	Contact Oxley, P.
	Telephone 44 234 752751
	Fax 44 234 750875

2032 TELAID:	Telematic Applications for the Integration of Drivers with Special Needs
	TRUTH (GR)
	Contact — Naniopoulos, A.
	Telephone — 30 31 991560/2636
	Fax — 30 31 991564

2045 ROSES:	Road Safety Enhancement System
	TNO (NL)
	Contact — Pauwelussen, J.
	Telephone — 31 15 696412
	Fax — 31 15 620766

2046 ACCEPT:	Alert Concerted Cooperation in European Pilots for TMC
	ROBERT BOSCH (D)
	Contact — Heinzelmann, A.
	Telephone — 49 5121 492170
	Fax — 49 5121 492520

Glossary

AI	Artificial Intelligence
AID	Automatic Incident Detection
CD-ROM	Compact Disc Read only Memory
CEC	Commission of the European Communities
CEN	Comité Européen de Normalisation—standards body
CENELEC	CEN standards body for electrical systems
CEPT	European Telecommunications and Posts Administration's technical group
DRCO	DRIVE Central Office
EC	European Community
EDRM	European Digital Road Map
EFTA	European Free Trade Association
ENP	Electronic Number Plates
ERP	Electronic Road Pricing
ERTP	Eureka Road Transport Monitoring Group
ESPRIT	European Strategic Programme for Research and Development in Information Technology
GIS	Geographic Information Systems
GSM	Groupe Special Mobile—digital pan-European cellular system planned for the early 1990s
IBC	Integrated Broadband Communications
IBCN	Integrated Broadband Communications Network
IRTE	Integrated Road Transport Environment—the ultimate outcome of the work underway in DRIVE, the emphasis being on **Integrated**
ISO	International Standards Organisation
IT	Information Technology

LCD	Liquid Crystal Display
LED	Light Emitting Diode
MMI	Man Machine Interface or Man Machine Interaction
OECD	Organisation for Economic Cooperation and Development
PROMETHEUS	PROgraMme for a European Traffic system with Highest Efficiency and Unprecedented Safety
PTT	Posts, Telegraph and Telecommunication Authority
RACE	Research in Advanced Communications for Europe—a Community R&D programme
RDS	Radio Data System defined by the European Broadcasting Union
RDS-TMC	Radio Data System Traffic Message Channel
RTI	Road Transport Informatics
SECFO	Systems Engineering and Consensus Formation Office—a DRIVE project
VDU	Visual Display Unit
VMS	Variable Message Sign
VSCS	Vehicle Scheduling and Control Systems.

Index

Page numbers in bold denote chapters primarily concerned with that subject